# Advances in Intelligent Systems and Computing

Volume 245

*Series Editor*

Janusz Kacprzyk, Warsaw, Poland

For further volumes:
http://www.springer.com/series/11156

Van-Nam Huynh · Thierry Denœux
Dang Hung Tran · Anh Cuong Le
Son Bao Pham

Editors

# Knowledge and Systems Engineering

Proceedings of the Fifth International
Conference KSE 2013, Volume 2

 Springer

*Editors*
Van-Nam Huynh
School of Knowledge Science
Japan Advanced Institute of Science
    and Technology
Ishikawa
Japan

Thierry Denœux
Universite de Technologie de Compiegne
Compiegne Cedex
France

Dang Hung Tran
Faculty of Information Technology
Hanoi National University of Education
Hanoi
Vietnam

Anh Cuong Le
Faculty of Information Technology
University of Engineering and
    Technology - VNU Hanoi
Hanoi
Vietnam

Son Bao Pham
Faculty of Information Technology
University of Engineering and
    Technology - VNU Hanoi
Hanoi
Vietnam

ISSN 2194-5357          ISSN 2194-5365    (electronic)
ISBN 978-3-319-02820-0    ISBN 978-3-319-02821-7    (eBook)
DOI 10.1007/978-3-319-02821-7
Springer Cham Heidelberg New York Dordrecht London

Library of Congress Control Number: 2013950935

Printed on acid-free paper

Springer is part of Springer Science+Business Media (www.springer.com)

# Preface

This volume contains papers presented at the Fifth International Conference on Knowledge and Systems Engineering (KSE 2013), which was held in Hanoi, Vietnam, during 17–19 October, 2013. The conference was jointly organized by Hanoi National University of Education and the University of Engineering and Technology, Vietnam National University. The principal aim of KSE Conference is to bring together researchers, academics, practitioners and students in order to not only share research results and practical applications but also to foster collaboration in research and education in Knowledge and Systems Engineering.

This year we received a total of 124 submissions. Each of which was peer reviewed by at least two members of the Program Committee. Finally, 68 papers were chosen for presentation at KSE 2013 and publication in the proceedings. Besides the main track, the conference featured six special sessions focusing on specific topics of interest as well as included one workshop, two tutorials and three invited speeches. The kind cooperation of Yasuo Kudo, Tetsuya Murai, Yasunori Endo, Sadaaki Miyamoto, Akira Shimazu, Minh L. Nguyen, Tzung-Pei Hong, Bay Vo, Bac H. Le, Benjamin Quost, Sébastien Destercke, Marie-Hélène Abel, Claude Moulin, Marie-Christine Ho Ba Tho, Sabine Bensamoun, Tien-Tuan Dao, Lam Thu Bui and Tran Dinh Khang in organizing these special sessions and workshop is highly appreciated.

As a follow-up of the Conference, two special issues of the Journal of *Data & Knowledge Engineering* and *International Journal of Approximate Reasoning* will be organized to publish a small number of extended papers selected from the Conference as well as other relevant contributions received in response to subsequent calls. These journal submissions will go through a fresh round of reviews in accordance with the journals' guidelines.

We would like to express our appreciation to all the members of the Program Committee for their support and cooperation in this publication. We would also like to thank Janusz Kacprzyk (Series Editor) and Thomas Ditzinger (Senior Editor, Engineering/Applied Sciences) for their support and cooperation in this publication.

Last, but not the least, we wish to thank all the authors and participants for their contributions and fruitful discussions that made this conference a success.

Hanoi, Vietnam                                                              Van-Nam Huynh
October 2013                                                                 Thierry Denœux
                                                                             Dang Hung Tran
                                                                             Anh Cuong Le
                                                                             Son Bao Pham

# Organization

## Honorary Chairs

Van Minh Nguyen – Hanoi National University of Education, Vietnam
Ngoc Binh Nguyen – VNU University of Engineering and Technology, Vietnam

## General Chairs

Cam Ha Ho – Hanoi National University of Education, Vietnam
Anh Cuong Le – VNU University of Engineering and Technology, Vietnam

## Program Chairs

Van-Nam Huynh – Japan Advanced Institute of Science and Technology, Japan
Thierry Denœux – Université de Technologie de Compiègne, France
Dang Hung Tran – Hanoi National University of Education, Vietnam

## Program Committee

Akira Shimazu, Japan
Azeddine Beghdadi, France
Son Bao Pham, Vietnam
Benjamin Quost, France
Bernadette Bouchon-Meunier, France
Binh Thanh Huynh, Vietnam
Bay Vo, Vietnam
Cao H, Tru, Vietnam
Churn-Jung Liau, Taiwan
Dinh Dien, Vietnam
Claude Moulin, France

Cuong Nguyen, Vietnam
Dritan Nace, France
Duc Tran, USA
Duc Dung Nguyen, Vietnam
Enrique Herrera-Viedma, Spain
Gabriele Kern-Isberner, Germany
Hiromitsu Hattori, Japan
Hoang Truong, Vietnam
Hung V. Dang, Vietnam
Hung Son Nguyen, Poland
Jean Daniel Zucker, France

# Contents

# Part I
# Workshop Invited Talks

# The Place of Causal Analysis in the Analysis of Simulation Data

Ladislav Hluch

**Abstract.** This talk briefly reviews selected basic concepts and principles of structural approach to causal analysis, and outlines how they could be harnessed for analyzing and summarizing the data from simulations of complex dynamic systems, and for exploratory analysis of simulation models through machine learning. We illustrate the proposed method in the context of human behaviour modeling on a sample scenario from the EDA project A-0938-RT-GC EUSAS. The method revolves around the twin concepts of a causal partition of a variable of interest, and a causal summary of a simulation run. We broadly define a causal summary as a partition of the significant values of the analyzed variables (in our case the simulated motives fear and anger of human beings) into separate contributions by various causing factors, such as social influence or external events. We demonstrate that such causal summaries can be processed by machine learning techniques (e.g. clustering and classification) and facilitate meaningful interpretations of the emergent behaviours of complex agent-based models.

**Acknowledgement.** This work was supported by the European Defence Agency project A-0938-RT-GC EUSAS, by the Slovak Research and Development Agency under the contract No. APVV-0233-10, and by the project VEGA No. 2/0054/12.

Ladislav Hluch
Institute of Informatics, Slovak Academy of Sciences

V.-N. Huynh et al. (eds.), *Knowledge and Systems Engineering, Volume 2*,
Advances in Intelligent Systems and Computing 245,
DOI: 10.1007/978-3-319-02821-7_1, © Springer International Publishing Switzerland 2014

# Evolutionary Computation in the Real World: Successes and Challenges

Graham Kendall

**Abstract.** Evolutionary Computation has the potential to address many problems which may seem intractable to some of the methodologies that are available today. After briefly describing what evolutionary computation is (and what it is not), I will outline some of the success stories before moving onto the challenges we face in having these algorithms adopted by the industrial community at large.Some of the areas I will draw upon include Checkers and Chess, Scheduling and Timetabling, Hyper-heuristics and Meta-heuristics, as well some other problems drawn from the Operational Research literature.

Graham Kendall
The University of Nottingham Malaysia Campus,
Selangor Darul Ehsan, Malaysia

V.-N. Huynh et al. (eds.), *Knowledge and Systems Engineering, Volume 2*,
Advances in Intelligent Systems and Computing 245,
DOI: 10.1007/978-3-319-02821-7_2, © Springer International Publishing Switzerland 2014

# Part II
# KSE 2013 Special Sessions and Workshop

# A Method of Two-Stage Clustering with Constraints Using Agglomerative Hierarchical Algorithm and One-Pass $k$-Means++

Yusuke Tamura, Nobuhiro Obara, and Sadaaki Miyamoto

**Abstract.** The aim of this paper is to propose a two-stage method of clustering in which the first stage uses one-pass $k$-means++ and the second stage uses an agglomerative hierarchical algorithm. This method outperforms a foregoing two-stage algorithm by replacing the ordinary one-pass $k$-means by one-pass $k$-means++ in the first stage. Pairwise constraints are also taken into consideration in order to improve its performance. Effectiveness of the proposed method is shown by numerical examples.

## 1 Introduction

Clustering techniques [7, 9] has recently been becoming more and more popular, as huge data on the web should be handled. Such data are frequently unclassified in contrast to those in traditional pattern classification problems where most data have classification labels [5]. Not only methods of unsupervised classification but also those of semi-supervised classification [6] and constrained clustering [2, 3] have been developed to handle such data.

Clustering techniques in general can be divided into two categories of hierarchical clustering and non-hierarchical clustering. Best-known methods in the first category are agglomerative hierarchical clustering, while that in the second category is the method of $k$-means [8]. Most methods of semi-supervised classification and constrained clustering are non-hierarchical, but agglomerative hierarchical clustering is at least as useful as non-hierarchical techniques in various applications. A drawback in agglomerative hierarchical clustering is that larger computation is needed when compared with simple non-hierarchical methods such as the $k$-means.

Yusuke Tamura · Nobuhiro Obara
Master's Program in Risk Engineering, University of Tsukuba, Ibaraki 305-8573, Japan

Sadaaki Miyamoto
Department of Risk Engineering, University of Tsukuba, Ibaraki 305-8573, Japan
e-mail: miyamoto@risk.tsukuba.ac.jp

V.-N. Huynh et al. (eds.), *Knowledge and Systems Engineering, Volume 2,*
Advances in Intelligent Systems and Computing 245,
DOI: 10.1007/978-3-319-02821-7_3, © Springer International Publishing Switzerland 2014

Here is a question: how can we develop a method of agglomerative hierarchical clustering that can handle large amount of data with semi-supervision or constraints? We have partly answered this question by developing a method of agglomerative hierarchical clustering in which pairwise constraints can be handled using penalties in the agglomerative clustering algorithm [11]. Moreover a two-stage clustering has been suggested in which the first-stage uses $k$-means and the second stage is a class of agglomerative hierarchical clustering [10]. However, performance of the two-stage algorithm should still be improved.

In this paper we introduce a variation of the algorithm presented in [10]. In short, we use one-pass $k$-means++[1] in the first stage and show an improved two stage clustering algorithm with pairwise constraints. Several numerical examples are shown to observe the usefulness of the proposed method.

The rest of this paper is organized as follows. Section 2 provides preliminaries, then Section 3 shows the two-stage algorithm herein. Section 4 shows effectiveness and efficiency of the proposed algorithm using a number of numerical examples. Finally, Section 5 concludes the paper.

## 2   Preliminary Consideration

We begin with notations. Let the set of objects be $X = \{x_1, \cdots, x_n\}$. Each object $x_k$ is a point in the $p$-dimensional Euclidean space $\boldsymbol{R}^p$: $x_i = (x_{i1}, \cdots, x_{ip}) \in \boldsymbol{R}^p$

Clusters are denoted by $G_1, G_2, \cdots, G_C$, and the collection of clusters is given by $\mathscr{G} = \{G_1, G_2, \cdots, G_C\}$. Clusters are partition of $X$:

$$\bigcup_{i=1}^{C} G_i = X, \; G_i \cap G_j = \emptyset \; (i \neq j) \tag{1}$$

### 2.1   Agglomerative Hierarchical Clustering

Assume that $d(G, G')$ is a dissimilarity measure defined between two clusters; calculation formula of $d(G, G')$ will be given after the following general algorithm of agglomerative hierarchical clustering, abbreviated **AHC** in which **AHC 1** and **AHC 2** are the steps of this algorithm.

AHC1:      Let initial clusters given by objects.
   $G_i = \{x_i\}, (i = 1, \cdots, n)$
   $C = n$, ($C$ is the number of clusters and $n$ is the number of objects)
   Calculate $d(G, G')$ for all pairs $G, G' \in \mathscr{G} = \{G_1, G_2, \cdots, G_C\}$.
AHC2:      Merge the pair of clusters of minimum dissimilarity:

$$d(G_q, G_r) = \arg \min_{G, G' \in \mathscr{G}} d(G, G') \tag{2}$$

Add $\hat{G} = G_q \cup G_r$ to $\mathcal{G}$ and remove $G_q, G_r$ from $\mathcal{G}$.
$C = C - 1$.
If $C = 1$, then output the process of merge of clusters as a dendrogram and stop.
AHC3:     Calculate $d(\hat{G}, G')$ for $\hat{G}$ and all other $G' \in \mathcal{G}$. go to **AHC2**.

We assume that the dissimilarity between two objects is given by the squared Euclidean distance:

$$d(x_k, x_l) = \|x_k - x_l\|^2 = \sum_{j=1}^{p} (x_{kj} - x_{lj})^2.$$

Moreover the centroid method is used here, which calculate $d(\hat{G}, G')$ as follows.

Centroid method:

Let $M(G)$ be the centroid (the center of gravity) of $G$:

$$M(G) = (M_1(G), \cdots, M_p(G))^T,$$

where

$$M_j(G) = \frac{1}{|G|} \sum_{x_k \in G} x_{kj}, \ (j = 1, \cdots, p) \tag{3}$$

and let

$$d(G, G') = \|M(G) - M(G')\|^2 \tag{4}$$

## 2.2   k-Means and k-Means++

The method of $k$-means repeats the calculation of centroids of clusters and nearest centroid allocation of each object until convergence [4]. It has been known that the result is strongly dependent on the choice of initial values.

The method of $k$-means++ [1] improves such dependence on initial clusters by using probabilistic selection of initial centers. To describe $k$-means++, let $v_i$ be the $i$-th cluster center and $D(x)$ be the Euclidean distance between object $x$ and the already selected centers nearest to $x$. The algorithm is as follows [1].

1a:     Let the first cluster center $v_1$ be a randomly selected object from $X$.
1b:     Let a new center $v_i$ be selected from $X$ with probability $\frac{D(x)^2}{\sum_{x \in X} D(x)^2}$.
1c:     Repeat **1b** until $k$ cluster centers are selected.
2:     Carry out the ordinary $k$-means algorithm.

Step **1b** is called "$D^2$ weighting", whereby a new cluster center that have larger distance from already selected centers will have larger probability to be selected.

## 2.3 Pairwise Constraints

Two sets $ML$ and $CL$ of constraints are used in constrained clustering [2, 3]. A set $ML = \{(x_i, x_j)\} \subset X \times X$ consists of *must-link* pairs so that $x_i$ and $x_j$ should be in a same cluster, while another set $CL = \{(x_k, x_l)\} \subset X \times X$ consists of *cannot-link* pairs so that $x_i$ and $x_j$ should be in different clusters. $ML$ and $SL$ are assumed to be symmetric in the sense that if $(x_i, x_j) \in ML$ then $(x_j, x_i) \in ML$, and if $(x_k, x_l) \in CL$ then $(x_l, x_k) \in CL$.

Note that $ML$ is regarded as an undirected graph in which nodes are objects appeared in $ML$, and an undirected edge is $(x_i, x_j) \in ML$.

Introduction of the pairwise constraints to $k$-means has been done by Wagstaff et al. [12]. The developed algorithm is called COP $k$-means.

## 3 A Two-Stage Algorithm

A two-stage algorithm of clustering for large-scale data is proposed, in which the first stage uses one-pass $k$-means++ to have a medium number of cluster centers and the second stage uses the centroid method. Pairwise constraints are taken into account in both stages.

### 3.1 One-Pass COP $k$-Means++

One pass $k$-means implies that the algorithm does not iterate the calculation of the centroid and the nearest center allocation: it first generates initial cluster centers, then each object is allocated to the cluster of the nearest center. After the allocation, new cluster centers are calculated as the centroids (3). Then the algorithm stops without further iteration.

#### Pairwise Constraints in the First Stage

Moreover the one-pass algorithm must take pairwise constraints into account. $ML$ (must-link) is handled as the initial set of objects, as $ML$ defines a connected components of a graph. Then the centroid of the connected components is used instead of the objects in the components. On the other hand, $CL$ (cannot-link) is handled in the algorithm.

Thus the algorithm in the first stage is called one-pass COP $k$-means++, which is as follows.

One-Pass COP $k$-means++ in the first stage
1:    Let initial clusters be generated by using the $D^2$ weighting.
2:    Each object $x \in X$ is allocated to the cluster of the nearest center that does not break the given pairwise constraints $CL$. If $x$ cannot be allocated to any cluster due to the constraints, stop with flag **FAILURE**.
3:    Cluster centers are updated as the centroids (3).

4:      Stop. (Note that this step is replaced by 'repeat steps 2 and 3 until convergence'
      if the one-pass condition is removed.)
End of One-Pass COP $k$-means++.

## 3.2 Agglomerative Algorithm in the Second Stage

Information of the centroids $M(G_i)$ and the number of elements $|G_i|$ in cluster $G_i$
$(i = 1, 2, \ldots, c)$ is passed to the second stage. Note that information concerning every
object $x \in X$ is not required to generate clusters by AHC.

Different sets of $M(G_i)$ are obtained from the first stage. To have better clusters
in the second stage, a number of different trials of the first stage are made and those
centroids with the minimum value of

$$J = \sum_{i=1}^{C} \sum_{x \in G_i} \|x - M(G_i)\|^2 \tag{5}$$

is taken for the second stage.

### Pairwise Constraints in the Second Stage

Although must-link constraints is already handled in the first stage, cannot-link constraints still exist in the second stage. Hence $CL$ is handled by a penalty term in the
following algorithm.

### Penalized Agglomerative Hierarchical Clustering Algorithm (P-AHC)

P-AHC1:    For initial clusters derived from the first stage, calculate $d(G, G')$ for
      all $G, G' \in \mathscr{G}$.
P-AHC2:

$$d(G_q, G_r) = \arg \min_{G, G' \in \mathscr{G}} \left\{ d(G, G') + \sum_{x_k \in G, x_l \in G'} \omega_{kl} \right\}$$

using the penalty term with $\omega_{kl}$:
if $(x_k, x_l) \in CL$, $\omega_{kl} > 0$; if $(x_k, x_l) \notin CL$, $\omega_{kl} = 0$.
Let $\bar{G} = G_q \cup G_r$.
Add $\bar{G}$ to $\mathscr{G}$ and delete $G_q, G_r$ from $\mathscr{G}$.
$C = C - 1$. If $C = 1$, stop.
P-AHC3:    Calculate $d(\bar{G}, G')$ for all other $G' \in \mathscr{G}$. Go to **P-AHC2**.

Note that $\omega$ is taken to be sufficient large, i.e., we assume hard constraints.

## 4 Numerical Examples

Two data sets were used for evaluating the present method with other methods already proposed elsewhere. One is an artificial data set on the plane, while the second is a real data set from a data repository [1].

As for the methods, the following abbreviated symbols are used:

- PAHC: penalized AHC algorithm;
- COPKPP: one-pass COP $k$-means++ ;
- COPK: ordinary one-pass COP $k$-means ;
- COPKPP($n$): one-pass COP $k$-means++ with $n$ different initial values;
- COPK($n$): one-pass COP $k$-means with $n$ different initial values.

The computation environment is as follows.

CPU:     Intel(R) Core(TM) i5-3470 CPU @ 3.20GHz - 3.60GHz
Memory:     8.00 GB
OS:     Windows 7 Professional 64bit
Programming Language:     C

**Two Circles**

First data is shown In Fig. 1. The objective is to separate the outer circle having 700 points and the inner circle with 9,300 points. Note that the two clusters are 'unbalanced' in the sense that the numbers of objects are very different.

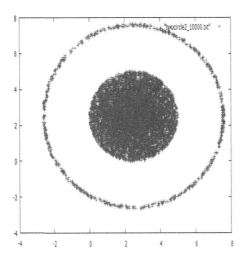

**Fig. 1** Data of 'two circles'

**Shuttle Data Set**

The Shuttle data set downloaded from [1] has 9 dimensions that can be divided into seven classes. About 80% of points belong to Class 1. We divide this data set into two clusters: one cluster is Class 1 and another cluster should be other six classes, since to detect small six clusters in 20% of points and one large cluster of 80% of points directly is generally a difficult task.

**Evaluation Criteria**

The evaluation has been done using three criteria: objective function values, the Rand index, and the run time.

Note that *CL* alone is used and *ML* is not used here, since *ML* was found to be not useful when compared with *CL* by preliminary tests on these data sets.

Pairs of objects in *CL* were randomly selected from the data set: one object from a cluster and another object from another cluster. For artificial data set the number in *CL* varies from 0 to 50; for the Shuttle data the number in *CL* varies from 0 to 500. The number of trials $n = 100$ (the number of trials in the first stage is 100) or $n = 10$ were used.

## 4.1  *Evaluation by Objective Function Value*

The averages of objective function values *J* are plotted in Figs. 2 and 3, respectively for the artificial data and the Shuttle data.

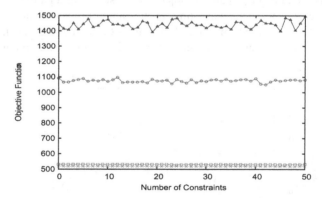

**Fig. 2** Objective function values with *CL* for artificial data. Red circles are for COPK(100)-PAHC. Green × are for COPKPP(100)-PAHC. Blue triangles are for COPK(10)-PAHC. Pink squares are for COPKPP(10)-PAHC.

From these figures it is clear that COPKPP-PAHC has less values of the objective function than COPK-PAHC.

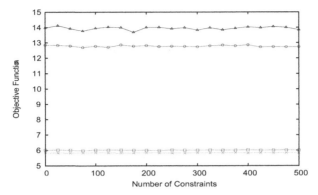

**Fig. 3** Objective function values with *CL* for the Shuttle data. Red circles are for COPK(100)-PAHC. Green × are for COPKPP(100)-PAHC. Blue triangles are for COPK(10)-PAHC. Pink squares are for COPKPP(10)-PAHC.

## 4.2   Evaluation by RandIndex

The Rand index has been used as a standard index to measure precision of classification [12]:

$$Rand(P_1, P_2) = \frac{|C_a| + |C_b|}{{}_nC_2} \tag{6}$$

where $P_1$ and $P_2$ means the precise classification and the actually obtained classification. $|C_a|$ is the number of pairs of objects in $C_a$ such that a pair in $C_a$ is in the same precise class and at the same time in the same cluster obtained by the experiment; $|C_b|$ is the number of pairs of objects in $C_b$ such that a pair in $C_a$ is in different precise classes and at the same time in different clusters obtained by the

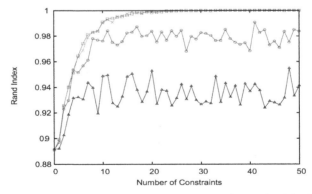

**Fig. 4** Rand index values with *CL* for artificial data. Red circles are for COPK(100)-PAHC. Green × are for COPKPP(100)-PAHC. Blue triangles are for COPK(10)-PAHC. Pink squares are for COPKPP(10)-PAHC.

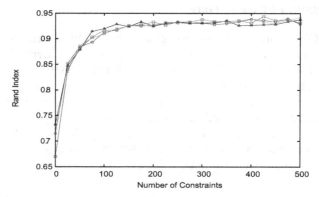

**Fig. 5** Rand index values with $CL$ for the Shuttle data. Red circles are for COPK(100)-PAHC. Green × are for COPKPP(100)-PAHC. Blue triangles are for COPK(10)-PAHC. Pink squares are for COPKPP(10)-PAHC.

experiment. If the resulting clusters precisely coincide with the precise classes, then $Rand(P_1, P_2) = 1$, and vice versa.

The Rand index with $n = 100$ has been calculated and the results are shown in Figs. 4 and 5, respectively for the artificial data and the Shuttle data. The former figure shows advantage of COPKPP, while the effect of K-means++ is not clear in the second example.

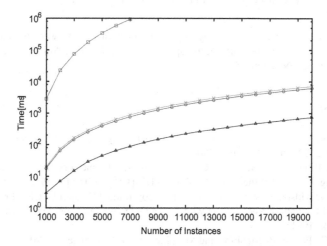

**Fig. 6** Relation between the number of objects in artificial data and the CPU time. Red circles are for COPK(100)-PAHC. Green × are for COPKPP(100)-PAHC. Blue triangles are for COPKPP(10)-PAHC. Pink squares are for PAHC.

## 4.3  Evaluation by CPU Time

How total CPU time varies by using one-pass COP $k$-means++ or one-pass COP $k$-means was investigated. The used methods were COPK(100)-PAHCCCOPKPP(100)-PAHCCCOPKPP(10)-PAHCC and PAHC (without the first stage). Ten trials with $n$ objects and their average CPU time was measured with $n = 1,000 - 20,000$. In the first stage the number of objects was reduced to 1% and the second stage AHC was carried out. The result is shown in Fig. 6.

Fig. 6 shows that CPU time was reduced to 0.1% by introducing the two-stage method. When COPK(100)-PAHC and COPKPP(100)-PAHC are comparted, the latter needs more time, but the difference is not notable.

## 5  Conclusion

This paper proposed a two-stage algorithm in which the first stage uses one-pass $k$-means++ and the second stage uses the centroid method of agglomerative hierarchical clustering. Pairwise constraints were moreover introduced in the algorithm. It has been shown by numerical examples that one-pass $k$-means++ is effective when compared with one-pass $k$-means in the first stage. Thus the dependence on initial values was greatly improved. Moreover the use of cannot-links was effective in the numerical examples. This inclination is in accordance with other studies, e.g., [11].

The two-stage procedure could handle relatively large-scale data sets. However, more tests on larger real data should be done as a future work in order to show the usefulness of the proposed method in a variety of applications.

**Acknowledgment.** The authors greatly appreciate anonymous reviewers for their useful comments. This study has partially been supported by the Grant-in-Aid for Scientific Research, JSPS, Japan, No.23500269.

## References

1. Arthur, D., Vassilvitskii, S.: k-means++: The Advantages of Careful Seeding. In: Proc. of SODA 2007, pp. 1027–1035 (2007)
2. Basu, S., Bilenko, M., Mooney, R.J.: A Probabilistic Framework for Semi-Supervised Clustering. In: Proc. of the Tenth ACM SIGKDD (KDD 2004), pp. 59–68 (2004)
3. Basu, S., Davidson, I., Wagstaff, K.L. (eds.): Constrained Clustering. CRC Press (2009)
4. Bezdek, J.C.: Pattern Recognition with Fuzzy Objective Function Algorithms. Plenum, New York (1981)
5. Bishop, C.: Pattern Recognition and Machine Learning. Springer (2006)
6. Chapelle, O., Schölkopf, B., Zien, A. (eds.): Semi-Supervised Learning. MIT Press (2006)
7. Everitt, B.S.: Cluster Analysis, 3rd edn., Arnold (1993)
8. MacQueen, J.B.: Some methods of classification and analysis of multivariate observations. In: Proc. of 5th Berkeley Symposium on Math. Stat. and Prob., pp. 281–297 (1967)
9. Miyamoto, S.: Introduction to Cluster Analysis. Morikita-shuppan (1999) (in Japanese)

10. Obara, N., Miyamoto, C.S.: A Method of Two-Stage Clustering with Constraints Using Agglomerative Hierarchical Algorithm and One-Pass $K$-Means. In: Proc. of SCIS-ISIS 2012, pp. 1540–1544 (2012)
11. Terami, A., Miyamoto, S.: Constrained Agglomerative Hierarchical Clustering Algorithms with Penalties. In: Proc. of FUZZ-IEEE 2011, pp. 422–427 (2011)
12. Wagstaff, N., Cardie, C., Rogers, S., Schroedl, S.: Constrained K-means Clustering with Background Knowledge. In: Proc. of ICML 2001, pp. 577–584 (2001)
13. http://archive.ics.uci.edu/ml/

# An Algorithm Combining Spectral Clustering and DBSCAN for Core Points

So Miyahara, Yoshiyuki Komazaki, and Sadaaki Miyamoto

**Abstract.** The method of spectral clustering is based on the graph Laplacian, and outputs good results for well-separated groups of points even when they have non-linear boundaries. However, it is generally difficult to classify a large amount of data by this technique because computational complexity is large. We propose an algorithm using the concept of core points in DBSCAN. This algorithm first applies DBSCAN for core points and performs spectral clustering for each cluster obtained from the first step. Simulation examples are used to show performance of the proposed algorithm.

## 1 Introduction

Many researchers are now working on analysis of huge data on the web. In accordance with this, many methods of data analysis have been developed. Data clustering is not exceptional: nowadays a variety of new algorithms of clustering is being applied to large-scale data sets. Special attention has been paid to spectral clustering [4, 2, 3] which is based on a weighted graph model and uses the graph Laplacian. It has been known that this method works well even when clusters have strongly nonlinear boundaries between clusters, as far as they are well-separated.

In spite of its usefulness, the spectral clustering has a drawback: it has a relatively large computation when compared with a simple algorithm of the $K$-means [4, 5]. The latter can be applied to huge data, since the algorithm is very simple, but the former uses eigenvalues and eigenvectors which needs much more computation.

This paper proposes a method combining the spectral clustering and the idea in a simple graph-theoretical method based on DBSCAN [6]. The both methods are

So Miyahara · Yoshiyuki Komazaki
Master's Program in Risk Engineering, University of Tsukuba, Ibaraki 305-8573, Japan

Sadaaki Miyamoto
Department of Risk Engineering, University of Tsukuba, Ibaraki 305-8573, Japan
e-mail: miyamoto@risk.tsukuba.ac.jp

V.-N. Huynh et al. (eds.), *Knowledge and Systems Engineering, Volume 2*,
Advances in Intelligent Systems and Computing 245,
DOI: 10.1007/978-3-319-02821-7_4, © Springer International Publishing Switzerland 2014

well-known, but their combination with a simple modification leads a new algorithm. A related study has been done by Yan et al. [7] in which $K$-means is first used and the centers from $K$-means are clustered using the spectral clustering. The present study is different from [7], since the original objects are made into clusters by the spectral clustering by the method herein, whereas the $K$-means centers are clustered in [7]. A key point is that only core-points are used for clustering, and other 'noise points' are allocated to clusters using a simple technique of supervised classification. Moreover, these two methods of the spectral clustering and DBSCAN has a common theoretical feature that is useful for reducing computation, and hence the combination proposed here has a theoretical basis, as we will see later. Such a feature cannot be found between $K$-means and the spectral clustering.

The rest of this paper is organized as follows. Section 2 gives preliminaries, and then Section 3 proposes a new algorithm using the spectral clustering and DBSCAN for core points. Section 4 shows illustrative examples and a real example. Finally, Section 5 concludes the paper.

## 2  Preliminary Consideration

This section discusses the well-known methods of the spectral clustering and DBSCAN.

### 2.1  Spectral Clustering

The spectral clustering, written as SC here, uses a partition of a graph of objects $D = \{1, 2, \ldots, n\}$ for clustering. The optimality of the partition is discussed in [3] but omitted here.

Assume that the number of clusters is fixed and given by $c$. A similarity matrix $S = (s_{ij})$ is generated using a dissimilarity $d(i, j)$ between $i$ and $j$. We assume that $d(i, j)$ is the Euclidean distance in this paper, although many other dissimilarity can also be used for the same purpose.

$$S = [s_{ij}], \; s_{ij} = \exp\left(-\frac{d(i,j)}{(2\sigma^2)}\right)$$

where $\sigma$ is a positive constant. When the $\varepsilon$-neighborhood graph should be used, then those $s_{ij}$ with $d(i, j) > \varepsilon$ should be set to zero. We then calculate

$$D = diag(d_1, \cdots, d_n), \; d_i = \sum_{j=1}^{n} s_{ij}$$

and the graph Laplacian $L$:

$$L = D^{-\frac{1}{2}}(D - S)D^{-\frac{1}{2}}$$

Minimum $c$ eigenvalues are taken and the corresponding eigenvectors are assumed to be $u_1, \cdots, u_c$. A matrix

$$U = (u_1, \cdots, u_c)$$

is then defined. Each component of the eigenvalues has correspondence to an object. Then $K$-means clustering of each rows with $c$ clusters will give the results of clustering by SC [3]. Concretely, suppose row vectors of $U$ are $u_1^\top, \ldots, u_n^\top$: $U = (u_1, \ldots, u_n)^\top$, then $K$-means algorithm is applied to objects $u_1, \ldots, u_n$, where $u_j$ $(j = 1, \ldots, n)$ is a $c$-vector [3].

## 2.2 DBSCAN-CORE

DBSCAN proposed by Ester et al. [6] generates clustering based on density of objects using two parameters Eps and MinPts. For given Eps and MinPts, the Eps-neighborhood of $p \in D$ is given by

$$N_{\text{Eps}}(p) = \{q \in D \mid d(p,q) \leq \text{Eps}\}$$

When an object $p$ satisfies $|N_{\text{Eps}}(p)| \geq \text{MinPts}$, then $p$ is called a core-point (note: $|N_{\text{Eps}}(p)|$ is the number of elements in $N_{\text{Eps}}(p)$).

If the next two conditions are satisfied, then $p$ is called *directly density-reachable from q*:

1. $p \in N_{\text{Eps}}(q)$, and
2. $|N_{\text{Eps}}(q)| \geq \text{MinPts}$ ($q$ is a core-point).

A variation of the DBSCAN algorithm used here starts from a core-point called seed, and then collects all *core points* that are directly density-reachable from the seed. Then they form a cluster. Then the algorithm repeats the same procedure until no more cluster is obtained. The remaining objects are left unclassified. In other words, this algorithm searches the connected components of the graph generated from core points with the edges of direct reachability, and defines clusters as the connected components.

This algorithm is simpler than the original DBSCAN in that only core-points are made into clusters, while non-core points are included in clusters by the original DBSCAN. Therefore the present algorithm is called DBSCAN-CORE in this paper. Specifically, The set $D$ is first divided into $C$ of core points and $N$ of non-core points:

$$D = C \cup N, \qquad C \cap N = \emptyset.$$

Clusters $C_1, \ldots, C_l$ generated by DBSCAN-CORE is a partition of $C$:

$$\bigcup_{i=1}^{l} C_i = C, \qquad C_i \cap C_j = \emptyset \ (i \neq j).$$

How to decide appropriate values of the parameters is given in [6], but omitted here.

## 3   Combining DBSCAN-CORE and Spectral Clustering

A method proposed here first generates clusters of core-points using DBSCAN-CORE and then each clusters are subdivided by the spectral clustering. We assume that Eps-neighborhood graph is used for the both method, i.e., the same value of Eps is applied: $s_{ij} = 0$ iff $d(i,j) \geq Eps$ in the spectral clustering and $N_{\mathrm{Eps}}$ is used for DBSCAN-CORE.

We then have the next proposition.

**Proposition 1.** *Let $G_1, \ldots, G_K$ be clusters of set $C$ of core-points generated by the spectral clustering. Then, for arbitrary $G_i$, there exists $C_j$ such that $G_i \subseteq C_j$.*

The proof is based on the fact that no cluster by the spectral clustering connects different connected components of graph $C$ [3].

Note that DBSCAN-CORE has a fast algorithm similar to generation of spanning trees. Thus the complexity is $O(n)$, which is less than the complexity of the spectral clustering. We hence have the following simple algorithm combining DBSCAN-CORE and the spectral clustering.

**Algorithm DBSCAN-CORE-SC:**

1. Define core points and carry out DBSCAN-CORE. Let $C_1, \ldots, C_l$ be clusters of $C$.
2. Generate subclusters of $C_i$ for all $i = 1, 2, \ldots, l$ by the spectral clustering.

### 3.1   *Clusters of Data Set* D

The above procedure generates clusters of $C$, the set of core points, but the non-core points will remain as noises. When we wish to classify noises to one of the clusters of $C$, a simple supervised classification algorithm can be used. A typical algorithm is the $k$ nearest neighbor method ($k$NN) [4]: Let $x \in N$ should be allocated to some cluster. Suppose $y_1, \ldots, y_k \in C$ be $k$ nearest neighbors of $x$ in $C$. Then the class $h$ is determined by the following:

$$h = \arg \max_{1 \leq j \leq l} |\{y_1, \ldots, y_k\} \cap C_j|.$$

When $k = 1$, the above is reduced to the nearest neighbor allocation:

$$h = \arg \min_{1 \leq j \leq l} d(x, C_j),$$

where $d(x, C_j) = \min_{y \in C_j} d(x, y)$. The nearest neighbor allocation is used for numerical examples below.

We thus have an algorithm to generate clusters of $D$ by first generating clusters of $C$ using DBSCAN-CORE-SC and then allocate other points. We moreover use a particular option that only those points in $N_{\mathrm{Eps}}(q)$ for some core point $q$ should be allocated using $k$NN, but those points $p' \notin N_{\mathrm{Eps}}(q')$ for all $q' \in C$ should be left as *noise points*. This algorithm is called DBSCAN-CORE-SC-$k$NN in this paper.

## 3.2 Other Related Algorithms

Although we propose DBSCAN-CORE-SC and DBSCAN-CORE-SC-$k$NN here, there are other algorithms that should be compared with the proposed algorithms.

**Algorithm SC-CORE**

Step 1.    Select core points by the same procedure as the DBSCAN-CORE.

Step 2.    Generate clusters by the spectral clustering for the core points without using DBSCAN-CORE.

**End of SC-CORE**

Thus SC-CORE generates clusters of $C$. Accordingly, we can define SC-CORE-$k$NN by using the $k$NN after applying SC-CORE.

## 4    Numerical Examples

Algorithms of DBSCANCSCCSC-CORE-$k$NNCDBSCAN-CORE-SC-$k$NN, and SC-CORE-$k$NN have been done by using the following computational environment.

–    Hardware: Deginnos Series
–    OS: Ubuntu 12.10 i64 bit OSj
–    CPU: Intel(R) Core(TM) i7-3630QM CPU @ 2.40GHz
–    Memory: 16.00 GB
–    Language: Python 2.7
–    Eigenvalue solver: linalg.eig in Numpy library

In order to reduce the effect of initial values in the $K$-means used in the spectral clustering, 50 trials with different random initial values were used and the clusters with minimum objective function values were selected.

The used parameters were the same values for all methods: The nearest neighbor allocation: $k = 1$ and the neighborhood graph with $\sigma = 1.0$ were used. Eps were determined by using the sorted 4-dist graph given in [6]. Thus MinPts $= 4$. The value of Eps is thus different according to the examples. First example uses Eps $= 0.0015$, Eps $= 0.0006$ for the second, and Eps $= 0.18$ for the third.

Noise points in the following figures are shown by black $*$, while clusters are shown by $+$ and $\bigcirc$ with different colors.

## 4.1    Results for Artificial Data Sets

Two artificial data sets on the plane were used. First data shown in Fig. 1 called *test data 1* has $2,650$ objects with 100 noise points. Second data shown in Fig. 2 called *test data 2* has $5,030$ objects with 50 noise points. Figures 3 and 4 show the results from SC-CORE-$k$NN and DBSCAN-CORE-SC-$k$NN for test data 1, respectively; Figures 5 and 6 show the results from SC-CORE-$k$NN and DBSCAN-CORE-SC-$k$NN for test data 2, respectively.

In the both examples DBSCAN-CORE divided the set of core points into two clusters: upper cluster and lower cluster in test data 1 and inner cluster and outer cluster in test data 2.

CPU times for SC, SC-CORE-$k$NN, and DBSCAN-CORE-SC-$k$NN are compared in Table 1 (Note that the time for preprocessing to calculate similarity values is not included in Table 1 and Table 3). The four figures show that good and same clusters are obtained by the two methods, and Table shows that run time is effectively reduced by the proposed method.

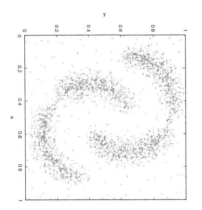　　　　　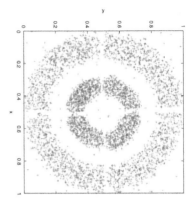

**Fig. 1** Test data 1　　　　　　　　　**Fig. 2** Test data 2

**Table 1** CPU time for artificial data with different methods

|  | Time(s) | |
|---|---|---|
| Method | test data1 | test data2 |
| SC | 85.99019 | 510.94347 |
| SC-CORE-$k$NN | 84.04765 | 495.55304 |
| DBSCAN-CORE-SC-$k$NN | 29.05077 | 179.54790 |

## 4.2  The Iris Data Set

The well-known *iris* data has been handled by the different methods. As shown in Table 2, the same classification results were obtained from the different methods of SC-CORE-$k$NN and DBSCAN-CORE-SC-$k$NN. DBSCAN-CORE generated two well-separated clusters in *iris*. Then SC generated two subclusters from the larger cluster by DBSCAN-CORE.

The CPU time is again reduced by using DBSCAN-CORE-SC-$k$NN, as shown in Table 3.

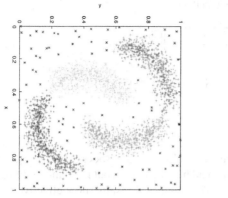 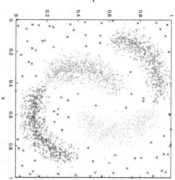

**Fig. 3** Clusters generated by SC-CORE-kNN for test data 1

**Fig. 4** Clusters generated by DBSCAN-CORE-SC-kNN for test data 1

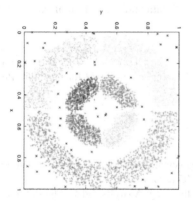 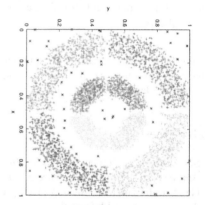

**Fig. 5** Clusters generated by SC-CORE-kNN for test data 2

**Fig. 6** Clusters generated by DBSCAN-CORE-SC-kNN for test data 2

**Table 2** The results for iris data from different methods, where the Rand index is used

| Method | Rand Index |
|---|---|
| HCM | 0.87374 |
| SC (complete graph) | 0.87373 |
| SC ($\varepsilon$-neighborhood graph) | 0.85682 |
| SC-CORE-kNN | 0.85682 |
| DBSCAN-CORE-SC-kNN | 0.85682 |

**Table 3** CPU time for *iris* data with different methods

| Method | Time[s] |
|---|---|
| SC | 0.36957 |
| SC-CORE-$k$NN | 0.35769 |
| DBSCAN-CORE-SC-$k$NN | 0.20951 |

## 5    Conclusion

The combination of DBSCAN with core points alone and the spectral clustering has been discussed. This combination is not an ad hoc technique, but has a methodological consistency shown in Proposition 1. The numerical results show effectiveness and efficiency of the proposed method. In the numerical examples, the values of the parameters greatly affects the results, and hence how good values of the parameters can be found should be an important subject of future study.

A fundamental problem is that no definite method to determine the number of clusters beforehand in DBSCAN-CORE-SC proposed here, which needs further research. More experiments for huge amount of real data and evaluation of the results should also be done.

**Acknowledgment.** The authors greatly appreciate anonymous reviewers for their useful comments. This study has partially been supported by the Grant-in-Aid for Scientific Research, JSPS, Japan, No.23500269.

## References

1. Shi, J., Malik, J.: Normalized Cuts and Image Segmentation. IEEE Transactions on Pattern Analysis and Machine Intelligence 22(8), 888–905 (2000)
2. Ng, A.Y., Jordan, M.I., Weiss, Y.: On Spectral Clustering: Analysis and an Algorithm. In: Advances in Neural Information Processing System, pp. 849–856 (2001)
3. von Luxburg, U.: A Tutorial on Spectral Clustering. Statistics and Computing 17(4), 395–416 (2007)
4. Duda, R.O., Hart, P.E.: Pattern Classification and Scene Analysis. Wiley, Chichester (1973)
5. Miyamoto, S., Ichihashi, H., Honda, K.: Algorithms for Fuzzy Clustering. Springer, Berlin (2008)
6. Ester, M., Kriegel, H.-P., Sander, J., Xu, X.: A Density-Based Algorithm for Discovering Clusters in Large Spatial Databases with Noise. In: Proceedings of 2nd International Conference on Knowledge Discovery and Data Mining, pp. 226–231 (1996)
7. Yan, D., Huang, L., Jordan, M.I.: Fast Approximate Spectral Clustering. In: Proceedings of the 15th ACM SIGKDD International Conference on Knowledge Discovery and Data Mining, pp. 907–916 (2009)

# Relational Fuzzy $c$-Means and Kernel Fuzzy $c$-Means Using a Quadratic Programming-Based Object-Wise $\beta$-Spread Transformation

Yuchi Kanzawa

**Abstract.** Clustering methods of relational data are often based on the assumption that a given set of relational data is Euclidean, and kernelized clustering methods are often based on the assumption that a given kernel is positive semidefinite. In practice, non-Euclidean relational data and an indefinite kernel may arise, and a $\beta$-spread transformation was proposed for such cases, which modified a given set of relational data or a given a kernel Gram matrix such that the modified $\beta$ value is common to all objects.

In this paper, we propose a quadratic programming-based object-wise $\beta$-spread transformation for use in both relational and kernelized fuzzy $c$-means clustering. The proposed system retains the given data better than conventional methods, and numerical examples show that our method is efficient for both relational and kernel fuzzy $c$-means.

## 1 Introduction

Fuzzy $c$-means (FCM) [2] is a well-known clustering method for vectorial data. In contrast, relational fuzzy $c$-means (RFCM) [2] clusters relational data. However, RFCM is not always able to cluster non-Euclidean relational data, because the membership cannot always be calculated. To overcome this limitation, a non-Euclidean RFCM (NERFCM) has been proposed [3]. NERFCM modifies the given data so that the memberships can be calculated, and this modification is called a $\beta$-spread transformation.

In order to cluster data with nonlinear borders, an algorithm that converts the original pattern space to a higher-dimensional feature space has been proposed [4]. This algorithm, known as kernel FCM (K-FCM), uses a nonlinear transformation defined by kernel functions in the support vector machine (SVM) [5]. In kernel

Yuchi Kanzawa
Shibaura Institute of Technology, Koto 135-8548 Tokyo, Japan
e-mail: kanzawa@sic.shibaura-it.ac.jp

V.-N. Huynh et al. (eds.), *Knowledge and Systems Engineering, Volume 2,*    29
Advances in Intelligent Systems and Computing 245,
DOI: 10.1007/978-3-319-02821-7_5, © Springer International Publishing Switzerland 2014

data analysis, it is not necessary to know the explicit mapping of the feature space; however, its inner product must be known. Despite this, an explicit mapping has been reported and this was used to describe the appearance of clusters in a high-dimensional space [6], [7].

K-FCM fails for indefinite kernel matrices when the magnitude of the negative eigenvalues is extremely large, because the memberships cannot be calculated if the dissimilarity between a datum and a cluster center is updated to become a negative value. Although indefinite kernel matrices can be transformed to positive-definite ones by subtracting the minimal eigenvalue from their diagonal components, or by replacing negative eigenvalues with 0, these procedures result in over-transformation of the matrix. Although the clustering can still be executed, the risk is that the memberships can become extremely fuzzy and worsen the clustering result. Therefore, an indefinite-kernel FCM (IK-FCM) method has been developed [8]; this adopts a $\beta$-spread transformation and is similar to the derivation of NERFCM from RFCM.

In the conventional $\beta$-spread transformation for NERFCM or IK-FCM, the modified $\beta$ value is common to all objects in the given relational data matrix or kernel Gram matrix. In this paper, we propose that a different value is added to each object in the given matrices. We refer to this as an object-wise $\beta$-spread transformation, and it allows clustering to be performed while retaining the original relational data matrix or kernel Gram matrix to the maximum possible extent. Because $\beta$ is vector valued, we cannot determine its minimal value such that the dissimilarities between elements in the data set and cluster centers would be non-negative. Hence, we consider determining this vector for the case where the dissimilarities are non-negative, minimizing the squared Frobenius norms of the difference between the original matrix and the object-wise $\beta$-spread transformed matrix, which can be achieved by solving a quadratic programming problem. The proposed methods retain the given data better than previous methods, and so we expect them to produce better clustering results. Numerical examples show that this is the case.

The remainder of this paper is organized as follows. In Section 2, we introduce some conventional FCM methods. In Section 3, we propose two clustering algorithms: RFCM using a quadratic programming-based object-wise $\beta$-spread transformation (qO-NERFCM) and K-FCM using a quadratic programming-based object-wise $\beta$-spread transformation (qO-IK-FCM). In Section 4, we present some numerical examples, and conclude this paper in Section 5.

## 2 Preliminaries

In this section, we introduce RFCM, NERFCM, K-FCM, and IK-FCM. RFCM and K-FCM provide the basic methodology for NERFCM and IK-FCM, which apply a $\beta$-spread transformation to non-Euclidean relational data and indefinite kernel Gram matrices, respectively.

## 2.1 RFCM and NERFCM

For a given data set $X = \{x_k \mid k \in \{1,\ldots,N\}\}$, the dissimilarity $R_{k,j}$ between $x_k$ and $x_j$ is given. Here, $R$ is a matrix whose $(k,j)$-th element is $R_{k,j}$. Let $C$ denote the cluster number. The goal of RFCM and NERFCM is obtaining the membership by which the datum $x_k$ belongs to the $i$-th cluster, denoted by $u_{i,k}$, from $R$. $u \in \mathbb{R}^{C \times N}$ is referred to as the partition matrix.

RFCM is obtained by solving the optimization problem

$$\text{minimize}_u \sum_{i=1}^{C} \frac{\sum_{k=1}^{N} \sum_{j=1}^{N} u_{i,k}^m u_{j,k}^m R_{k,j}}{2 \sum_{t=1}^{N} u_{i,t}^m}, \tag{1}$$

$$\text{subject to} \sum_{i=1}^{C} u_{i,k} = 1, \tag{2}$$

where $m > 1$ is a fuzzifier parameter. The RFCM procedure is as follows.

---

**1**

STEP 1. Fix $m > 1$ and assume an initial partition matrix $u$.
STEP 2. Update $v_i \in \mathbb{R}^N$ as

$$v_i = \left( u_{i,1}^m, \cdots, u_{i,N}^m \right)^{\mathsf{T}} / \sum_{k=1}^{N} u_{i,k}^m. \tag{3}$$

STEP 3. Update $d_{i,k}$ as

$$d_{i,k} = (Rv_i)_k - v_i^{\mathsf{T}} Rv_i / 2. \tag{4}$$

STEP 4. Update the membership as

$$u_{i,k} = 1 / \sum_{j=1}^{C} \left( d_{i,k}/d_{j,k} \right)^{1/(m-1)}. \tag{5}$$

STEP 5. If the stopping criterion is satisfied, terminate this algorithm. Otherwise, return to STEP 2.

---

We say that a matrix $R \in \mathbb{R}^{N \times N}$ is *Euclidean* if there exists a set of points $\{y_1, \cdots, y_N\} \in \mathbb{R}^{N-1}$ such that $R_{k,j} = \|y_k - y_j\|_2^2$, and *non-Euclidean* if no such set of points exists. $R$ is Euclidean if and only if $HRH$ is negative semi-definite for $H = E - \mathbf{1}\mathbf{1}^{\mathsf{T}}/N$, where $E$ is the $N$-dimensional unit matrix, and $\mathbf{1}$ is an $N$-dimensional vector whose elements are all 1. For a non-Euclidean $R$, RFCM only works when the positive eigenvalues of $HRH$ are not particularly large. However, RFCM fails for a non-Euclidean $R$ when the positive eigenvalues of $HRH$ are extremely large because the membership cannot be calculated after the value of $d_{i,k}$ is updated to a negative value.

In order to overcome this limitation, the following modification of $R$, called the $\beta$-spread transformation, has been developed [3]:

$$R_\beta = R + \beta(\mathbf{1}\mathbf{1}^{\mathsf{T}} - E), \tag{6}$$

where $\beta$ is a positive scalar value. With this $\beta$-spread transformation, NERFCM is given by the following algorithm.

1

STEP 1. Fix $m > 1$ and assume an initial partition matrix $u$. Set $\beta = 0$.

STEP 2. Execute STEP 2 of Algorithm 1.

STEP 3. Update $d_{i,k}$ as

$$d_{i,k} = \left(R_\beta v_i\right)_k - v_i^{\mathsf{T}} R_\beta v_i / 2. \tag{7}$$

STEP 4. If $d_{i,k} < 0$, update $\Delta\beta, d_{i,k}$, and $\beta$ as

$$\Delta\beta = \max\{-2d_{i,k}/\|e_k - v_i\|^2\}, \tag{8}$$
$$d_{i,k} \leftarrow d_{i,k} + \Delta\beta/2\|e_k - v_i\|^2, \tag{9}$$
$$\beta \leftarrow \beta + \Delta\beta. \tag{10}$$

STEP 5. Execute STEP 4 of Algorithm 1.

STEP 6. If the stopping criterion is satisfied, terminate this algorithm. Otherwise, return to STEP 2.

---

Another option for tackling non-Euclidean relational data is to apply RFCM to a set of Euclidean relational data $R'$, that has been modified from $R$ in the following ways. The first $R'$ is obtained by:

$$R'_{k,j} = K'_{k,k} - 2K'_{k,j} + K'_{j,j}, \tag{11}$$

where $K'$ is the positive semi-definite matrix obtained from $K = -(1/2)HRH$ by subtracting the scaled identity matrix with its minimal eigenvalue if it is negative, that is,

$$K' = K - \lambda_{\min}E \quad (\lambda_{\min} < 0), \tag{12}$$

where $\lambda_{\min}$ is the minimal eigenvalue of $K$. In this paper, we refer to this revision as "diagonal shift" (DS), and its application to RFCM as RFCM-DS. The second $R'$ is obtained by Eq. (11), when $K'$ is the positive semi-definite matrix formed from $K = -(1/2)HRH$ by setting all the negative eigenvalues to zero. We refer to this modification as "nearest positive semi-definite" (nPSD), and thus, its application to RFCM is denoted as RFCM-nPSD.

In the NERFCM algorithm, $\beta$ is adaptively determined at STEP 4; hence, the modification from $R$ to $R_\beta$ is suppressed to a minimum such that the algorithm execution can continue, whereas DS and nPSD may cause an over-transformation, only allowing the execution of RFCM. Indeed, it has been reported that RFCM-DS causes the memberships to become extremely fuzzy [3].

## 2.2 K-FCM and IK-FCM

For a given data set $X = \{x_k \mid k \in \{1,\ldots,N\}\}$, K-FCM assumes that the kernel matrix $K \in \mathbb{R}^{N \times N}$ is given. Let $\mathbb{H}$ be a higher-dimensional feature space, $\Phi : X \to \mathbb{H}$ be a map from the data set $X$ to the feature space $\mathbb{H}$, and $W = \{W_i \in \mathbb{H} \mid i \in \{1,\cdots,C\}\}$ be a set of cluster centers in the feature space.

K-FCM is obtained by solving the following optimization problem:

$$\text{minimize}_{u,W} \sum_{i=1}^{C} \sum_{k=1}^{N} u_{i,k}^{m} \| \Phi(x_k) - W_i \|_{\mathbb{H}}^{2} \tag{13}$$

subject to Eq. (2). Generally, $\Phi$ cannot be given explicitly, so the K-FCM algorithm assumes that a kernel function $\mathcal{K} : x \times x \to \mathbb{R}$ is given. This function describes the inner product value of the pairs of elements in the data set of the feature space as $\mathcal{K}(x_k, x_j) = \langle \Phi(x_k), \Phi(x_j) \rangle$. However, it can be interpreted that $\Phi$ is given explicitly by allowing $\mathbb{H} = \mathbb{R}^{N}$, $\Phi(x_k) = e_k$, where $e_k$ is the $N$-dimensional unit vector whose $\ell$-th element is the Kronecker delta $\delta_{k,\ell}$, and by introducing $K \in \mathbb{R}^{N \times N}$ such that

$$K_{k,j} = \langle \Phi(x_k), \Phi(x_j) \rangle. \tag{14}$$

According to this discussion, K-FCM is given as follows.

---

1

STEP 1. Fix $m > 1$. Assume a kernel matrix $K \in \mathbb{R}^{N \times N}$ and an initial partition matrix $u$.

STEP 2. Update cluster centers as

$$W_i = \left(u_{i,1}^{m}, \cdots, u_{i,N}^{m}\right)^{\mathsf{T}} / \sum_{k=1}^{N} u_{i,k}^{m}. \tag{15}$$

STEP 3. Update the dissimilarity between each element in the data set and the cluster center as

$$d_{i,k} = (e_k - W_i)^{\mathsf{T}} K (e_k - W_i). \tag{16}$$

STEP 4. Update the membership as

$$u_{i,k} = 1 / \sum_{j=1}^{C} \left(d_{i,k}/d_{j,k}\right)^{1/(m-1)} \tag{17}$$

STEP 5. If $(u, d, W)$ converge, terminate this algorithm. Otherwise, return to STEP 2.

---

K-FCM is constructed based on Eq. (14), i.e., $K$ is positive semi-definite. Even so, $K$ is sometimes introduced without the existence of $\Phi$ being guaranteed. In this case, $K$ is not always positive semi-definite. Similar to RFCM, K-FCM works for an indefinite $K$ when the magnitude of negative eigenvalues is not particularly large. However, K-FCM fails for indefinite $K$ when the magnitude of negative eigenvalues is extremely large, because the memberships cannot be calculated after the dissimilarity between a datum and a cluster center is updated as a negative value. In order to overcome this limitation, the following $\beta$-spread transformation of $K$ has been developed [8]:

$$K_{\beta} = K + \beta E. \tag{18}$$

With this $\beta$-spread transformation, IK-FCM is given by the following algorithm.

---

1

STEP 1. Fix $m > 1$ for K-FCM. Assume a kernel matrix $K \in \mathbb{R}^{N \times N}$ and an initial partition matrix $u$. Set $\beta = 0$ and $K_0 = K$.
STEP 2. Execute STEP 2 of Algorithm 1.
STEP 3. Update $d_{i,k}$ as

$$d_{i,k} = (e_k - W_i)^\mathsf{T} K_\beta (e_k - W_i). \tag{19}$$

STEP 4. If $d_{i,k} < 0$, update $\Delta\beta, d_{i,k}, \beta$, and $K_\beta$ as:

$$\Delta\beta = \max\{-d_{i,k}/\|e_k - W_i\|_2^2\}, \tag{20}$$
$$d_{i,k} \leftarrow d_{i,k} + \Delta\beta \|e_k - W_i\|^2, \tag{21}$$
$$\beta \leftarrow \beta + \Delta\beta, \tag{22}$$
$$K_\beta \leftarrow K_\beta + \Delta\beta E. \tag{23}$$

STEP 5. Execute STEP 4 of Algorithm 1.
STEP 6. If the stopping criterion is satisfied, terminate this algorithm. Otherwise, return to STEP 2.

---

Another option for handling indefinite kernel data is to apply K-FCM to a positive semi-definite matrix $K'$, which is modified from $K$ in the following two ways. The first $K'$ is obtained from $K$ by adding the scaled identity matrix with its minimal eigenvalue if it is negative, that is,

$$K' = K + \lambda_{\min} E \quad (\lambda_{\min} < 0), \tag{24}$$

where $\lambda_{\min}$ is the minimal eigenvalue of $K$. As for RFCM, we refer to this revision as "diagonal shift" (DS), and its application to K-FCM is thus K-FCM-DS. The second $K'$ is obtained from $K$ by setting all the negative eigenvalues to zero, and thus K-FCM becomes K-FCM-nPSD.

In the IK-FCM algorithm, $\beta$ is adaptively determined at STEP 4; hence, the modification from $K$ to $K_\beta$ is suppressed to a minimum such that the algorithm execution can continue, whereas DS and nPSD may cause an over-transformation, only allowing the execution of K-FCM.

## 3 Quadratic Programming-Based Object-Wise $\beta$-Spread Fuzzy Clustering

### 3.1 Concept of the Proposed Algorithms

In the conventional $\beta$-spread transformation given by Eq. (6) for NERFCM or Eq. (18) for IK-FCM, the modified $\beta$ value is common to all objects in the given relational data matrix or kernel Gram matrix. In this paper, we propose that a different value is added to each object in the given matrices. We refer to this as an object-wise $\beta$-spread transformation, and it allows clustering to be performed while retaining the original relational data matrix or kernel Gram matrix to the maximum possible extent. The object-wise $\beta$-spread transformation for RFCM is

$$R_\beta = R + \frac{1}{2}\beta\mathbf{1}^\mathsf{T} + \frac{1}{2}\mathbf{1}\beta^\mathsf{T} - \mathrm{diag}(\beta), \tag{25}$$

and that for K-FCM is

$$K_\beta = K + \mathrm{diag}(\beta), \tag{26}$$

where $\beta \in \mathbb{R}_+^N$. If all the elements of $\beta$ are the same, then the object-wise $\beta$-spread transformation is identical to that in NERFCM and IK-FCM.

Because $\beta$ is vector valued, we cannot determine its minimal value such that the dissimilarities between elements in the data set and cluster centers would be non-negative. Therefore, we consider determining $\beta$ for the case where the dissimilarities are non-negative, minimizing the squared Frobenius norms $\|R_\beta - R\|_\mathrm{F}^2$ and $\|K_\beta - K\|_\mathrm{F}^2$, which can be achieved by solving a quadratic programming problem.

## 3.2 RFCM Using a Quadratic Programming-Based Object-Wise $\beta$-Spread Transformation

Using RFCM with an object-wise $\beta$-spread transformation, the following condition must be satisfied in order for the dissimilarities between the elements in the data set and cluster centers to be non-negative:

$$-\frac{1}{2}(e_k - v_i)^\mathsf{T} R_\beta (e_k - v_i) \geq 0 \tag{27}$$

$$\Leftrightarrow -\frac{1}{2}(e_k - v_i)^\mathsf{T} R_0 (e_k - v_i) - \frac{1}{4}(e_k - v_i)^\mathsf{T}\beta\mathbf{1}^\mathsf{T}(e_k - v_i)$$
$$- \frac{1}{4}(e_k - v_i)^\mathsf{T}\mathbf{1}\beta^\mathsf{T}(e_k - v_i)$$
$$+ \frac{1}{2}(e_k - v_i)^\mathsf{T}\mathrm{diag}(\beta)(e_k - v_i) \geq 0 \tag{28}$$

$$\Leftrightarrow d_{i,k} + \frac{1}{2}\sum_{\ell=1}^N (e_k^{(\ell)} - v_i^{(\ell)})^2 \beta_\ell \geq 0, \tag{29}$$

where $e_k^{(\ell)}$ and $v_i^{(\ell)}$ are the $\ell$-th element of $e_k$ and $v_i$, respectively. Under this condition, the value of $\beta$ that minimizes $\|R_\beta - R\|_\mathrm{F}^2$ can be obtained by solving the following quadratic programming problem:

$$\mathrm{minimize}_\beta \frac{1}{2}\beta^\mathsf{T} A \beta \tag{30}$$

$$\text{subject to } d_{i,k} + \frac{1}{2}\sum_{\ell=1}^N (e_k^{(\ell)} - v_i^{(\ell)})^2 \beta_\ell \geq 0 \quad (k \in \{1, \cdots, N\}, i \in \{1, \cdots, C\}), \tag{31}$$

where

$$A_{k,j} = \begin{cases} N - 1 & (k = j), \\ 1 & (k \neq j). \end{cases} \tag{32}$$

Using the obtained value of $\beta$, we can describe the dissimilarity between the datum $x_k$ and the cluster center $v_i$ as

$$d_{i,k}(\boldsymbol{\beta}) = d_{i,k}(0) + \frac{1}{2} \sum_{\ell=1}^{N} (e_k^{(\ell)} - v_i^{(\ell)})^2 \boldsymbol{\beta}_{\ell}. \tag{33}$$

If we set a tentative value of $\boldsymbol{\beta}$, and obtain the modified value of $\boldsymbol{\beta} + \Delta\boldsymbol{\beta}$ satisfying the above constraint, we need only solve the following quadratic programming problem for $\Delta\boldsymbol{\beta}$.

$$\text{minimize}_{\Delta\boldsymbol{\beta}} \frac{1}{2} \Delta\boldsymbol{\beta}^{\mathsf{T}} A \Delta\boldsymbol{\beta} \tag{34}$$

$$\text{subject to } d_{i,k}(\boldsymbol{\beta}) + \frac{1}{2} \sum_{\ell=1}^{N} (e_k^{(\ell)} - v_i^{(\ell)})^2 \Delta\boldsymbol{\beta}_{\ell} \geq 0$$

$$(k \in \{1, \cdots, N\}, i \in \{1, \cdots, C\}). \tag{35}$$

Hence, we set $\boldsymbol{\beta}$ to 0 at the beginning of the algorithm and then modify $\boldsymbol{\beta}$ by the value of $\Delta\boldsymbol{\beta}$ obtained from the above programming problem, provided that at least one of dissimilarities between a datum and a cluster center is non-negative while the algorithm execution continues. On the basis of the above, we modify the NERFCM algorithm to the following quadratic programming-based object-wise $\beta$-spread NERFCM (qO-NERFCM).

---

1

STEP 1. Fix $m > 1$ and assume an initial partition matrix $u$. Set $\boldsymbol{\beta} = \Delta\boldsymbol{\beta} = 0$.

STEP 2. Update the cluster center $v_i \in \mathbb{R}^N$ as

$$v_i = \left(u_{i,1}^m, \cdots, u_{i,N}^m\right)^{\mathsf{T}} / \sum_{k=1}^{N} u_{i,k}^m. \tag{36}$$

STEP 3. Update the dissimilarity between data and cluster centers $d_{i,k}$ as

$$d_{i,k} = \left(R_\beta v_i\right)_k - v_i^{\mathsf{T}} R_\beta v_i / 2. \tag{37}$$

STEP 4. If $d_{i,k} < 0$, solve the quadratic programming problem for $\Delta\boldsymbol{\beta}$

$$\text{minimize}_{\boldsymbol{\beta}} \frac{1}{2} \Delta\boldsymbol{\beta}^{\mathsf{T}} A \Delta\boldsymbol{\beta} \tag{38}$$

$$\text{subject to } d_{i,k}(\boldsymbol{\beta}) - \frac{1}{2} \sum_{\ell=1}^{N} (e_k^{(\ell)} - v_i^{(\ell)})^2 \Delta\boldsymbol{\beta}_{\ell} \geq 0$$

$$(k \in \{1, \cdots, N\}, i \in \{1, \cdots, C\}) \tag{39}$$

and update $d_{i,k}$ and $\boldsymbol{\beta}$ as

$$d_{i,k} \leftarrow d_{i,k} + \frac{1}{2} \|e_k - v_i\|_{\boldsymbol{\beta}}^2, \tag{40}$$

$$\boldsymbol{\beta} \leftarrow \boldsymbol{\beta} + \Delta\boldsymbol{\beta}. \tag{41}$$

STEP 5. Update the membership $u_{i,k}$ as

$$u_{i,k} = 1 / \sum_{j=1}^{C} (d_{i,k}/d_{j,k})^{1/(m-1)}. \tag{42}$$

STEP 6. If the stopping criterion is satisfied, terminate this algorithm. Otherwise, return to STEP 2.

---

Determining $\Delta\boldsymbol{\beta}$ in conventional NERFCM is identical to solving the quadratic programming problem given by Eqs. (38) and (39) with the additional constraint $\boldsymbol{\beta}_k = \boldsymbol{\beta}_j$ $(k \neq j)$, because the objective function $\boldsymbol{\beta}^\mathsf{T} A \boldsymbol{\beta}$ becomes $\frac{1}{2}\mathbf{1}^\mathsf{T} A \mathbf{1} \beta^2$, resulting in the expression given in Eq. (8). The constraints for $\beta$ in Eqs. (30), (31) are more relaxed in qO-NERFCM than in conventional NERFCM, and hence qO-NERFCM achieves a lower objective function value than conventional NERFCM.

## 3.3 K-FCM Using Quadratic Programming-Based Object-Wise $\beta$-Spread

Using a quadratic programming-based object-wise $\beta$-spread transformation in K-FCM, the following condition must be satisfied in order for the dissimilarities between data and cluster centers to be non-negative:

$$(e_k - v_i)^\mathsf{T} K_{\boldsymbol{\beta}}(e_k - v_i) \geq 0 \tag{43}$$
$$\Leftrightarrow (e_k - v_i)^\mathsf{T} K_0(e_k - v_i) + (e_k - v_i)^\mathsf{T} \text{diag}(\boldsymbol{\beta})(e_k - v_i) \geq 0 \tag{44}$$
$$\Leftrightarrow d_{i,k} + \sum_{\ell=1}^{N} (e_k^{(\ell)} - v_i^{(\ell)})^2 \boldsymbol{\beta}_\ell \geq 0. \tag{45}$$

Under this condition, the value of $\boldsymbol{\beta}$ that minimizes $\|K_\beta - K\|_F^2$ can be obtained by solving the following quadratic programming problem.

$$\text{minimize}_{\boldsymbol{\beta}} \, \boldsymbol{\beta}^\mathsf{T} \boldsymbol{\beta} \tag{46}$$
$$\text{subject to } d_{i,k} + \sum_{\ell=1}^{N} (e_k^{(\ell)} - v_i^{(\ell)})^2 \boldsymbol{\beta}_\ell \geq 0 \quad (k \in \{1, \cdots, N\}, i \in \{1, \cdots, C\}) \tag{47}$$

Using the obtained value of $\boldsymbol{\beta}$, we can describe the dissimilarity between the datum $x_k$ and the cluster center $v_i$ as

$$d_{i,k}(\boldsymbol{\beta}) = d_{i,k}(0) + \sum_{\ell=1}^{N} (e_k^{(\ell)} - v_i^{(\ell)})^2 \boldsymbol{\beta}_\ell \tag{48}$$

If we set a tentative value of $\boldsymbol{\beta}$, and obtain the modified value of $\boldsymbol{\beta} + \Delta\boldsymbol{\beta}$ satisfying the above constraint, we need only solve the following quadratic programming problem for $\Delta\boldsymbol{\beta}$.

$$\text{minimize}_{\Delta\boldsymbol{\beta}} \, \Delta\boldsymbol{\beta}^\mathsf{T} \Delta\boldsymbol{\beta} \tag{49}$$
$$\text{subject to } d_{i,k}(\boldsymbol{\beta}) + \frac{1}{2} \sum_{\ell=1}^{N} (e_k^{(\ell)} - v_i^{(\ell)})^2 \Delta\boldsymbol{\beta}_\ell \geq 0$$
$$(k \in \{1, \cdots, N\}, i \in \{1, \cdots, C\}) \tag{50}$$

Hence, we set $\boldsymbol{\beta}$ to 0 at the beginning of the algorithm and then modify $\boldsymbol{\beta}$ using the value of $\Delta\boldsymbol{\beta}$ obtained from the above programming problem, provided that at least

one of the dissimilarities between a datum and a cluster center is non-negative while
algorithm execution continues. On the basis of these discussions, we modify IK-
FCM to obtain the following algorithm of the quadratic programming-based object-
wise $\beta$-spread IK-FCM (qO-IK-FCM).

---

1

STEP 1. Fix $m > 1$ and assume an initial partition matrix $u$. Set
$\beta = \Delta\beta = 0$.

STEP 2. Update the cluster center $v_i \in \mathbb{R}^N$ as

$$W_i = \left(u_{i,1}^m, \cdots, u_{i,N}^m\right)^{\mathsf{T}} / \sum_{k=1}^{N} u_{i,k}^m. \tag{51}$$

STEP 3. Update the dissimilarity $d_{i,k}$ between the datum $x_k$ and the
cluster centers $W_i$ as

$$d_{i,k} = (e_k - W_i)^{\mathsf{T}} K_\beta (e_k - W_i) \tag{52}$$

STEP 4. If $d_{i,k} < 0$, solve the following quadratic programming problem
for $\Delta\beta$:

$$\text{minimize}_{\Delta\beta}\, \Delta\beta^{\mathsf{T}} \Delta\beta \tag{53}$$

$$\text{subject to } d_{i,k}(\beta) + \sum_{\ell=1}^{N} (e_k^{(\ell)} - v_i^{(\ell)})^2 \Delta\beta_\ell \geq 0 \quad (k \in \{1, \cdots, N\}, i \in \{1, \cdots, C\}) \tag{54}$$

and update $d_{i,k}$ and $\beta$ as

$$d_{i,k} \leftarrow d_{i,k} + \|e_k - v_i\|_\beta^2, \tag{55}$$

$$\beta \leftarrow \beta + \Delta\beta. \tag{56}$$

STEP 5. Update the membership $u_{i,k}$ as

$$u_{i,k} = 1 / \sum_{j=1}^{C} (d_{i,k}/d_{j,k})^{1/(m-1)}. \tag{57}$$

STEP 6. If the stopping criterion is satisfied, terminate this algorithm.
Otherwise, return to STEP 2.

---

Determining $\Delta\beta$ in conventional IK-FCM is identical to solving the quadratic
programming problem in Eqs. (53) and (54) with the additional constraint $\beta_k =
\beta_j$ $(k \neq j)$, because the objective function $\beta^{\mathsf{T}}\beta$ becomes $N\beta^2$, resulting in the ex-
pression given by Eq. (20). The constraints for $\beta$ given by the programming problem
in Eqs. (46), (47) are more relaxed in qO-IK-FCM than in conventional IK-FCM,
and hence qO-IK-FCM achieves a lower objective function value than conventional
IK-FCM.

## 3.4 Relation between RFCM and K-FCM, NERFCM and IK-FCM, and qO-NERFCM and qO-IK-FCM

RFCM and K-FCM are identical when $R$ and $K$ are constructed as $R_{k,j} = \|x_k - x_j\|_2^2$ and $K_{k,j} = (x_k - \bar{x})^\mathsf{T}(x_j - \bar{x})$, where $\bar{x} = (\sum_{k=1}^N x_k)/N$, from the same vector-data $\{x_k\}_{k=1}^N$. This identity can been shown by the Young–Householder transformation:

$$K = -HRH/2. \tag{58}$$

Furthermore, NERFCM and IK-FCM are identical when $K$ and $R$ satisfy Eq. (58), i.e.,

$$R_{k,j} = K_{k,k} - 2K_{k,j} + K_{j,j}. \tag{59}$$

Similarly, qO-NERFCM and qO-IK-FCM are identical when $K$ and $R$ satisfy Eq. (58) or (59).

## 4 Numerical Examples

In this section, we numerically compare the proposed methods with conventional methods for clustering data sets in order to show the efficiency of the proposed methods compared to others which are the most popular ones for the task.

The qO-NERFCM, NERFCM, RFCM-DS, and RFCM-nPSD methods for clustering relational data are applied to the following three data sets. The first is obtained from the vector data of two ball-shaped clusters, in which each ball is composed of 50 points in a two-dimensional space, as shown in Fig. 1, with $R_{k,j} = \|x_k - x_j\|_1^2$, which is known to be non-Euclidean. We call this data set "L1 2-balls" and set $N = 100$ and $C = 2$. This is the most commonly used way (as in [3] and [8]) to obtain artificial non-Euclidean relational data from vectorial data. The second data set, "Tamura" [9], shows the relation between 16 people from three families; thus, $N = 16$ and $C = 3$ for this data set. The third data set is called "cat-cortex" [10], and shows the relation between 65 cortical areas of a cat with four functional regions; thus, $N = 65$ and $C = 4$. It is important to experimentally test the second and the third data sets because they are not only from the real world but also are essentially non-Euclidean relational data than other non-Euclidean relational data obtained by artificially measuring dissimilarity between objects in vectorial data. Thus, the validity of the proposed methods can be shown by comparing the proposed methods with the conventional methods using such non-Euclidean relational data. We use each data set with two settings: one without any changes, and the other after subtracting the minimal value of the off-diagonal elements, which makes the set less Euclidean. Hence, we apply six data sets to each clustering method.

Three indices are used to compare the clustering results. The first is the normalized mutual information (NMI) [11], which is used for evaluating clustering results and for which a higher value is preferred. The second is the partition coefficients (PC) [12]. This evaluates the hardness of clustering, and again a higher value is preferred. Finally, the squared Frobenius norm $\|R_\beta - R\|_\mathrm{F}^2$, is used to evaluate the

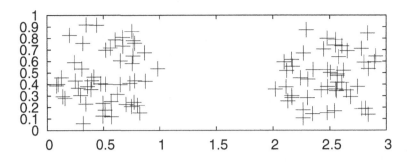

**Fig. 1** L1 2-ball data set

ability to retain the given data in relational clustering, and a lower value of this index is preferred.

For kernel data, we compared the qO-IK-FCM, IK-FCM, K-FCM-DS, and K-FCM-nPSD clustering methods. We apply these to the six sets of kernel data obtained from the relational data sets described above using $K = -1/2HRH$. From the relations described in the previous section, RFCM and K-FCM produce the same clustering results when $K = -1/2HRH$; similarly, NERFCM and IK-FCM, and qO-NERFCM and qO-IK-FCM, will give the same results. We use the squared Frobenius norm $\|K_\beta - K\|_F^2$ as an index for evaluating retention of the given data in the kernelized clusterings, where a lower value is again preferred.

In each method, and for each data set, we selected the fuzzifier parameter $m \in \{2, 1.2\}$, performed 100 trials with different initial membership values, and selected the solution with the minimal objective function value. The results for each data set are summarized in Tables 1–6. In each table, the four indices described above are listed for eight methods with two fuzzifier parameters.

From Tables 1–6, we can see that the proposed methods produce better results than the previous methods. In RFCM-DS and K-FCM-DS, $\|R_\beta - R\|_F^2$ and $\|K_\beta - K\|_F^2$ are much larger than in other methods, and hence the partition coefficients tend to be fuzzy. Sometimes, all the membership values are equal to $1/C$, i.e., the NMI values are 0. This confirms the claim in [3],[8] that extremely large $\beta$-spread transformation values produce very fuzzy clustering results. For the other six methods, the NMI values are similar; however, in RFCM-nPSD and K-FCM-nPSD, $\|R_\beta - R\|_F^2$ and $\|K_\beta - K\|_F^2$ are considerably larger than in the other four methods, which shows that these techniques cannot retain the given data. Upon comparing the proposed qO-NERFCM and qO-IK-FCM methods with NERFCM and IK-FCM, respectively, we find that the performance indices are similar, although those of the proposed methods are always better. In particular, the NMI index of the proposed methods with $m = 1.2$ for the modified cat-cortex data set is better, as shown in Table 6, because, in our opinion, the proposed methods can retain the given data better than the other methods.

**Table 1** Results for L1 2-ball data set

| | $m$ | RFCM-DS or K-FCM-DS | RFCM-nPSD or K-FCM-nPSD | NERFCM or IK-FCM | qO-NERFCM or qO-IK-FCM |
|---|---|---|---|---|---|
| NMI | 2.0 | 0.00 | **1.00** | **1.00** | **1.00** |
| | 1.2 | **1.00** | **1.00** | **1.00** | **1.00** |
| PC | 2.0 | 0.50 | 0.86 | 0.93 | 0.93 |
| | 1.2 | 0.62 | **0.99** | **0.99** | **0.99** |
| $\|R_\beta - R\|_F^2$ | 2.0 | $2.54 \times 10^3$ | $5.99 \times 10^2$ | **0.00** | **0.00** |
| | 1.2 | $2.54 \times 10^3$ | $5.99 \times 10^2$ | **0.00** | **0.00** |
| $\|K_\beta - K\|_F^2$ | 2.0 | $1.23 \times 10^2$ | $1.31 \times 10^1$ | **0.00** | **0.00** |
| | 1.2 | $1.23 \times 10^2$ | $1.31 \times 10^1$ | **0.00** | **0.00** |

**Table 2** Results for modified L1 2-ball data set

| | $m$ | RFCM-DS or K-FCM-DS | RFCM-nPSD or K-FCM-nPSD | NERFCM or IK-FCM | qO-NERFCM or qO-IK-FCM |
|---|---|---|---|---|---|
| NMI | 2.0 | 0.00 | **1.00** | **1.00** | **1.00** |
| | 1.2 | **1.00** | **1.00** | **1.00** | **1.00** |
| PC | 2.0 | 0.50 | 0.86 | 0.93 | 0.93 |
| | 1.2 | 0.62 | **0.99** | **0.99** | **0.99** |
| $\|R_\beta - R\|_F^2$ | 2.0 | $2.54 \times 10^3$ | $5.99 \times 10^2$ | **0.00** | **0.00** |
| | 1.2 | $2.54 \times 10^3$ | $5.99 \times 10^2$ | **0.00** | **0.00** |
| $\|K_\beta - K\|_F^2$ | 2.0 | $1.23 \times 10^2$ | $1.31 \times 10^1$ | **0.00** | **0.00** |
| | 1.2 | $1.23 \times 10^2$ | $1.31 \times 10^1$ | **0.00** | **0.00** |

**Table 3** Results for Tamura data set

| | $m$ | RFCM-DS or K-FCM-DS | RFCM-nPSD or K-FCM-nPSD | NERFCM or IK-FCM | qO-NERFCM or qO-IK-FCM |
|---|---|---|---|---|---|
| NMI | 2.0 | 0.00 | **1.00** | **1.00** | **1.00** |
| | 1.2 | **1.00** | **1.00** | **1.00** | **1.00** |
| PC | 2.0 | 0.33 | 0.43 | 0.48 | 0.48 |
| | 1.2 | 0.93 | **0.96** | **0.96** | **0.96** |
| $\|R_\beta - R\|_F^2$ | 2.0 | 6.91 | 0.88 | **0.00** | **0.00** |
| | 1.2 | 6.91 | 0.88 | **0.00** | **0.00** |
| $\|K_\beta - K\|_F^2$ | 2.0 | 0.90 | 0.25 | **0.00** | **0.00** |
| | 1.2 | 0.90 | 0.25 | **0.00** | **0.00** |

**Table 4** Results for modified Tamura data set

| | $m$ | RFCM-DS or K-FCM-DS | RFCM-nPSD or K-FCM-nPSD | NERFCM or IK-FCM | qO-NERFCM or qO-IK-FCM |
|---|---|---|---|---|---|
| NMI | 2.0 | 0.00 | **1.00** | **1.00** | **1.00** |
| | 1.2 | **1.00** | **1.00** | **1.00** | **1.00** |
| PC | 2.0 | 0.33 | 0.50 | 0.56 | 0.61 |
| | 1.2 | 0.93 | 0.96 | **0.97** | **0.97** |
| $\|R_\beta - R\|_F^2$ | 2.0 | $1.00 \times 10^1$ | 1.67 | 1.56 | $\mathbf{2.16 \times 10^{-1}}$ |
| | 1.2 | $1.00 \times 10^1$ | 1.67 | 1.56 | $\mathbf{1.44 \times 10^{-1}}$ |
| $\|K_\beta - K\|_F^2$ | 2.0 | 1.30 | $3.87 \times 10^{-1}$ | $2.72 \times 10^{-2}$ | $\mathbf{1.17 \times 10^{-2}}$ |
| | 1.2 | 1.30 | $3.87 \times 10^{-1}$ | $2.72 \times 10^{-2}$ | $\mathbf{2.53 \times 10^{-3}}$ |

**Table 5** Results for cat-cortex data set

| | $m$ | RFCM-DS or K-FCM-DS | RFCM-nPSD or K-FCM-nPSD | NERFCM or IK-FCM | qO-NERFCM or qO-IK-FCM |
|---|---|---|---|---|---|
| NMI | 2.0 | 0.00 | 0.00 | 0.00 | 0.00 |
| | 1.2 | 0.00 | 0.70 | **0.73** | **0.73** |
| PC | 2.0 | 0.25 | 0.25 | 0.25 | 0.25 |
| | 1.2 | 0.25 | 0.59 | **0.66** | **0.66** |
| $\|R_\beta - R\|_F^2$ | 2.0 | $3.18 \times 10^2$ | $3.44 \times 10^1$ | **0.00** | **0.00** |
| | 1.2 | $3.18 \times 10^2$ | $3.44 \times 10^1$ | **0.00** | **0.00** |
| $\|K_\beta - K\|_F^2$ | 2.0 | $1.99 \times 10^1$ | $4.69 \times 10^0$ | **0.00** | **0.00** |
| | 1.2 | $1.99 \times 10^1$ | $4.69 \times 10^0$ | **0.00** | **0.00** |

**Table 6** Results for modified cat-cortex data set

| | $m$ | RFCM-DS or K-FCM-DS | RFCM-nPSD or K-FCM-nPSD | NERFCM or IK-FCM | qO-NERFCM or qO-IK-FCM |
|---|---|---|---|---|---|
| NMI | 2.0 | 0.00 | 0.00 | 0.00 | 0.00 |
| | 1.2 | 0.00 | 0.77 | 0.84 | **0.87** |
| PC | 2.0 | 0.25 | 0.25 | 0.25 | 0.25 |
| | 1.2 | 0.25 | 0.68 | 0.77 | **0.78** |
| $\|R_\beta - R\|_F^2$ | 2.0 | $3.82 \times 10^2$ | $5.28 \times 10^1$ | $2.65 \times 10^0$ | $\mathbf{5.47 \times 10^{-1}}$ |
| | 1.2 | $3.82 \times 10^2$ | $5.28 \times 10^1$ | $2.65 \times 10^0$ | $\mathbf{5.45 \times 10^{-1}}$ |
| $\|K_\beta - K\|_F^2$ | 2.0 | $2.39 \times 10^1$ | $6.39 \times 10^0$ | $3.03 \times 10^{-1}$ | $\mathbf{2.32 \times 10^{-2}}$ |
| | 1.2 | $2.39 \times 10^1$ | $6.39 \times 10^0$ | $3.03 \times 10^{-1}$ | $\mathbf{2.30 \times 10^{-2}}$ |

# 5 Conclusion

In this paper, we proposed a quadratic programming-based object-wise $\beta$-spread transformation for use in relational fuzzy $c$-means clustering and kernelized fuzzy $c$-means clustering. Our method differs from conventional methods (NERFCM and IK-FCM) in that the modified $\beta$ value is not common to all objects in the given relational data matrix or kernel Gram matrix. Comparing the squared Frobenius norm of the given data and the modified data as an evaluation index, we theoretically proved that the proposed methods retain the given data better than conventional methods. Furthermore, our numerical experiments established that the proposed methods produce the same, or sometimes better, results in terms of three evaluation indices.

**Acknowledgment.** This work has partly been supported by the Grant-in-Aid for Scientific Research, Japan Society for the Promotion of Science, No. 00298176.

# References

1. Bezdek, J.C.: Pattern Recognition with Fuzzy Objective Function Algorithms. Plenum, New York (1981)
2. Hathaway, R.J., Davenport, J.W., Bezdek, J.C.: Relational Duals of the $c$-Means Clustering Algorithms. Pattern Recognition 22(2), 205–212 (1989)
3. Hathaway, R.J., Bezdek, J.C.: NERF C-means: Non-Euclidean Relational Fuzzy Clustering. Pattern Recognition 27, 429–437 (1994)
4. Miyamoto, S., Suizu, D.: Fuzzy $c$-Means Clustering Using Kernel Functions in Support Vector Machines. J. Advanced Computational Intelligence and Intelligent Informatics 7(1), 25–30 (2003)
5. Vapnik, V.N.: Statistical Learning Theory. Wiley, New York (1998)
6. Miyamoto, S., Kawasaki, Y., Sawazaki, K.: An Explicit Mapping for Kernel Data Analysis and Application to Text Analysis. In: Proc. IFSA-EUSFLAT 2009, pp. 618–623 (2009)
7. Miyamoto, S., Sawazaki, K.: An Explicit Mapping for Kernel Data Analysis and Application to $c$-Means Clustering. In: Proc. NOLTA 2009, pp. 556–559 (2009)
8. Kanzawa, Y., Endo, Y., Miyamoto, S.: Indefinite Kernel Fuzzy $c$-Means Clustering Algorithms. In: Torra, V., Narukawa, Y., Daumas, M. (eds.) MDAI 2010. LNCS (LNAI), vol. 6408, pp. 116–128. Springer, Heidelberg (2010)
9. Tamura, S., Higuchi, S., Tanaka, K.: Pattern Classification Based on Fuzzy Relations. IEEE Trans. Syst. Man Cybern. 1(1), 61–66 (1971)
10. Scannell, J.W., Blakemore, C., Young, M.P.: Analysis of Connectivity in the Cat Cerebral Cortex. J. Neuroscience 15(2), 1463–1483 (1995)
11. Ghosh, G., Strehl, A., Merugu, S.: A Consensus Framework for Integrating Distributed Clusterings under Limited Knowledge Sharing. In: Proc. NSF Workshop on Next Generation Data Mining, pp. 99–108 (2002)
12. Bezdek, J.C., Keller, J., Krishnapuram, R., Pal, N.-R.: Fuzzy Models and Algorithms for Pattern Recognition and Image Processing. Kluwer Academic Publishing, Boston (1999)

# The Utilities of Imprecise Rules and Redundant Rules for Classifiers

Masahiro Inuiguchi and Takuya Hamakawa

**Abstract.** Rules inferring the memberships to single decision classes have been induced in rough set approaches and used to build a classifier system. Rules inferring the memberships to unions of multiple decision classes can be also induced in the same manner. In this paper, we show the classifier system with rules inferring the memberships to unions of multiple decision classes has an advantage in the accuracy of classification. We look into the reason of this advantage from the view point of robustness. The robustness in this paper implies the maintenance of the classification accuracy against missing values. Moreover, we examine the relationship between the accuracy and the robustness via numerical experiments. We demonstrate an advantage of the redundant rules. Finally, a stronger advantage of rules inferring the memberships to unions of multiple decision classes is reexamined.

## 1 Introduction

Rule induction is known as one of data mining techniques. There are several approaches to induce rules, association rule mining approach, decision tree learning approach and so on. Rough set approaches [1, 2] provide also rule induction techniques which are characterized by the minimality in the conditions of induced rules and/or the minimality in the number of rules to explain all given training data. Rough set approaches have been applied to various fields including medicine, engineering, management and economy.

In the conventional rough set approaches, rules inferring the memberships to single decision classes (simply called "precise rules") have been induced and used to build the classifier system. However, rules inferring the memberships to unions of multiple decision classes (simply called "imprecise rules") can be also induced

Masahiro Inuiguchi · Takuya Hamakawa
Graduate School of Engineering Science, Osaka University
Toyonaka, Osaka, 560-8531, Japan
http://www-inulab.sys.es.osaka-u.ac.jp/

V.-N. Huynh et al. (eds.), *Knowledge and Systems Engineering, Volume 2*,
Advances in Intelligent Systems and Computing 245,
DOI: 10.1007/978-3-319-02821-7_6, © Springer International Publishing Switzerland 2014

based on the rough set model. Each of such imprecise rules cannot give the conclusion univocally. Moreover, some imprecise rule may have sufficient condition to conclude a member of a decision class precisely but its conclusion is only a member of the union of the decision classes. Nevertheless, as demonstrated in this paper, the classifier system with imprecise rules has an advantage in the accuracy of classification over the conventional classifier system with precise rules.

Recently, the robustness measure of a rule has been proposed as a measure to evaluate to what extent the rule preserves the accuracy against the partial matching data or missing values (see [3]). It has been demonstrated that rules with high robustness usually classify new objects with higher accuracy than rules with low robustness. On the other hand, the redundancy in the condition part of a rule increases the robustness measure. Then the robustness measure may provide a contrasting perspective with the previous idea in mining minimal rules.

In this paper, we look into the reason of the advantage of the classifier system with imprecise rules in the accuracy from the view point of robustness. We define the robustness of the rule-based classifier system. This robustness measure is different from the robustness measure of a rule proposed by Ohki and Inuiguchi [3], although the concept of robustness is the same. Moreover, we examine the relationship between the accuracy and the robustness via numerical experiments using several data-sets available at public domain. We demonstrate an advantage of the redundant rules. Finally, a stronger advantage of imprecise rules is reexamined by numerical experiments.

This paper is organized as follows. In Section 2, we briefly review the rough set approaches to rule induction. We describe the idea of imprecise rule induction and the classifier system with imprecise rules in Section 3. In Section 4, we explain the numerical experiments and the results. We made three kinds of numerical experiments: (1) experiments to demonstrate the advantage of classifiers with imprecise rules, (2) experiments to examine the performance of redundant rules and (3) experiments to confirm the advantage of imprecise rules over redundant rules. Finally, concluding remarks are given in Section 5.

## 2   Rough Set Approach to Classification

Rough set theory is useful in the analysis of decision table. A decision table is defined by a 4-tuple $DT = \langle U, C \cup \{d\}, V, f \rangle$, where $U$ is a finite set of objects, $C$ is a finite set of condition attributes and $d$ is a decision attribute, $V = \bigcup_{a \in C \cup \{d\}} V_a$ with attribute value set $V_a$ of attribute $a \in C \cup \{d\}$ and $f : U \times C \cup \{d\} \to V$ is called an information function which is a total function. By decision attribute value $v_j^d \in V_d$, decision class $D_j \subseteq U$ is defined by $D_j = \{u \in U \mid f(u,d) = v_j^d\}$, $j = 1, 2, \ldots, p$. Using condition attributes in $A \subseteq C$, we define equivalence classes $[u]_A = \{x \in U \mid f(x,a) = f(u,a), \forall a \in A\}$.

The lower and upper approximations of an object set $X \subseteq U$ are defined by

$$A_*(X) = \{x \in U \mid [x]_A \subseteq X\}, \quad A^*(X) = \{x \in U \mid [x]_A \cap X \neq \emptyset\}. \tag{1}$$

Suppose that members of $X$ can be described by condition attributes in $A$, all elements in $[x]_A$ are classified either into $X$ or $U - X$. If $[x]_A \cap X \neq \emptyset$ and $[x]_A \cap (U - X) \neq \emptyset$ hold, the membership of $x$ to $X$ or $U - X$ is questionable. Form this point of view, each element of $A_*(X)$ is a consistent member of $X$ while each element of $A^*(X)$ is a possible member of $X$ (a consistent member or a questionable member). The pair $(A_*(X), A^*(X))$ is called the rough set of $X$ under $A \subseteq C$.

In rough set approaches, the attribute reduction, i.e., the minimal attribute set $A \subseteq C$ satisfying $A_*(D_j) = C_*(D_j)$, $j = 1, 2, \ldots, p$, and the minimal rule induction, i.e., inducing rules inferring the membership to $D_j$ with minimal conditions which can differ members of $C_*(D_j)$ from non-members, are investigated well. In this paper, we use minimal rule induction algorithms proposed in rough set approaches: one is LEM2-based algorithm [4] and the other is decision matrix method [5]. By the former algorithm, LEM2-based algorithm, we obtain minimal set of rules with minimal conditions which can explain all objects in lower approximations of $X$ of the given decision table. We use LEM2 algorithm and MLEM2 algorithm [4]. Those algorithms are different in their forms of condition parts of rules: by LEM2 algorithm, we obtain rules of the form of "if $f(u, a_1) = v_1$, $f(u, a_2) = v_2$, $\ldots$ and $f(u, a_p) = v_p$ then $u \in X$", while by MLEM2 algorithm, we obtain rules of the form of "if $v_1^L \leq f(u, a_1) \leq v_1^R$, $v_2^L \leq f(u, a_2) \leq v_2^R$, $\ldots$ and $v_p^L \leq f(u, a_p) \leq v_p^R$ then $u \in X$". Namely, MLEM2 algorithm is a generalized version of LEM2 algorithm to cope with numerical/ordinal condition attributes. On the other hand, by the latter method, decision matrix method, we obtain all rules inferring the membership of $X$ with minimal conditions.

For each decision class $D_i$ we induce rules inferring the membership of $D_i$. Using all those rules, we build a classifier system. To build the classifier system, we apply the idea of LERS [4]. The classification of an unseen object $u$ is as follows: When the condition attribute values of $u$ match to at least one of the elementary conditions of the rule, we calculate

$$S(D_i) = \sum_{\text{matching rules } r \text{ for } D_i} Strength(r) \times Specificity(r), \qquad (2)$$

where $r$ is called a *matching rule* if the condition part of $r$ is satisfied. $Strength(r)$ is the total number of objects in given decision table correctly classified by rule $r$. $Specificity(r)$ is the total number of condition attributes in the condition part of rule $r$. For convenience, when rules from a particular class $D_i$ are not matched by the object, we define $S(D_i) = 0$.

On the other hand, when the condition attribute values of $u$ do not match totally to any condition part of rule composing the classifier system. For each $D_i$, we calculate

$$M(D_i) = \sum_{\substack{\text{partially matching} \\ \text{rules } r \text{ for } D_i}} Matching\_factor(r) \times Strength(r) \times Specificity(r), \qquad (3)$$

where $r$ is called a *partially matching rule* if a part of the premise of $r$ is satisfied. $Matching\_factor(r)$ is the ratio of the number of matched conditions of rule $r$ to the total number of conditions of rule $r$.

The classification can be performed as follows: if there exists $D_j$ such that $S(D_j) > 0$, the class $D_i$ with the largest $S(D_i)$ is selected. Otherwise, the class $D_i$ with the largest $M(D_i)$ is selected.

## 3   Imprecise Rules and Robustness

### 3.1   Inducing Imprecise Rules and Classification

In the same way as the induction of rules inferring the membership to $D_i$, we can induce rules inferring the membership to the union of $D_i$'s. Namely, both a LEM2-based algorithm and a decision matrix method can be applied because the union of $D_i$'s is a set of objects. Inducing rules inferring the membership of the union of $D_i \cup D_j$ for all pairs $(D_i, D_j)$ such that $i \neq j$, we may build a classifier. Moreover, in the same way, we can build a classifier by induced rules inferring the membership to $\bigcup_{j=i_1, i_2, \ldots, i_l} D_j$ for all combinations of $l$ decision classes.

To do this, we should consider a classification method under rules inferring the membership to the union of decision classes. We examined several possible way which is an extension of LERS classifier described in the previous section. By several auxiliary numerical experiments, the following simple classification method performs best: the classification method simply replace $S(D_i)$ and $M(D_i)$, respectively, with

$$\hat{S}(D_i) = \sum_{\substack{\text{matching rules } r \text{ for } Z \subseteq D_i}} Strength(r) \times Specificity(r), \tag{4}$$

$$\hat{M}(D_i) = \sum_{\substack{\text{partially matching} \\ \text{rules } r \text{ for } Z \subseteq D_i}} Matching\_factor(r) \times Strength(r) \times Specificity(r),$$

$$\tag{5}$$

where $Z$ is a union of decision classes. Obviously, this classification method is reduced to the LERS classification method when $Z = D_i$.

### 3.2   Robustness

Ohki and Inuiguchi [3] has proposed the robustness measure of a rule. The robustness measure indicates to what extent the rule preserves the classification accuracy against partial matching data. In this paper, we extend this idea to a classifier. The robustness measure of a classifier is a measure showing the degree to what extent the classifier preserves the classification accuracy against partial matching data. In this paper, we use the training data-set as the checking data-set of the robustness by erasing all data of a condition attribute. Namely the checking data-set is composed of data in which values of the condition attribute are missing. Because there are $m$ condition attributes, we obtain $m$ checking data-sets, where we define $m = |C|$. After all rules are induced from the training data, the average accuracy in the classification

**Table 1** Data-sets

| Data-set(abbreviation) | $\|U\|$ | $\|C\|$ | $\|V_d\|$ | attribute type |
|---|---|---|---|---|
| car(C) | 1,728 | 6 | 4 | ordinal |
| ecoli(E) | 336 | 7 | 8 | numerical |
| glass(G) | 214 | 9 | 6 | numerical |
| iris(I) | 150 | 4 | 3 | numerical |
| hayes-roth(H) | 159 | 4 | 3 | nominal |
| damatology(D) | 358 | 34 | 6 | numerical |
| wine(W) | 178 | 13 | 3 | numerical |
| zoo(Z) | 101 | 16 | 7 | nominal |

of checking data is calculated for each of $m$ checking data-sets. The robustness is evaluated by the average and the standard deviation of $m$ average accuracy.

# 4 Numerical Experiments

## 4.1 Outline

In this paper, we first demonstrate the good performance of the classifier based on the imprecise rules by a numerical experiment using eight or five data-sets. Then we look into the reason of this advantage of imprecise rules from the view point of robustness. Then we examine the relationship between the accuracy and the robustness in the classifier based on precise rules. Finally, the advantage of imprecise rules is reexamined.

## 4.2 Data-Sets

In the numerical experiments, we use eight data-sets obtained from UCI Machine Learning Repository [1]. The eight data-sets are shown in Table 1. In Table 1, $\|U\|$, $\|C\|$ and $\|V_d\|$ means the number of objects in the given data table, the number of condition attributes and the number of decision classes.

MLEM2 algorithm is applied to all those data-sets because MLEM2 algorithm produces the same results as LEM2 when all condition attributes are nominal. On the other hand, decision matrix method cannot be applied to the three bottom data-sets in Table 1 due to the out of memory in our computer setting. Then only top five data-sets are used in the experiments by decision matrix method. In order to apply the decision matrix method to data-sets with numerical condition attributes, we discretize the numerical condition attributes by WEKA [7], namely entropy-based discretization with a threshold. To compare the performance with classifier systems with precise rules induced by the decision matrix method, we apply LEM2 algorithm to the discretized data-sets so that we induce imprecise rules. We note that although the five data are consistent, i.e., the lower approximations of decision

**Table 2** Performances of classifiers with imprecise rules induced by MLEM2

| $A(k)$ | No. rules | Accuracy | Robustness | $A(k)$ | No. rules | Accuracy | Robustness |
|---|---|---|---|---|---|---|---|
| C(1) | 57.22±1.12 | 98.67±0.01 | 66.84±11.64 | H(1) | 23.17±0.45 | 68.68±1.56 | 39.08±19.32 |
| C(2) | 128.02±2.37 | 99.15*±0.00 | 82.28*±8.17 | H(2) | 39.25±0.57 | 66.50*±1.49 | 44.39*±18.94 |
| C(3) | 69.55±0.85 | 99.68*±0.00 | 83.48*±7.37 | D(1) | 12.09±0.32 | 92.24±1.12 | 90.89±15.48 |
| E(1) | 35.90±0.56 | 77.38±1.75 | 56.33±31.75 | D(2) | 61.32±0.84 | 94.67*±1.00 | 98.76*±3.72 |
| E(2) | 220.87±2.07 | 82.94*±1.37 | 84.18*±16.80 | D(3) | 103.58±0.88 | 96.15*±0.64 | 99.33*±2.19 |
| E(3) | 566.07±5.09 | 84.40*±1.68 | 95.33*±6.89 | D(4) | 77.28±1.28 | 95.62*±0.84 | 98.43*±4.42 |
| E(4) | 781.78±8.14 | 84.43*±2.38 | 97.48*±4.12 | D(5) | 23.84±0.62 | 91.10±1.39 | 89.96*±18.24 |
| E(5) | 617.24±6.07 | 83.33*±2.57 | 96.75*±5.75 | W(1) | 4.65±0.14 | 93.15±1.77 | 75.15±20.14 |
| E(6) | 269.31±3.27 | 82.41*±1.90 | 91.31*±12.70 | W(2) | 7.31±0.11 | 88.78*±2.17 | 71.49*±26.72 |
| E(7) | 54.09±0.86 | 75.98±2.96 | 50.84*±27.98 | Z(1) | 9.67±0.14 | 96.33±1.75 | 91.38±10.85 |
| G(1) | 25.38±0.34 | 68.20±3.04 | 61.72±20.52 | Z(2) | 48.54±0.55 | 96.16±1.37 | 97.52*±5.37 |
| G(2) | 111.40±1.08 | 72.61*±1.65 | 87.55*±11.86 | Z(3) | 105.37±1.48 | 96.74±1.68 | 99.25*±2.39 |
| G(3) | 178.35±1.49 | 73.55*±1.31 | 95.95*±5.66 | Z(4) | 113.70±1.01 | 96.84±2.38 | 99.23*±2.40 |
| G(4) | 130.14±1.61 | 71.50*±1.63 | 92.92*±8.33 | Z(5) | 66.76±0.89 | 97.34±2.57 | 93.45*±15.89 |
| G(5) | 39.59±0.71 | 64.41*±2.81 | 46.51*±20.50 | Z(6) | 17.72±0.22 | 97.15±1.90 | 86.10*±19.63 |
| I(1) | 7.40±0.24 | 92.87±1.34 | 79.52±24.99 | | | | |
| I(2) | 8.52±0.28 | 93.14±1.31 | 78.41±21.86 | | | | |

classes equal to the original decision classes, the discretized data-set may be inconsistent. When the discretized data-set is inconsistent, we induce rules only to objects in the lower approximations of decision classes.

## 4.3  10-Fold Cross Validation

To obtain the accuracy, we apply the 10-fold cross validation method. Namely we divide the data-set into 10 subsets and data in 9 subsets are used for training data and data in the remaining subset is used for checking data. Changing the combination of 9 subsets, we obtain 10 different accuracies. Then we take the average. We execute this procedure 10 times with different divisions.

## 4.4  Performances of Imprecise Rules Induced by MLEM2

First, we demonstrate the good performance of the classifier with imprecise rules. Using training data, we induce rules inferring the membership to the union of $k$ decision classes by MLEM2 algorithm for all possible combination of $k$ decision classes. Then by the classifier based on induced imprecise rules, all of checking data are classified and the classification accuracy as well as the robustness are obtained. The results of this numerical experiment are shown in Table 2. In Table 2, the average number of induced rules, the average accuracy and the robustness are shown. Column $A(k)$ indicates the abbreviation of data-set by $A$ and the number of decision classes composing the union by $k$. Each entry in the other columns is

**Table 3** Effects of precisiation in the accuracy and robustness

| $A(k)$ | No Precisiation | | Precisiation | |
|---|---|---|---|---|
| | Accuracy | Robustness | Accuracy | Robustness |
| E(7) | $76.00_{\pm7.28}$ | $50.84_{\pm27.98}$ | $78.97_{\pm7.02}$ | $72.59_{\pm21.91}$ |
| D(5) | $91.06_{\pm4.85}$ | $89.96_{\pm18.24}$ | $91.87_{\pm4.40}$ | $94.92_{\pm8.55}$ |
| G(5) | $64.02_{\pm10.86}$ | $46.51_{\pm20.50}$ | $65.80_{\pm10.85}$ | $66.01_{\pm16.81}$ |
| Z(6) | $97.15_{\pm5.56}$ | $86.10_{\pm19.63}$ | $96.15_{\pm6.86}$ | $93.72_{\pm10.16}$ |

composed of the average $av$ and the standard deviation $st$ in the form of $av_{\pm st}$. Mark $*$ means the value is significantly different from the case of precise rules ($k = 1$) by the paired $t$-test with significance level $\alpha = 0.05$. For each data-set, the largest values are underlined.

As shown in Table 2, the classification accuracy sometimes attains its largest value around the middle value of $k$. We observe that the classifier with imprecise rules often performs better than the classifier with precise rules. Moreover, we observe the tendency to some extent that the larger the classification accuracy, the larger the number of induced rules and the larger the robustness.

*Remark 1.* As shown in Table 2, the accuracy of the classifier with respect to $A(k)$ with large $k$ is not always good especially when $|V_d|$ is large. This could be caused by the fact that some imprecise rules include superfluous decision classes in their conclusions. When rule $r$, "if $u$ is $E$ then $u$ is $H$", includes at least one superfluous decision class in $H$, we can precisiate $r$ to "if $u$ is $E$ then $u$ is $\check{H}$" with $\check{H} = \{f(u_i, d) \mid u_i \in ||H||\}$, where $||H||$ is a set of object $u_i \in U$ satisfying condition $H$.

In cases of E(7), G(5), D(5) and Z(6), let us examine the effect of the precisiation. The results are shown in Table 3. In this table, the average accuracy and robustness with their standard deviations are given. As shown in Table 3, the precisiation works well except Z(6). The robustness is improved much more than the accuracy. However, the improved accuracy and robustness are not always better than the best average accuracy shown in Table 2.

Let us check the tendency described above in the results by the classifiers with imprecise rules induced by the decision matrix method. We note that data-set is not always totally consistent by the discretization of numerical attributes. We note also that only to five data-sets we can apply the decision matrix method. The results are shown in Table 4. The format of this table is as same as Table 2. From the results in Table 2, the tendency is observed rather weakly but we cannot disaffirm this tendency. We note that except "hayes-roth", the accuracies in Table 4 become smaller than those in Table 2 due to the fact that the discretization leads to the inconsistencies in data-sets.

The relationship between the accuracy and the robustness can be understood from the fact that unseen data cannot always match the conditions of induced rules but partially and thus the classifier with high performance against the partial matching data may perform well for unseen data. The performance to unseen data is evaluated

**Table 4** Performances of classifiers with imprecise rules induced by decision matrix

| $A(k)$ | No. rules | Accuracy | Robustness |
|---|---|---|---|
| C(1) | $914.05_{\pm 65.00}$ | $92.14_{\pm 2.17}$ | $69.87_{\pm 12.49}$ |
| C(2) | $\underline{3271.14}_{\pm 175.97}$ | $95.80^*_{\pm 1.62}$ | $83.84^*_{\pm 7.33}$ |
| C(3) | $2781.70_{\pm 182.04}$ | $\underline{96.70^*}_{\pm 1.47}$ | $\underline{84.18}^*_{\pm 6.96}$ |
| E(1) | $295.61_{\pm 115.61}$ | $63.30_{\pm 14.56}$ | $55.73_{\pm 16.15}$ |
| E(2) | $2181.28_{\pm 821.27}$ | $73.15^*_{\pm 13.33}$ | $68.15^*_{\pm 15.33}$ |
| E(3) | $6789.54_{\pm 2498.91}$ | $\underline{73.42^*}_{\pm 12.92}$ | $68.95^*_{\pm 14.94}$ |
| E(4) | $\underline{11332.19}_{\pm 4114.79}$ | $73.09^*_{\pm 12.94}$ | $\underline{68.99}^*_{\pm 14.91}$ |
| E(5) | $10557.15_{\pm 3810.27}$ | $73.15^*_{\pm 13.00}$ | $\underline{68.99}^*_{\pm 14.91}$ |
| E(6) | $4961.50_{\pm 1760.08}$ | $73.09^*_{\pm 12.84}$ | $\underline{68.99}^*_{\pm 14.91}$ |
| E(7) | $775.87_{\pm 249.79}$ | $64.85_{\pm 25.00}$ | $55.86_{\pm 29.98}$ |
| G(1) | $747.27_{\pm 129.64}$ | $51.90_{\pm 11.38}$ | $50.80_{\pm 7.25}$ |
| G(2) | $4386.3_{\pm 742.15}$ | $62.67^*_{\pm 11.02}$ | $63.44^*_{\pm 4.77}$ |
| G(3) | $10452.51_{\pm 1781.22}$ | $62.62^*_{\pm 10.82}$ | $\underline{63.45}^*_{\pm 4.77}$ |
| G(4) | $\underline{12139.00}_{\pm 2142.02}$ | $62.81^*_{\pm 10.91}$ | $\underline{63.45}^*_{\pm 4.76}$ |
| G(5) | $6529.13_{\pm 1218.04}$ | $\underline{62.86^*}_{\pm 10.92}$ | $\underline{63.45}^*_{\pm 4.77}$ |
| I(1) | $29.72_{\pm 8.34}$ | $94.60_{\pm 6.84}$ | $\underline{91.71}_{\pm 8.49}$ |
| I(2) | $\underline{53.81}_{\pm 11.06}$ | $\underline{94.73}_{\pm 10.47}$ | $\underline{91.71}_{\pm 8.49}$ |
| H(1) | $48.24_{\pm 4.76}$ | $\underline{71.00}_{\pm 9.82}$ | $46.36_{\pm 19.26}$ |
| H(2) | $\underline{170.97}_{\pm 6.57}$ | $68.81_{\pm 11.02}$ | $\underline{64.41}^*_{\pm 14.60}$ |

by the accuracy while the performance to partially matched data is evaluated by the robustness. Therefore, the classifier with high robustness would perform well in the accuracy. On the other hand, the relation between the number of induced rules and the robustness can be understood from the fact that the classifier with the more independent rules will perform better against data with missing values because some other rule could be fired even if a rule could not be fired by the missing value.

## 4.5   The Performance of Classifier with Redundant Rules

By the syllogism, we may expect that the classifier with redundant rules will perform better. In this subsection, we examine this conjecture by increasing a set of minimal rules iteratively.

This experiment is explained as follows. We first induce all minimal precise rules by the decision matrix method. From those minimal rules we increase a subset of rules iteratively. The induced rules are totally accurate then rules with large coverage (also called recall) can be seen as good rules, where, for a rule "if $u$ satisfies $E$ then $u$ satisfies $H$", the accuracy (also called confidence) is defined by $acr(E \to H) = ||E \wedge H||/||E||$ while the coverage (also called recall) is defined by $cov(E \to H) = ||E \wedge H||/||H||$. $||E \wedge H||$, $||E||$ and $||H||$ show the number of objects in $U$ satisfying $E \wedge H$, $E$ and $H$, respectively. Let $||r|| = ||E \wedge H||$ for a rule $r$, "if $u$ satisfies $E$ then

**Table 5** Performances of classifiers with redundant precise rules

| | Classifier with no modification | | | | Classifier with modified strength | | |
|---|---|---|---|---|---|---|---|
| $A[q]$ | No. rules | Accuracy | Robustness | $A[q]$ | No. rules | Accuracy | Robustness |
| C[1] | 213.58±5.93 | 90.96±2.06 | 68.52±13.95 | C[1] | 213.58±5.93 | 90.96±2.06 | 68.52±13.95 |
| C[2] | 317.37±9.98 | 92.64±1.95 | 69.39±12.83 | C[2] | 444.18±14.57 | 92.89±1.9 | 69.42±12.81 |
| C[3] | 386.46±17.92 | <u>92.66</u>±1.90 | 69.59±12.72 | C[3] | 667.71±29.44 | 93.00±1.88 | 69.61±12.7 |
| C[4] | 450.86±25.93 | 92.59±1.91 | 69.69±12.65 | C[4] | 888.82±47.65 | <u>93.04</u>±1.87 | 69.73±12.51 |
| all | 914.05±65.00 | 92.23±2.06 | <u>69.87</u>±12.49 | all | 914.05±65.00 | 92.23±2.06 | <u>69.87</u>±12.49 |
| E[1] | 24.63±6.56 | 62.52±14.63 | 45.35±18.21 | E[1] | 24.62±6.55 | 62.48±14.67 | 45.37±18.15 |
| E[2] | 48.32±12.71 | 62.91±14.70 | 49.31*±17.40 | E[2] | 51.94±13.69 | 62.85±14.68 | 49.09*±17.52 |
| E[3] | 70.75±18.92 | 63.18±14.64 | 52.67*±16.74 | E[3] | 80.98±21.48 | 63.21±14.60 | 52.04*±16.91 |
| E[4] | 92.54±25.66 | <u>63.42</u>±14.60 | 54.02*±16.41 | E[4] | 110.17±29.25 | <u>63.36</u>±14.56 | 53.17*±16.68 |
| all | 295.61±115.61 | 63.30±14.56 | <u>55.73</u>*±16.15 | all | 295.61±115.61 | 63.30±14.56 | <u>55.73</u>*±16.15 |
| G[1] | 18.80±2.69 | 50.19±11.48 | 42.97±9.89 | G[1] | 18.80±2.69 | 50.19±11.69 | 42.99±9.87 |
| G[2] | 34.13±4.94 | 50.95±12.00 | 45.26*±8.84 | G[2] | 39.01±6.04 | 50.86±11.99 | 43.86±7.44 |
| G[3] | 50.10±6.28 | 51.10±11.60 | 47.38*±8.37 | G[3] | 60.52±8.63 | 51.14±11.66 | 46.70*±8.59 |
| G[4] | 66.21±8.02 | 51.38±11.73 | 48.40*±7.94 | G[4] | 82.46±11.43 | 51.48±11.66 | 47.90*±8.31 |
| all | 747.27±129.64 | <u>51.90</u>±11.38 | <u>50.80</u>*±7.25 | all | 747.27±129.64 | <u>51.90</u>±11.38 | <u>50.80</u>*±7.25 |
| I[1] | 4.90±1.33 | 94.73±6.69 | 73.42±24.68 | I[1] | 4.90±1.33 | 94.73±6.69 | 74.01±24.17 |
| I[2] | 10.07±2.16 | <u>94.93</u>±6.61 | 82.59*±20.35 | I[2] | 10.79±2.59 | <u>94.87</u>±6.59 | 82.11*±20.44 |
| I[3] | 14.35±3.04 | 94.87±6.59 | 90.57*±10.62 | I[3] | 16.19±3.92 | 94.73±6.62 | 89.47*±12.86 |
| I[4] | 19.15±4.14 | 94.80±6.64 | 91.45*±8.90 | I[4] | 22.11±5.41 | 94.67±6.67 | 89.91*±10.85 |
| all | 29.72±8.34 | 94.60±6.84 | <u>91.71</u>*±8.49 | all | 29.72±8.34 | 94.60±6.84 | <u>91.71</u>*±8.49 |
| H[1] | 23.63±1.47 | 68.56±9.46 | 39.85±18.15 | H[1] | 23.61±1.46 | 68.56±9.46 | 39.81±18.17 |
| H[2] | 41.75±2.61 | <u>71.50</u>±9.93 | 45.97±18.88 | H[2] | 48.99±3.01 | 72.13±9.45 | 42.73±20.90 |
| H[3] | 45.59±3.17 | 71.44±10.01 | 46.22±19.17 | H[3] | 72.70±4.41 | 72.25±9.2 | 42.79±20.93 |
| H[4] | 47.67±3.99 | 71.06±9.83 | <u>46.36</u>±19.26 | H[4] | 96.34±5.90 | <u>72.38</u>±9.36 | 42.83±20.94 |
| all | 48.24±4.76 | 71.00±9.82 | <u>46.36</u>±19.26 | all | 48.24±4.76 | 71.00±9.82 | <u>46.36</u>±19.26 |

$u$ satisfies $H$". For a subset $R_j$ of rules, we define $||R_j|| = \bigcup\{||r|| \mid r \in R_j\}$. The rule selection procedure is as follows:

**A1.** Let $q$ be the number of minimal subsets of rules covering all data in $U$. Let $F$ be initialized by a set of all induced rules and $S = \emptyset$. Let $R_s = \emptyset$, $s = 1, 2, \ldots, q$. Set $j = 1$.

**A2.** If $j > q$ terminate the procedure. $R = \bigcup_{s=1,2,\ldots,q} R_s$ is the set of selected rules.

**A3.** Select $r$ with largest coverage from $r \in F$ satisfying $||r|| \cap (U \setminus ||R_j||) \neq \emptyset$. If $r$ exists, update $R_j = R_j \cup \{r\}$, $S = S \cup \{r\}$ and go to A5.

**A4.** Select $r$ with largest coverage from $r \in S$ satisfying $||r|| \cap (U \setminus ||R_j||) \neq \emptyset$. Update $R_j = R_j \cup \{r\}$.

**A5.** For each $r \in R_j$, update $R_j = R_j \setminus \{r\}$ if $||R_j \setminus \{r\}|| = ||R_j||$.

**A6.** If $||R_j|| = U$ update $j = j + 1$.

**A7.** Return to A2.

Namely, we try to select $q$ minimal subsets of rules, $R_k$, $k = 1, 2, \ldots, q$ which are independent as much as possible. Then the subset $F$ of first candidate rules are composed of rules which have not yet selected earlier. If we cannot explain all data in $U$, we select rules from $S$, the subset of second candidate rules which have been already selected in the previous iterations. We say rule $r$ covers $u \in U$ if and only if both of the condition and decision attribute values match to those specified in $r$.

At the classification stage, we consider two classifiers. One is the same as the classifier described in the previous section. However, in this method, duplicatively selected rules are treated same as only once selected rules. Then we use the same classifier with modification of the strength of rule $r$, $Strength(r)$, by taking into account the number of duplication. Namely, we use the following new strength $Strength^{new}(r)$ instead of the previous strength $Strength^{old}(r)$ defined by the total number of objects in given decision table correctly classified by rule $r$:

$$Strength^{new}(r) = |\{R_k \mid r \in R_k\}| \times Strength^{old}(r). \tag{6}$$

For $q = 1, 2, 3, 4$, we executed the selection of rules. The results are shown in Table 5. The results of all rules obtained by decision matrix method are also shown. The format of Table 5 is almost same as that of Table 2 but Column $A[q]$ indicates the abbreviation of data-set by $A$ and the number of minimal subsets $R_s$ of precise rules by $q$. We do not underline the largest number of rules in each data-set because it obviously increases as $q$ increases and it attains the maximal when all rules are selected.

As shown in Table 5, in both classifiers, we neither say that the larger the number of rules, the larger the accuracy, nor that the larger the robustness, the larger the accuracy. However, we can observe that the larger the number of rules, the larger the robustness. Therefore, the hypothesis based on the syllogism is not supported strongly. Finding the reason is not an easy task because the results depend not only on rules but also on the classifier. However, we observe that classifiers with $q$ minimal subsets of precise rules ($q > 1$) are better than the classifier with only one minimal subset of precise rules ($q = 1$) in all cases although we cannot say which $q$ is the best. We cannot say which classifier is better between the classifiers with no modification and with modified strength. It depends on data-set: in data-sets, "car", "grass" and "hayes-roth" the classification with modified strength seems to be better but in the other data-sets, "ecoli" and "iris" the classification with no modification seems to be better.

## 4.6   Effects of Imprecision in Rule Conclusions

In the previous subsection, we neither observed that the larger the number of rules, the larger the accuracy, nor that the larger the robustness, the larger the accuracy. In this subsection, we investigate the effects of imprecision in rule conclusions. To this end, we compare classifiers with $p_1$ precise rules and classifiers of $p_2$ imprecise rules when $p_1$ is approximately equal to $p_2$ ($p_1 \approx .p_2$). To this end, we induce imprecise rules by LEM2 from five data-sets with discretization of numerical attributes and evaluate the performances of the classifiers with the induced imprecise rules. The results are shown in the left half part of Table 6. The format of this table is the same as previous tables. By the comparison of entries in those columns of Table 6

**Table 6** Performances of classifiers with imprecise rules induced by LEM2 for discretized data

| \multicolumn{4}{Classifiers with imprecise rules} | | | | Classifiers with redundant precise rules | | | |
|---|---|---|---|---|---|---|---|
| $A(k)$ | No. rules | Accuracy | Robustness | $A[q]$ | No. rules | Accuracy | Robustness |
| C(1) | 210.10±5.82 | 91.18±2.01 | 68.51±13.04 | C[1] | 213.58±5.93 | 90.96±2.06 | 68.52±13.95 |
| C(2) | 372.11±13.78 | 97.92*±0.18 | 82.82*±7.96 | C[3] | 386.46±17.92 | 92.66±1.90 | 69.59±12.72 |
| C(3) | 147.90±3.41 | 99.41*±0.02 | 83.50*±7.36 | C[1] | 213.58±5.93 | 90.96±2.06 | 68.52±13.95 |
| E(1) | 25.69±6.31 | 62.48±14.55 | 46.20±18.51 | E[1] | 24.63±6.56 | 62.52±14.63 | 45.35±18.21 |
| E(2) | 158.78±39.24 | 72.97*±13.45 | 63.83*±16.15 | E[8] | 162.29±48.45 | 63.30±14.49 | 55.41±16.2 |
| E(3) | 426.39±97.49 | 73.09*±13.16 | 66.78*±15.50 | all | 295.61±115.61 | 63.30±14.56 | 55.73±16.15 |
| E(4) | 619.35±129.25 | 72.91*±12.95 | 67.35*±15.26 | all | 295.61±115.61 | 63.30±14.56 | 55.73±16.15 |
| E(5) | 517.61±100.68 | 72.82*±12.96 | 67.58*±15.11 | all | 295.61±115.61 | 63.30±14.56 | 55.73±16.15 |
| E(6) | 241.59±40.66 | 72.67*±13.03 | 67.05*±15.21 | E[12] | 207.98±66.40 | 63.27±14.57 | 55.65*±16.19 |
| E(7) | 52.88±8.31 | 72.27±12.89 | 53.45±18.50 | E[2] | 48.32±12.71 | 62.91±14.70 | 49.31*±17.40 |
| G(1) | 20.86±2.53 | 50.62±11.48 | 43.01±10.02 | G[1] | 18.80±2.69 | 50.19±11.48 | 42.97±9.89 |
| G(2) | 103.12±12.95 | 62.05*±10.89 | 61.27*±5.47 | G[6] | 96.70±10.40 | 51.81±11.65 | 49.51±7.57 |
| G(3) | 225.73±25.82 | 62.05*±10.46 | 62.71*±5.03 | G[12] | 183.73±20.47 | 51.81±11.53 | 50.55±7.32 |
| G(4) | 255.87±26.08 | 62.14*±10.68 | 62.90*±4.96 | G[12] | 183.73±20.47 | 51.81±11.53 | 50.55±7.32 |
| G(5) | 152.01±13.85 | 61.90*±11.02 | 62.66*±4.98 | G[10] | 155.45±16.73 | 51.86±11.54 | 50.33±7.41 |
| I(1) | 4.91±1.33 | 94.87*±6.66 | 79.81±19.33 | I[1] | 4.90±1.33 | 94.73±6.69 | 73.42±24.68 |
| I(2) | 8.35±1.44 | 94.80±9.38 | 75.56*±20.40 | I[2] | 10.07±2.16 | 94.93±6.61 | 82.59*±20.35 |
| H(1) | 23.17±1.41 | 68.88±9.23 | 39.08±19.32 | H[1] | 23.63±1.47 | 68.56±9.46 | 39.85±18.15 |
| H(2) | 39.25±2.20 | 66.50*±10.89 | 44.39*±18.94 | H[2] | 41.75±2.61 | 71.50±9.93 | 45.97±18.88 |

with corresponding entries in Table 2, except data-set "iris", we observe that rules induced by MLEM2 outperforms rules induced by LEM2 with the discretization. This is because MLEM2 finds suitable intervals of attribute values, i.e., a suitable discretization, during rule induction without creating conflicting data.

Moreover, we examine the performances of classifiers with $q$ subsets of precise rules from $q = 1$ to $q = 12$ in order to know the results of classifiers with sufficient large number of precise rules. Nevertheless, the number of imprecise rules composing a classifier sometimes larger than that of prepared classifiers of precise rules and that of all precise rules induced by the decision matrix method. In such cases, we use the results of the classifiers of precise rules with the nearest number of rules among the prepared ones. The performances of the corresponding classifiers with $p_1$ precise rules such that $p_1 \approx p_2$ are shown in the right half of Table 6. In Table 6, we can compare a subset of minimal rules induced by LEM2 and a subset of rules selected from all minimal rules based on the coverage, respectively, by $A(1)$ and $A[1]$ rows. The performances of those classifiers are similar. By the comparison of accuracy evaluations between $A(k)$ ($k > 1$) and the corresponding $A[q]$ ($q > 1$), generally speaking, classifiers with imprecise rules seem to be better than classifiers with precise rules. This is because classifiers with imprecise rules strongly outperform in the first three data-sets while they are weakly inferior to classifiers with precise rules in the other two data-sets. This implies that it is better to increase imprecise rules when we increase rules composing a classifier.

## 5   Concluding Remarks

In this paper, we first showed that the classifiers with minimal imprecise rules perform better than the classifiers with minimal precise rules. Then we tried to understand the reason of the good performance from the view point of robustness assuming that the good accuracy is related to the good robustness. The robustness is closely related to the number of rules. Then we examined the performances of the classifiers with redundant precise rules whether the redundancy improves the robustness and eventually the accuracy. It is confirmed that the redundancy improves the robustness. However, we did not observe the strong improvement of the accuracy by the redundancy. We only observed that the classifiers of redundant precise rules perform better than the classifiers of non-redundant precise rules. Finally, we compared the classifiers of imprecise rules to the classifiers of redundant rules in the settings that the numbers of rules in those classifiers are similar each other. As the results, we observed that the classifiers of imprecise rules outperform the classifiers of redundant rules.

In our future study we should investigate: the performances of the classifiers of redundant imprecise rules, the effects of the precisiation of imprecise rules, a guideline for the selections of imprecision degree ($k$) and redundancy degree ($q$), and so on.

## References

1. Pawlak, Z.: Rough Sets. International Journal of Computer and Information Sciences 11(5), 341–356 (1982)
2. Pawlak, Z.: Rough Sets: Theoretical Aspects of Reasoning About Data. Kluwer Academic Publishing, Dordrecht (1991)
3. Ohki, M., Inuiguchi, M.: Robustness Measure of Decision Rules (submitted for presentation 2013)
4. Grzymala-Busse, J.W.: MLEM2 - Discretization During Rule Induction. In: Proceedings of the IIPWM 2003, pp. 499–508 (2003)
5. Shan, N., Ziarko, W.: Data-based Acquisition and Incremental Modification of Classification Rules. Computational Intelligence 11, 357–370 (1995)
6. UCI Machine Learning Repository, http://archive.ics.uci.edu/ml/
7. Hall, M., Frank, E., Holmes, G., Pfahringer, B., Reutemann, P., Witten, I.H.: The WEKA Data Mining Software: An Update. SIGKDD Explorations 11(1), 1 (2009)

# On Cluster Extraction from Relational Data Using Entropy Based Relational Crisp Possibilistic Clustering

Yukihiro Hamasuna and Yasunori Endo

**Abstract.** The relational clustering is one of the clustering methods for relational data. The membership grade of each datum to each cluster is calculated directly from dissimilarities between datum and the cluster center which is referred to as representative of cluster is not used in relational clustering. This paper discusses a new possibilistic approach for relational clustering from the viewpoint of inducing the crispness. In the previous study, crisp possibilistic clustering and its variant has been proposed by using $L_1$-regularization. These crisp possibilistic clustering methods induce the crispness in the membership function. In this paper, entropy based crisp possibilistic relational clustering is proposed for handling relational data. Next, the way of sequential extraction is also discussed. Moreover, the effectiveness of proposed method is shown through numerical examples.

## 1 Introduction

Clustering is one of the data analysis method which divides a set of objects into some groups called clusters. Objects classified in same cluster are considered similar, while objects classified in different cluster are considered dissimilar. Fuzzy $c$-means clustering (FCM) is the most well-known clustering method [1, 2, 3]. Possibilistic clustering (PCM) is also well-known as one of the useful methods as well as FCM. PCM is well-known as robust clustering method for noises and outliers which negatively affect clustering results [4, 5]. A proposal of algorithms extracting

Yukihiro Hamasuna
Department of Informatics, School of Science and Engineering, Kinki University,
Kowakae 3-4-1, Higashi-osaka, Osaka, 577-8502, Japan
e-mail: yhama@info.kindai.ac.jp

Yasunori Endo
Faculty of Engineering, Information and Systems, University of Tsukuba, Tennodai 1-1-1,
Tsukuba, Ibaraki, 305-8573, Japan
e-mail: endo@risk.tsukuba.ac.jp

V.-N. Huynh et al. (eds.), *Knowledge and Systems Engineering, Volume 2,*
Advances in Intelligent Systems and Computing 245,
DOI: 10.1007/978-3-319-02821-7_7, © Springer International Publishing Switzerland 2014

"one cluster at a time" is based on the idea of noise clustering [7, 6]. Sequential extraction does not need to determine the number of clusters in advance. In order to detect high-density cluster, this advantage is quite important for handling massive and complex data.

Specifically, clustering methods such as FCM handles numerical data in the form of a set of vectors. The fuzzy non-metric model (FNM) is also one of the clustering method in which the membership grade of each datum to each cluster is calculated directly from dissimilarities between data [8]. The cluster center which is referred to as representative of cluster is not used in FNM. The data space need not necessarily be Euclidean space in FNM. Therefore, relational data is handled in FNM and other relational clustering methods such as Ref. [10]. The relational fuzzy $c$-means (RFCM) for euclidean relational data [11] and the one for non-euclidean relational data (NERFCM) [12] have been proposed for handling data described in relational form. Handling euclidean and non-euclidean relational data has been also important problem in the field of clustering [13].

$L_1$-regularization is well-known as useful techniques and applied to induce the sparseness, that is, small variables are calculated as zero [14]. In the field of clustering, sparse possibilistic clustering method has been proposed by introducing $L_1$-regularization [15]. $L_1$-regularization induce the sparseness with calculating the small membership grade as zero. This means that it induces the sparseness from the viewpoint of not being in the cluster. In the field of clustering, it should be also considered that the sparseness for being in the cluster. The crisp possibilistic $c$-means clustering (CPCM) have been proposed and describes its classification function for inducing sparseness from the viewpoint of being in the cluster [16]. The way of sequential extraction by CPCM has also been proposed as the same way for previous studies [5, 6].

In this paper, we will propose entropy based relational crisp possibilistic clustering (ERCPCM) for handling relational data and constructing sequential extraction algorithm. This paper is organized as follows: In section 2, we introduce some symbols, FCM and RFCM. In section 3, we propose entropy based relational crisp possibilistic clustering (ERCPCM). In section 4, we show the sequential extraction algorithm based on ERCPCM . In section 5, we show the effectiveness of sequential extraction through numerical examples. In section 6, we conclude this paper.

## 2  Preparation

A set of objects to be clustered is given and denoted by $X = \{x_1, \ldots, x_n\}$ in which $x_k\,(k = 1, \ldots, n)$ is an object. In most cases, $x_1, \ldots, x_n$ are $p$-dimensional vectors $\Re^p$, that is, a datum $x_k \in \Re^p$. A cluster is denoted as $C_i\,(i = 1, \ldots, c)$. A membership grade of $x_k$ belonging to $C_i$ and a partition matrix is also denoted as $u_{ki}$ and $U = (u_{ki})_{1 \leq k \leq n,\ 1 \leq i \leq c}$.

## 2.1  Entropy Based Fuzzy c-Means Clustering

Entropy Based Fuzzy c-means clustering (EFCM) is based on optimizing an objective function under the constraint for membership grade.

We consider following objective function $J_e$ for EFCM:

$$J_e(U,V) = \sum_{k=1}^{n} \sum_{i=1}^{c} \left\{ u_{ki} \| x_k - v_i \|^2 + \lambda u_{ki} \log u_{ki} \right\}.$$

here, $\lambda > 0$ is fuzzification parameters.

$J_e$ is an entropy based fuzzy c-means clustering (EFCM) [2].

Probabilistic constraint for FCM is as follows:

$$\mathcal{U}_f = \left\{ (u_{ki}) : u_{ki} \in [0,1], \sum_{i=1}^{c} u_{ki} = 1, \; ^\forall k \right\}.$$

## 2.2  Entropy Based Relational Fuzzy c-Means Clustering

The squared $L_2$-norm between $x_k$ and $x_t$ is denoted as $r_{kt} = \| x_k - x_t \|^2$ in entropy based relational fuzzy c-means clustering (ERFCM). $R$ is the matrix whose $(k,t)$-th element is $r_{kt}$. The objective function and constraint for $u_{ki}$ are the same as conventional EFCM. However, the dissimilarity between $x_k$ and $i$-th cluster denoted as $d_{ki}$ is calculated by using $r_{kt}$. This procedure is derived by substituting the optimal solution of cluster center $v_i$ for $\| x_k - v_i \|^2$ [11].

The algorithm of ERFCM is described as Algorithm 1.

---

**Algorithm 1.** Algorithm of ERFCM

---

**STEP1**  Set initial values $u_{ki}$ and parameter $\lambda$.

**STEP2**  Calculate the intermediate variable $v_i$ for ERFCM as follows:

$$v_i = \frac{(u_{1i}, \ldots, u_{ni})^T}{\sum_{k=1}^{n} u_{ki}}.$$

**STEP3**  Calculate $d_{ki}$ for ERFCM as follows:

$$d_{ki} = (Rv_i)_k - \frac{1}{2} v_i^T R v_i.$$

**STEP4**  Calculate $u_{ki}$ for ERFCM as follows:

$$u_{ki} = \frac{\exp\left(-\lambda^{-1} d_{ki}\right)}{\sum_{l=1}^{c} \exp\left(-\lambda^{-1} d_{kl}\right)}.$$

**STEP5**  If convergence criterion is satisfied, stop. Otherwise go back to **STEP2**.

---

## 3   Entropy Based Relational Crisp Possibilistic Clustering

### 3.1   Objective Function and Optimal Solution

We consider the following objective function for entropy based relational crisp possibilistic clustering (ERCPCM):

$$J_{ecp}(U) = \frac{\sum_{i=1}^{c} \sum_{k=1}^{n} \sum_{t=1}^{n} u_{ki} u_{ti} r_{kt}}{2 \sum_{l=1}^{n} u_{li}} + \lambda \sum_{i=1}^{c} \sum_{k=1}^{n} u_{ki} (\log u_{ki} - 1) + \gamma \sum_{i=1}^{c} \sum_{k=1}^{n} |1 - u_{ki}|,$$

here, $\lambda > 0.0$ and $\gamma > 0.0$ are the parameters of ERCPCM. The condition $\mathscr{U}_p$ for ERCPCM is written as follows:

$$\mathscr{U}_p = \left\{ (u_{ki}) \; : \; u_{ki} \in [0,1], \quad ^{\forall}k \right\}, \tag{1}$$

where, we have omitted original constraint $0 < \sum_{k=1}^{n} u_{ki} \le n$. $d_{ki}$ is as follows:

$$d_{ki} = (Rv_i)_k - \frac{1}{2} v_i^T R v_i,$$

here, $v_i$ is as follows:

$$v_i = \frac{(u_{1i}, \ldots, u_{ni})^T}{\sum_{k=1}^{n} u_{ki}}.$$

The main problem of constructing the algorithm of ERCPCM is how to derive the optimal solution of $u_{ki}$. Each membership $u_{ki}$ could be solved separately in ERCPCM procedure because of the condition $\mathscr{U}_p$. First, we will consider the following semi-objective function:

$$J_{ecp}^{ki}(u_{ki}) = u_{ki} d_{ki} + \lambda u_{ki} (\log u_{ki} - 1) + \gamma |1 - u_{ki}|.$$

We will decompose $1 - u_{ki} = \xi^+ - \xi^-$, in order to obtain partial derivatives with respect to $u_{ki}$ where all element of $\xi^+$ and $\xi^-$ are nonnegative. The semi-objective function is rewritten by using decomposition method as follows:

$$J_{ecp}^{ki}(u_{ki}) = (u_{ki})^m d_{ki} + \lambda u_{ki} (\log u_{ki} - 1) + \gamma \left( \xi^+ + \xi^- \right).$$

Constraints are as follows:

$$1 - u_{ki} \leq \xi^+, \quad 1 - u_{ki} \geq -\xi^-, \quad \xi^+, \xi^- \geq 0.$$

Introducing the Lagrange multipliers $\beta^+$, $\beta^-$, $\psi^+$, and $\psi^- \geq 0$, Lagrangian $L^{ecp}$ is as follows:

$$
\begin{aligned}
L^{ecp} = {} & u_{ki}d_{ki} + \lambda u_{ki}(\log u_{ki} - 1) + \gamma(\xi^+ + \xi^-) \\
& + \beta^+(1 - u_{ki} - \xi^+) + \beta^-(-1 + u_{ki} - \xi^-) - \psi^+\xi^+ - \psi^-\xi^-.
\end{aligned}
\tag{2}
$$

From $\frac{\partial L^{ecp}}{\partial \xi^+} = 0$ and $\frac{\partial L^{ecp}}{\partial \xi^-} = 0$,

$$\gamma - \beta^+ - \psi^+ = 0, \quad \gamma - \beta^- - \psi^- = 0. \tag{3}$$

Since $\psi^+$, $\psi^- \geq 0$, conditions $0 \leq \beta^+ \leq \gamma$ and $0 \leq \beta^- \leq \gamma$ are obtained from (3). Substituting (3) into (2), the Lagrangian $L^{ecp}$ is simplified as follows:

$$L^{ecp} = u_{ki}d_{ki} + \lambda u_{ki}(\log u_{ki} - 1) + \beta(1 - u_{ki}). \tag{4}$$

Here, $\beta = \beta^+ - \beta^-$ and satisfies $-\gamma \leq \beta \leq \gamma$.
From $\frac{\partial L^{ecp}}{\partial u_{ki}} = 0$,

$$u_{ki} = \exp\left(-\frac{d_{ki} - \beta}{\lambda}\right) \tag{5}$$

Substituting (5) to (4), the Lagrangian dual problem is written as follows:

$$L_d^{ecp} = \beta - \lambda \exp\left(-\frac{d_{ki} - \beta}{\lambda}\right)$$

From $\frac{\partial L_d^{cp}}{\partial \beta} = 0$, this dual problem is solved as,

$$\beta = d_{ki}. \tag{6}$$

The optimal solution of primal problem is derived by considering (5), (6) and $-\gamma \leq \beta \leq \gamma$. In the case of $\beta < 0$ is not realized since $d_{ki}$ is always positive. Second, the case of $0 \leq \beta \leq \gamma$, the optimal solution is $u_{ki} = 1$ since $\beta = d_{ki}$. Third, the case of $\gamma < \beta$, the optimal solution is $u_{ki} = \exp\left(-\frac{d_{ki} - \beta}{\lambda}\right)$. Finally, the optimal solution for $u_{ki}$ of proposed method is derived as follows:

$$u_{ki} = \begin{cases} 1 & 0 \leq d_{ki} \leq \gamma \\ \exp\left(-\frac{d_{ki} - \beta}{\lambda}\right) & \gamma < d_{ki} \end{cases} \tag{7}$$

## 3.2 Algorithm of Proposed Method

The algorithm of ERCPCM is described as Algorithm 2.

---

**Algorithm 2.** Algorithm of ERCPCM

---

**STEP1** Set initial values $u_{ki}$ and parameters $\lambda$ and $\gamma$.

**STEP2** Calculate the intermediate variable $v_i$ for ERCPCM as follows:

$$v_i = \frac{(u_{1i}, \ldots, u_{ni})^T}{\sum_{k=1}^{n} u_{ki}}.$$

**STEP3** Calculate $d_{ki}$ for ERCPCM as follows:

$$d_{ki} = (Rv_i)_k - \frac{1}{2} v_i^T R v_i.$$

**STEP4** Calculate $u_{ki}$ for ERCPCM as follows:

$$u_{ki} = \begin{cases} 1 & 0 \leq d_{ki} \leq \gamma \\ \exp\left(-\frac{d_{ki}-\beta}{\lambda}\right) & \gamma < d_{ki} \end{cases}$$

**STEP5** If convergence criterion is satisfied, stop. Otherwise go back to **STEP2**.

---

## 4 Sequential Extraction Algorithm

The objective function of PCM can be minimized separately because probabilistic constraint used in FCM is not considered [5]. This implies the drawback that the cluster centers calculated by PCM would be completely the same. The sequential extraction procedure is constructed by considering this drawback. The basis of this procedure has been already proposed [5] and discussed [6]. We consider the sequential extraction algorithm by ERCPCM as well as conventional methods. The membership grade of the datum which has small $d_{ki}$ could be calculated as $u_{ki} = 1$. These data should be considered in one crisp cluster. Therefore, ERCPCM can extract crisp one cluster at a time by minimizing the objective function in the case of $c = 1$. The algorithm of sequential extraction by ERCPCM is described as Algorithm 3.

---

**Algorithm 3.** Sequential extraction algorithm based on ERCPCM

---

**STEP 1** Give $X$, initial values $u_{ki}$ and parameters $\lambda$ and $\gamma$.

**STEP 2** Repeat ERCPCM algorithm with $c = 1$ until convergence criterion is satisfied.

**STEP 3** Extract $\{x_k \mid u_{ki} = 1\}$ from $X$.

**STEP 4** If $X = \emptyset$ or convergence criterion is satisfied, stop. Otherwise, give initial values and go back to **STEP2**.

---

# 5  Numerical Examples

## 5.1  Sequential Extraction with Artificial Data Set

We show the numerical examples of sequential extraction with polaris data set and three artificial data set described in Figs. 1, 5, 7 and 9. The polaris data set consists of 51 data point and should be classified into three clusters. Artificial data 1 consists of 100 data point and should be classified into two clusters. Artificial data 2 consists of 120 data point and should be classified into three clusters. Artificial data 3 consists of 120 data point and should be classified into four clusters.

We set the $\lambda = 0.5$ used in ERCPCM. First, we show the results of polaris data set with $\gamma = 1.0$, $\gamma = 2.0$, and $\gamma = 3.0$ described Figs. 2, 3, and 4. These results shows that the large $\gamma$ induce strong crispness. It is shown that sequential extraction algorithm with large $\gamma$ extract large crisp clusters from these results.

Next, we show the results of three artificial data set by sequential extraction. Figs. 6, 8, and 10 are the results of sequential extraction with $\gamma = 0.8$, $\gamma = 1.8$, and $\gamma = 1.0$, respectively. The value displayed on each data point means the order of extracting clusters. These results show that the large $\lambda$ induces strong crispness and shows the broad crisp area and extracts a few clusters.

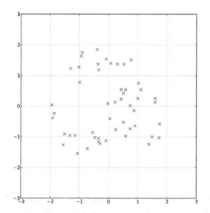

**Fig. 1** Polaris data ($n = 51$, $p = 2$, $c = 3$)

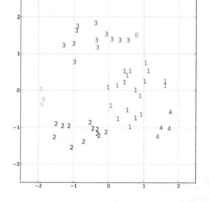

**Fig. 2** Sequential extraction by ERCPCM with $\lambda = 0.5$, $\gamma = 1.0$, number of extracted cluster is 6

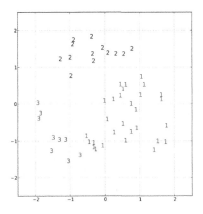

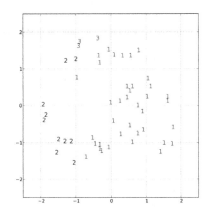

**Fig. 3** Sequential extraction by ERCPCM with $\lambda = 0.5$, $\gamma = 2.0$, number of extracted cluster is 3

**Fig. 4** Sequential extraction by ERCPCM with $\lambda = 0.5$, $\gamma = 3.0$, number of extracted cluster is 3

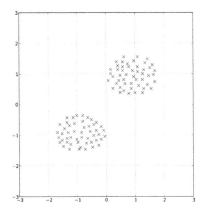

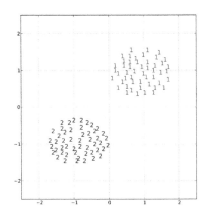

**Fig. 5** Artificial data 1 ($n = 100$, $p = 2$, $c = 2$)

**Fig. 6** Sequential extraction by ERCPCM with $\lambda = 0.5$, $\gamma = 0.8$, number of extracted cluster is 2

## 5.2 Sequential Extraction with Iris Data Set

Next, we show the results of sequential extraction by different $\gamma$ with Iris data set which consists of 150 data point and has four attributes. This data set should be classified into three clusters.

We evaluate the effectiveness of sequential extraction by ERCPCM with the results of average of Rand Index (RI) [17] and number of extracted clusters out of 100

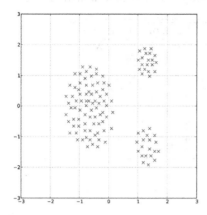

**Fig. 7** Artificial data 2
($n = 120$, $p = 2$, $c = 3$)

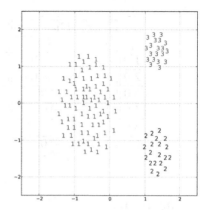

**Fig. 8** Sequential extraction by ERCPCM with $\lambda = 0.5$, $\gamma = 1.8$, number of extracted cluster is 3

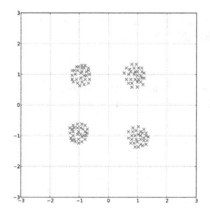

**Fig. 9** Artificial data 1
($n = 120$, $p = 2$, $c = 4$)

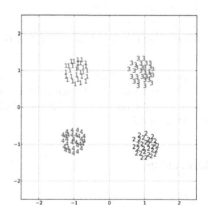

**Fig. 10** Sequential extraction by ERCPCM with $\lambda = 0.5$, $\gamma = 1.0$, number of extracted cluster is 4

trials described in Table 1. It can be shown that the larger $\gamma$ is, the smaller the number of extracted clusters is. In these tables, Ave SD Max and Min means average, standard variation, maximal value and minimal value, respectively. It can be shown that large $\gamma$ induce broad crisp area These tables show that ERCPCM are robust for initial values because the value of SD is quite small. The parameter $\gamma$ further influence the classification result and sequential extraction.

**Table 1** The results of 100 trials of sequential extraction by ERCPCM with Iris data set

| | Rand Index | | Num. of clusters | | | |
|---|---|---|---|---|---|---|
| $\gamma$ | Ave | SD | Ave | SD | Max | Min |
| 1.0 | 0.757 | $7.26 \times 10^{-4}$ | 15.66 | 0.62 | 17 | 14 |
| 2.0 | 0.768 | $3.90 \times 10^{-5}$ | 7.95 | 0.22 | 8 | 7 |
| 3.0 | 0.767 | 0.00 | 6 | 0 | 6 | 6 |
| 4.0 | 0.771 | 0.00 | 5 | 0 | 5 | 5 |
| 5.0 | 0.768 | $1.32 \times 10^{-4}$ | 5.58 | 0.49 | 6 | 5 |
| 6.0 | 0.772 | 0.00 | 4 | 0 | 4 | 4 |

## 6 Conclusions

In this paper, we have proposed entropy based relational crisp possibilistic clustering (ERCPCM) and sequential extraction algorithm. We have moreover shown the results of sequential extraction algorithm based on proposed method. We have verified that sequential extraction algorithm by proposed method could be useful for euclidean relational data.

As possible future works, we will consider sequential extraction algorithm for non-euclidean relational data based on NERFCM [12]

**Acknowledgments.** This study is partly supported by Research Fellowships of the Japan Society for the Promotion of Science for Young Scientists and the Grant-in-Aid for Scientific Research (C) (Project No.21500212) from the Ministry of Education, Culture, Sports, Science and Technology, Japan.

## References

1. Bezdek, J.C.: Pattern Recognition with Fuzzy Objective Function Algorithms. Plenum Press, New York (1981)
2. Miyamoto, S., Mukaidono, M.: Fuzzy c-means as a regularization and maximum entropy approach. In: Proc. of the 7th International Fuzzy Systems Association World Congress, IFSA 1997, vol. 2, pp. 86–92 (1997)
3. Miyamoto, S., Ichihashi, H., Honda, K.: Algorithms for Fuzzy Clustering. Springer, Heidelberg (2008)
4. Krishnapuram, R., Keller, J.M.: A possibilistic approach to clustering. IEEE Transactions on Fuzzy Systems 1(2), 98–110 (1993)
5. Davé, R.N., Krishnapuram, R.: Robust clustering methods: A unified view. IEEE Transactions on Fuzzy Systems 5(2), 270–293 (1997)
6. Miyamoto, S., Kuroda, Y., Arai, K.: Algorithms for Sequential Extraction of Clusters by Possibilistic Method and Comparison with Mountain Clustering. Journal of Advanced Computational Intelligence and Intelligent Informatics, JACIII 12(5), 448–453 (2008)
7. Davé, R.N.: Characterization and detection of noise in clustering. Pattern Recognition Letters 12(11), 657–664 (1991)

8. Roubens, M.: Pattern classification problems and fuzzy sets. Fuzzy Sets and Systems 1, 239–253 (1978)
9. Endo, Y.: On Entropy Based Fuzzy Non Metric Model – Proposal, Kernelization and Pairwise Constraints –. Journal of Advanced Computational Intelligence and Intelligent Informatics, JACIII 16(1), 169–173 (2012)
10. Ruspini, E.: Numerical methods for fuzzy clustering. Information Science 2(3), 319–350 (1970)
11. Hathaway, R.J., Davenport, J.W., Bezdek, J.C.: Relational Duals of the $c$-Means Clustering Algorithms. Pattern Recognition 22(2), 205–212 (1989)
12. Hathaway, R.J., Bezdek, J.C.: Nerf $c$-Means: Non-Euclidean Relational Fuzzy Clustering. Pattern Recognition 27(3), 429–437 (1994)
13. Filippone, M.: Dealing with non-metric dissimilarities in fuzzy central clustering algorithms. Journal International Journal of Approximate Reasoning 50(2), 363–384 (2009)
14. Tibshirani, R.: Regression shrinkage and selection via the lasso. Journal of the Royal Statistical Society Series B 58(1), 267–288 (1996)
15. Inokuchi, R., Miyamoto, S.: Sparse Possibilistic Clustering with L1 Regularization. In: Proc. of The 2007 IEEE International Conference on Granular Computing (GrC 2007), pp. 442–445 (2007)
16. Hamasuna, Y., Endo, Y.: On Sparse Possibilistic Clustering with Crispness – Classification Function and Sequential Extraction –. In: Joint 6th International Conference on Soft Computing and Intelligent Systems and 12th International Symposium on Advanced Intelligent Systems, SCIS & ISIS 2012, pp. 1801–1806 (2012)
17. Rand, W.M.: Objective criteria for the evaluation of clustering methods. Journal of the American Statistical Association 66(336), 846–850 (1971)

# EM-Based Clustering Algorithm for Uncertain Data

Naohiko Kinoshita and Yasunori Endo

**Abstract.** In recent years, advanced data analysis techniques to get valuable knowledge from data using computing power of today are required. Clustering is one of the unsupervised classification technique of the data analysis. Information on a real space is transformed to data in a pattern space and analyzed in clustering. However, the data should be often represented not by a point but by a set because of uncertainty of the data, e.g., measurement error margin, data that cannot be regarded as one point, and missing values in data.

These uncertainties of data have been represented as interval range and many clustering algorithms for these interval ranges of data have been constructed. However, the guideline to select an available distance in each case has not been shown so that this selection problem is difficult. Therefore, methods to calculate the dissimilarity between such uncertain data without introducing a particular distance, e.g., nearest neighbor one and so on, have been strongly desired. From this viewpoint, we proposed a concept of tolerance. The concept represents a uncertain data not as an interval but as a point with a tolerance vector. However, the distribution of uncertainty which represents the tolerance is uniform distribution and it it difficult to handle other distributions of uncertainty in the framework of tolerance, e.g., the Gaussian distribution, with HCM or FCM.

In this paper, we try to construct an clustering algorithm based on the EM algorithm which handles uncertain data which are represented by the Gaussian distributions through solving the optimization problem. Moreover, effectiveness of the proposed algorithm will be verified.

Naohiko Kinoshita
Graduate School of Systems and Information Engineering, University of Tsukuba,
Tennodai 1-1-1, Tsukuba, Ibaraki, 305-8573, Japan
e-mail: s1220594@u.tsukuba.ac.jp

Yasunori Endo
Faculty of Engineering, Information and Systems, University of Tsukuba, Tennodai 1-1-1,
Tsukuba, Ibaraki, 305-8573, Japan
e-mail: endo@risk.tsukuba.ac.jp

V.-N. Huynh et al. (eds.), *Knowledge and Systems Engineering, Volume 2,*
Advances in Intelligent Systems and Computing 245,
DOI: 10.1007/978-3-319-02821-7_8, © Springer International Publishing Switzerland 2014

# 1 Introduction

In recent years, data from many natural and social phenomena are accumulated into huge databases in the world wide network of computers. However, it is difficult to analyze by hand since data are massive and complex. Thus, advanced data analysis techniques to get valuable knowledge from data using computing power of today are required.

There are roughly supervised classification and unsupervised classification in the classification technique of the data analysis. Data clustering is the most popular unsupervised classification method. Clustering is to divide a set of objects into some categories which are referred to as clusters. Clustering techniques are also classified into non-parametric methods and parametric methods.

The method of hard $c$-means clustering [1] (HCM) is the most popular non-parametric method. HCM classifies a set of objects into $c$ clusters using a simple procedure. Assignments to clusters are deterministic, that is, an object $x$ is assigned to one cluster exclusively. There are many extensions of HCM. Fuzzy $c$-means clustering [2] (FCM) has been remarked by many researchers. It is an extension of HCM using the concept of fuzzy sets. It should be noted that FCM does not require the details of fuzzy set theory.

EM algorithm [3] is the most popular parametric method. The EM algorithm has two steps, an expectation (E) step and a maximization (M) step. The E step calculates the expectation of the log-likelihood using the current estimation for the parameters. The M step computes parameters maximizing the expected log-likelihood which is found in the E step. The EM algorithm finds maximum likelihood in statistical models, which depend on unobserved latent variables, by iteration of those steps.

By the way, each data on a real space is generally transformed to a point in a pattern space and analyzed in clustering. However, the data should be often represented not by a point but by a set because of uncertainty of the data. The uncertainty of data have been represented as interval range and many clustering algorithms for these interval ranges of data have been constructed [4, 5, 6]. In these algorithms, distances between intervals, e.g., nearest neighbor distance, furthest neighbor distance or Hausdorff distance, have been used to calculate the dissimilarity between the target data in clustering. However, there has been no guideline to select the available distance between intervals in each case so that this selection problem is difficult. Therefore, methods to calculate the dissimilarity between such uncertain data without introducing the distance between intervals, e.g., nearest neighbor one and so on, have been strongly desired.

We introduced the new concept of tolerance, which includes the above-mentioned uncertainties of data and proposed some clustering algorithms [7, 8]. The basic concept of tolerance is as follows: Now, let's assume that a datum $x \in \Re^p$ with uncertainty is represented as an interval $[\kappa, \overline{\kappa}]$. We have two ways to represent the data, one is $[x + \kappa, x + \overline{\kappa}]$ and the other is $x + (\overline{\kappa} + \kappa)/2 + \varepsilon$ $(\varepsilon = (\varepsilon_1, \ldots, \varepsilon_p)^T)$ under constraints $|\varepsilon_j| \leq (\overline{\kappa}_j - \kappa_j)/2$ $(j = 1, \ldots, p)$. The latter way corresponds to the basic concept of the tolerance and $\varepsilon$ is called tolerance vector. The concept is convenient to handle uncertain data. However, the distribution of uncertainty which represents the tolerance is uniform distribution and it it difficult to handle other distributions of uncertainty in the framework of tolerance, e.g., the Gaussian distribution, with HCM or FCM.

In this paper, we try to construct an clustering algorithm based on the EM algorithm which handles uncertain data which are represented by the Gaussian distributions. First, we will consider the expectation of log-likelihood for the uncertain data by the Gaussian distributions as an objective function. Next, we will derive the optimal solutions to parameters which maximizes the objective function and construct new EM-based clustering algorithm for uncertain data by Gaussian distributions. Moreover, effectiveness of the proposed algorithm will be verified.

## 2 Main Results

Conventional EM-based clustering algorithm classifies certain data, not uncertain data. In this section, we construct a new EM-based clustering algorithm which can classify uncertain data of which uncertainty is represented by the Gaussian distribution.

First, we consider the expectation of log-likelihood for uncertain data as an objective function. Next, we derive the optimal solutions which maximize the objective function. Third, we construct an EM-based clustering algorithm for uncertain data by using the optimal solutions. We discuss two cases separably, one-dimensional distribution and multi-dimensional distribution.

### 2.1 Objective Function

Let $g_k(x|\phi_{g_k})$ and $p_i(x|\phi_{p_i})$ be a probability density function which represents an uncertainty of an object $x_k$ and a probability density function which corresponds a class $G_i$, respectively. Here $\phi_{g_k}$ and $\phi_{p_i}$ are parameter vectors in the above density functions. In case of one-dimensional Gaussian distribution, $\phi_{g_k} = (\mu_{g_k}, \sigma_{g_k}^2)$ and $\phi_{p_i} = (\mu_{p_i}, \sigma_{p_i}^2)$. $\mu_{g_k}$ and $\sigma_{g_k}^2$ mean an average and a variance of Gaussian distribution $g_k(x|\phi_{g_k})$, respectively.

The value of $x_k$ in the mixture distribution is as follows:

$$q(x_k|\Phi, \phi_{g_k}) = \sum_{i=1}^{c} \alpha_i q_i(x_k|\phi_{p_i}, \phi_{g_k}).$$

$\alpha_i$ means a mixture ratio and $\sum_{i=1}^{c} \alpha_i = 1$, and

$$q_i(x_k|\phi_{p_i}, \phi_{g_k}) = \int_{-\infty}^{\infty} p_i(x|\phi_{p_i}) g_k(x|\phi_{g_k}) dx.$$

The expectation of the log-likelihood $Q$ is given as follows:

$$Q(\Phi|\Phi') = \sum_{i=1}^{c} \sum_{k=1}^{n} \log[\alpha_i q_i(x_k|\phi_{p_i}, \phi_{g_k})] \frac{\alpha_i' q_i(x_k|\phi_{p_i}', \phi_{g_k})}{q(x_k|\Phi, \phi_{g_k})}. \tag{1}$$

Here

$$\Phi = (\alpha_1, \ldots, \alpha_c, \phi_{p_1}, \ldots, \phi_{p_c}).$$

$\alpha_i'$ is an estimation of $\alpha_i$, $\Phi'$ is an estimation of $\Phi$ and $\sum_{i=1}^{c} \alpha_i' q_i(x_k|\phi_{p_i}', \phi_{g_k}) = q(x_k|\Phi', \phi_{g_k})$. By introducing the notations $\Psi_i$ and $\psi_{ki}$:

$$\Psi_i = \sum_{k=1}^{n} \frac{\alpha_i' q_i(x_k|\phi_{p_i}', \phi_{g_k})}{q(x_k|\Phi', \phi_{g_k})} = \sum_{k=1}^{n} \psi_{ki},$$

(1) is rewritten as

$$Q(\Phi|\Phi') = \sum_{i=1}^{c} \Psi_i \log \alpha_i + \sum_{i=1}^{c} \sum_{k=1}^{n} \psi_{ki} \log q_i(x_k|\phi_{p_i}, \phi_{g_k}) \tag{2}$$

from $\sum_{i=1}^{c} \Psi_i = n$.

From here, we assume Gaussian mixture distribution as the probability distribution. First, we consider one-dimensional distribution. Next, we extend the discussion into multi-dimensional distribution.

## 2.2 Optimal Solutions of One-Dimensional Distribution

First, we derive the optimal solution to the mixture ratio which maximizes (2) by using the method of Lagrange multiplier as follows:

$$L = Q(\Phi|\Phi') - \lambda(\sum_{i=1}^{c} \alpha_i - 1).$$

From partially differentiating $L$ by $\alpha_i$ as follows:

$$\frac{\partial L}{\partial \alpha_i} = \frac{\Psi_i}{\alpha_i} - \lambda = 0,$$

we get $\lambda = n$. Therefore, the optimal solution to $\alpha_i$ is as follows:

$$\alpha_i = \frac{\Psi_i}{n} = \frac{1}{n} \sum_{k=1}^{n} \frac{\alpha_i' q_i(x_k|\phi_{p_i}')}{q(x_k|\Phi', \phi_{g_k})} \tag{3}$$

Next, we consider the optimal solution to the average $\mu_{p_i}$ which maximizes (2). Let one-dimensional Gaussian distribution be

$$p_i(x|\phi_{p_i}) = \frac{1}{\sqrt{2\pi\sigma_{p_i}^2}}\exp\left\{-\frac{(x-\mu_{p_i})^2}{2\sigma_{p_i}^2}\right\}.$$

Here $\phi_{p_i} = (\mu_{p_i}, \sigma_{p_i}^2)$. $\mu_{p_i}$ and $\sigma_{p_i}^2$ mean the average and the variance of $C_i$, respectively. We assume that each object $x_k$ with uncertainty is represented by the following probability density function:

$$g_k(x|\phi_{g_k}) = \frac{1}{\sqrt{2\pi\sigma_{g_k}^2}}\exp\left\{-\frac{(x-\mu_{g_k})^2}{2\sigma_{g_k}^2}\right\}$$

Here $\phi_{g_k} = (\mu_{g_k}, \sigma_{g_k}^2)$. $\mu_{g_k}$ and $\sigma_{g_k}^2$ mean the average and the variance of $x_k$, respectively. Therefore, $q_i(x_k|\phi_{p_i}, \phi_{g_k})$ is represented as follows:

$$\begin{aligned}
q_i(x_k|\phi_{p_i}, \phi_{g_k}) &= \int_{-\infty}^{\infty} p_i(x|\phi_{p_i})g_k(x|\phi_{g_k})dx \\
&= \int_{-\infty}^{\infty} \frac{1}{2\pi\sigma_{p_i}\sigma_{g_k}}\exp\left\{-\frac{\sigma_{g_k}^2(x-\mu_{p_i})^2 + \sigma_{p_i}^2(x-\mu_{g_k})^2}{2\sigma_{p_i}^2\sigma_{g_k}^2}\right\}dx \\
&= \frac{1}{2\pi\sigma_{p_i}\sigma_{g_k}}\exp\left\{-\frac{(\mu_{g_k}-\mu_{p_i})^2}{2(\sigma_{p_i}^2+\sigma_{g_k}^2)}\right\} \\
&\quad \cdot \int_{-\infty}^{\infty}\exp\left\{-\frac{(\sigma_{p_i}^2+\sigma_{g_k}^2)}{2\sigma_{p_i}^2\sigma_{g_k}^2}\left(x-\frac{\sigma_{g_k}^2\mu_{p_i}+\sigma_{p_i}^2\mu_{g_k}}{\sigma_{p_i}^2+\sigma_{g_k}^2}\right)^2\right\}dx \\
&= \frac{1}{2\pi\sigma_{p_i}\sigma_{g_k}}\sqrt{\frac{2\pi\sigma_{o_i}^2\sigma_{g_k}^2}{(\sigma_{p_i}^2+\sigma_{g_k}^2)}}\exp\left\{-\frac{(\mu_{g_k}-\mu_{p_i})^2}{2(\sigma_{p_i}^2+\sigma_{g_k}^2)}\right\} \\
&= \frac{1}{\sqrt{2\pi(\sigma_{p_i}^2+\sigma_{g_k}^2)}}\exp\left\{-\frac{(\mu_{g_k}-\mu_{p_i})^2}{2(\sigma_{p_i}^2+\sigma_{g_k}^2)}\right\}.
\end{aligned}$$

Therefore, (2) is rewritten as follows:

$$Q(\Phi|\Phi') = \sum_{i=1}^{c}\Psi_i\log\alpha_i + \sum_{i=1}^{c}\sum_{k=1}^{n}\psi_{ki}\log\left(\frac{1}{\sqrt{2\pi(\sigma_{p_i}^2+\sigma_{g_k}^2)}}\exp\left\{-\frac{(\mu_{g_K}-\mu_{p_i})^2}{2(\sigma_{p_i}^2+\sigma_{g_k}^2)}\right\}\right).$$

We partially differentiate (2) for deriving $\mu_{p_i}$ which maximizes (2). The first term of (2) is also a function for $\mu_{p_i}$. However we now regard it as a constant term and partially differentiate it as follows:

$$\frac{\partial Q}{\partial \mu_{p_i}} = \sum_{k=1}^{n}\psi_{ki}\frac{\mu_{g_k}-\mu_{p_i}}{(\sigma_{p_i}^2+\sigma_{g_k}^2)} = 0.$$

Hence, the optimal solution to the average $\mu_{p_i}$ which maximizes $Q$ is as follows:

$$\mu_{p_i} = \frac{1}{\Psi_i} \sum_{k=1}^{n} \psi_{ki} \mu_{g_k}. \tag{4}$$

Last, we derive the optimal solution to the variance $\sigma_{p_i}^2$ which maximizes (2) similar to the average as follows:

$$\frac{\partial Q}{\partial \sigma_{p_i}^2} = \frac{1}{2} \sum_{k=1}^{n} \psi_{ki} \left( -\frac{1}{\sigma_{p_i}^2 + \sigma_{g_k}^2} + \frac{(\mu_{g_k} - \mu_{p_i})^2}{(\sigma_{p_i}^2 + \sigma_{g_k}^2)^2} \right) = 0 \tag{5}$$

There are many methods to solve (5). We use Newton's method in this paper.

$$f(\sigma_{p_i}^2) = \sum_{k=1}^{n} \psi_{ki} \frac{-(\sigma_{p_i}^2 + \sigma_{g_k}^2) + (\mu_{g_k} - \mu_{p_i})^2}{(\sigma_{p_i}^2 + \sigma_{g_k}^2)^2} = 0,$$

$$\sigma_{p_i}^{2\,(t+1)} = \sigma_{p_i}^{2\,(t)} - \frac{f(\sigma_{p_i}^{2\,(t)})}{f'(\sigma_{p_i}^{2\,(t)})}.$$

Hence, the approximation of the optimal solution to the variance $\sigma_{p_i}^2$ is as follows:

$$\sigma_{p_i}^{2\,(t+1)} = \sigma_{p_i}^{2\,(t)} - \frac{\sum_{k=1}^{n} \psi_{ki} \dfrac{-(\sigma_{p_i}^{2\,(t)} + \sigma_{g_k}^2) + (\mu_{g_k} - \mu_{p_i})^2}{(\sigma_{p_i}^{2\,(t)} + \sigma_{g_k}^2)^2}}{\sum_{k=1}^{n} \psi_{ki} \dfrac{(\sigma_{p_i}^{2\,(t)} + \sigma_{g_k}^2) - 2(\mu_{g_k} - \mu_{p_i})^2}{(\sigma_{p_i}^{2\,(t)} + \sigma_{g_k}^2)^3}} \tag{6}$$

### 2.3  Optimal Solutions of Multi-dimensional Distribution

First, we derive the optimal solution to the mixture ratio which maximizes $Q$. Similar to one-dimensional distribution, the optimal solution is given as (3).

Next, we consider the optimal solution to the average $\mu_{p_i}$ which maximizes $Q$. Let multi-dimensional Gaussian distribution be

$$p_i(x|\phi_{p_i}) = \frac{1}{(2\pi)^{\frac{p}{2}} |\Sigma_{p_i}|^{\frac{1}{2}}} \exp\left\{ -\frac{1}{2}(x - \mu_{p_i})^T \Sigma_{p_i}^{-1}(x - \mu_{p_i}) \right\}.$$

Here $\phi_{p_i} = (\mu_{p_i}, \Sigma_{p_i})$. $\mu_{p_i} = (\mu_{p_i}^1, \dots, \mu_{p_i}^p)^T$ and $\Sigma_{p_i} = (\sigma_{p_i}^{jl})$ $(1 \le j, l \le p)$ mean the average and the variance-covariance matrix of $C_i$, respectively. $|\Sigma_{p_i}|$ is the determinant of $\Sigma_{p_i}$. We assume that each object $x_k$ with uncertainty is represented by the following probability density function:

$$g_k(x|\phi_{g_k}) = \frac{1}{(2\pi)^{\frac{p}{2}} |\Sigma_{g_k}|^{\frac{1}{2}}} \exp\left\{ -\frac{1}{2}(x - \mu_{g_k})^T \Sigma_{g_k}^{-1}(x - \mu_{g_k}) \right\}.$$

Here $\phi_{g_k} = (\mu_{g_k}, \Sigma_{g_k})$. $\mu_{g_k} = (\mu_{g_k}^1, \ldots, \mu_{g_k}^p)^T$ and $\Sigma_{g_k} = (\sigma_{g_k}^{jl})$ $(1 \leq j, l \leq p)$ mean the average and the variance-covariance matrix of $x_k$, respectively. Therefore, $q_i(x_k|\phi_{p_i}, \phi_{g_k})$ is represented as follows:

$$
\begin{aligned}
&q_i(x_k|\phi_{p_i}, \phi_{g_k}) \\
&= \int_{-\infty}^{\infty} p_i(x_k|\phi_{p_i}) g_k(x|\phi_{g_k}) dx \\
&= \int_{-\infty}^{\infty} \frac{1}{(2\pi)^p |\Sigma_{p_i}|^{\frac{1}{2}} |\Sigma_{g_k}|^{\frac{1}{2}}} \\
&\quad \cdot \exp\left\{ -\frac{(x-\mu_{p_i})^T \Sigma_{p_i}^{-1}(x-\mu_{p_i}) + (x-\mu_{g_k})^T \Sigma_{g_k}^{-1}(x-\mu_{g_k})}{2} \right\} dx. \quad (7)
\end{aligned}
$$

Here

$$
\begin{aligned}
&(x-\mu_{p_i})^T \Sigma_{p_i}^{-1}(x-\mu_{p_i}) + (x-\mu_{g_k})^T \Sigma_{g_k}^{-1}(x-\mu_{g_k}) \\
&= x^T \Sigma_{p_i}^{-1} x - 2\mu_{p_i}^T \Sigma_{p_i}^{-1} x + \mu_{p_i}^T \Sigma_{p_i}^{-1} \mu_{p_i} + x^T \Sigma_{g_k}^{-1} x - 2\mu_{g_k}^T \Sigma_{g_k}^{-1} x + \mu_{g_k}^T \Sigma_{g_k}^{-1} \mu_{g_k} \\
&= x^T (\Sigma_{p_i}^{-1} + \Sigma_{g_k}^{-1}) x - 2(\mu_{p_i}^T \Sigma_{p_i}^{-1} + \mu_{g_k}^T \Sigma_{g_k}^{-1}) x + \mu_{p_i}^T \Sigma_{p_i}^{-1} \mu_{p_i} + \mu_{g_k}^T \Sigma_{g_k}^{-1} \mu_{g_k}. \quad (8)
\end{aligned}
$$

By substituting (8) in (7),

$$
\begin{aligned}
q_i(x_k|\phi_{p_i}, \phi_{g_k}) &= \int_{\infty}^{\infty} \frac{1}{(2\pi)^p |\Sigma_{p_i}|^{\frac{1}{2}} |\Sigma_{g_k}|^{\frac{1}{2}}} \\
&\quad \exp\left\{ -\frac{(x-\mu_{p_i})^T \Sigma_{p_i}^{-1}(x-\mu_{p_i}) + (x-\mu_{g_k})^T \Sigma_{g_k}^{-1}(x-\mu_{g_k})}{2} \right\} dx \\
&= \frac{1}{(2\pi)^{\frac{p}{2}} |\Sigma_{p_i}|^{\frac{1}{2}} |\Sigma_{g_k}|^{\frac{1}{2}} |\Sigma_{p_i}^{-1} + \Sigma_{g_k}^{-1}|^{\frac{1}{2}}} \exp\left\{ -\frac{\mu_{p_i}^T \Sigma_{p_i}^{-1} \mu_{p_i} + \mu_{g_k}^T \Sigma_{g_k}^{-1} \mu_{g_k}}{2} \right. \\
&\quad \left. + \frac{(\mu_{p_i}^T \Sigma_{p_i}^{-1} + \mu_{g_k}^T \Sigma_{g_k}^{-1})(\Sigma_{p_i}^{-1} + \Sigma_{g_k}^{-1})^{-1}(\mu_{p_i}^T \Sigma_{p_i}^{-1} + \mu_{g_k}^T \Sigma_{g_k}^{-1})^T}{2} \right\}.
\end{aligned}
$$

Therefore, (2) is rewritten as follows:

$$
\begin{aligned}
Q(\Phi|\Phi') &= \sum_{i=1}^{c} \Psi_i \log \alpha_i + \sum_{i=1}^{c} \sum_{k=1}^{n} \psi_{ki} \log q_i(x_k|\phi_{p_i}, \phi_{g_k}) \\
&= \sum_{i=1}^{c} \Psi_i \log \alpha_i + \sum_{i=1}^{c} \sum_{k=1}^{n} \psi_{ki} \log \frac{1}{(2\pi)^{\frac{p}{2}} |\Sigma_{p_i}|^{\frac{1}{2}} |\Sigma_{g_k}|^{\frac{1}{2}} |\Sigma_{p_i}^{-1} + \Sigma_{g_k}^{-1}|^{\frac{1}{2}}} \\
&\quad + \sum_{i=1}^{c} \sum_{k=1}^{n} \psi_{ki} \left\{ -\frac{\mu_{p_i}^T \Sigma_{p_i}^{-1} \mu_{p_i} + \mu_{g_k}^T \Sigma_{g_k}^{-1} \mu_{g_k}}{2} \right. \\
&\quad \left. + \frac{(\mu_{p_i}^T \Sigma_{p_i}^{-1} + \mu_{g_k}^T \Sigma_{g_k}^{-1})(\Sigma_{p_i}^{-1} + \Sigma_{g_k}^{-1})^{-1}(\mu_{p_i}^T \Sigma_{p_i}^{-1} + \mu_{g_k}^T \Sigma_{g_k}^{-1})^T}{2} \right\}. \quad (9)
\end{aligned}
$$

We partially differentiate (2) for deriving $\mu_{p_i}$ which maximizes (2). The first term of (2) is also a function for $\mu_{p_i}$. However we now regard it as a constant term and partially differentiate (9) with respect to $\mu_{p_i}^j$ as follows:

$$\frac{\partial Q}{\partial \mu_{p_i}^j} = [\sigma_{p_i}^{1j}, \ldots, \sigma_{p_i}^{jj}, \ldots, \sigma_{p_i}^{pj}] \sum_{k=1}^n -\psi_{ki}(\Sigma_{g_k}^{-1}\mu_{p_i} - \Sigma_{g_k}^{-1}\mu_{g_k})$$

Hence, the optimal solution to $\mu_{p_i}$ is given as follows:

$$\mu_{p_i} = \frac{1}{\Psi_i} \sum_{k=1}^n \psi_{ki}\mu_{g_k} \tag{10}$$

Last, we derive the optimal solution to the variance-covariance matrix $\Sigma_{p_i}$ similar to the average. We partially differentiate (9) with respect to $\sigma_{p_i}^{jl}$ as follows:

$$2\frac{\partial Q}{\partial \sigma_{p_i}^{jl}} = -\sum_{k=1}^n \psi_{ki}\left\{ \frac{\partial}{\partial \sigma_{p_i}^{jl}}\left(\log|\Sigma_{p_i} + \Sigma_{g_k}|\right) + \frac{\partial}{\partial \sigma_{p_i}^{jl}}(\mu_{p_i}^T \Sigma_{p_i}^{-1}\mu_{p_i}) \right.$$
$$\left. -\frac{\partial}{\partial \sigma_{p_i}^{jl}}\left((\mu_{p_i}^T \Sigma_{p_i}^{-1} + \mu_{g_k}^T \Sigma_{g_k}^{-1})(\Sigma_{p_i}^{-1} + \Sigma_{g_k}^{-1})^{-1}(\mu_{p_i}^T \Sigma_{p_i}^{-1} + \mu_{g_k}^T \Sigma_{g_k}^{-1})^T\right) \right\}$$
$$= 0. \tag{11}$$

First, we partially differentiate the first term in the right side of (11) as follows:

$$\left[\frac{\partial}{\partial \sigma_{p_i}^{jl}}\left(\log|\Sigma_{p_i} + \Sigma_{g_k}|\right)\right] = (\Sigma_{p_i} + \Sigma_{g_k})^{-1}. \tag{12}$$

Second, we partially differentiate the second term in the right side of (11) as follows:

$$\left[\frac{\partial}{\partial \sigma_{p_i}^{jl}}\Sigma_{p_i}^{-1}\right] = -\left[\Sigma_{p_i}^{-1} E^{jl} \Sigma_{p_i}^{-1}\right].$$

Here $E^{jl}$ is a matrix in which the $jl$-th element is 1 and otherwise 0. Therefore, we get

$$\left[\frac{\partial}{\partial \sigma_{p_i}^{jl}}(\mu_{p_i}^T \Sigma_{p_i}^{-1}\mu_{pi})\right] = -\left[(\Sigma_{p_i}^{-1}\mu_{p_i})^T E^{jl} \Sigma_{p_i}^{-1}\mu_{p_i}\right]$$
$$= -\Sigma_{p_i}^{-1}\mu_{p_i}\mu_{p_i}^T \Sigma_{p_i}^{-1}. \tag{13}$$

Third, we partially differentiate the third term in the right side of (11) as follows:

$$\left[\frac{\partial}{\partial \sigma_{p_i}^{jl}}(\mu_{p_i}^T \Sigma_{p_i}^{-1} + \mu_{g_k}^T \Sigma_{g_k}^{-1})(\Sigma_{p_i}^{-1} + \Sigma_{g_k}^{-1})^{-1}(\mu_{p_i}^T \Sigma_{p_i}^{-1} + \mu_{g_k}^T \Sigma_{g_k}^{-1})^T\right]$$

$$= -\Sigma_{p_i}^{-1}\mu_{p_i}(\mu_{p_i}^T \Sigma_{p_i}^{-1} + \mu_{g_k}^T \Sigma_{g_k}^{-1})(\Sigma_{p_i}^{-1} + \Sigma_{g_k}^{-1})^{-1}\Sigma_{p_i}^{-1}$$
$$+ \Sigma_{p_i}^{-1}(\Sigma_{p_i}^{-1} + \Sigma_{g_k}^{-1})^{-1}(\Sigma_{p_i}^{-1}\mu_{p_i} + \Sigma_{g_k}^{-1}\mu_{g_k})(\mu_{p_i}^T \Sigma_{p_i}^{-1} + \mu_{g_k}^T \Sigma_{g_k}^{-1})(\Sigma_{p_i}^{-1} + \Sigma_{g_k}^{-1})^{-1}\Sigma_{p_i}^{-1}$$
$$- \Sigma_{p_i}^{-1}(\Sigma_{p_i}^{-1} + \Sigma_{g_k}^{-1})^{-1}(\Sigma_{p_i}^{-1}\mu_{p_i} + \Sigma_{g_k}^{-1}\mu_{g_k})\mu_{p_i}^T \Sigma_{p_i}^{-1}$$
$$= \Sigma_{p_i}^{-1}(\mu_{p_i} - \beta)(\mu_{p_i}^T - \alpha)\Sigma_{p_i}^{-1} - \Sigma_{p_i}^{-1}\mu_{p_i}\mu_{p_i}^T \Sigma_{p_i}^{-1}. \tag{14}$$

Here

$$\beta_{ki} = (\mu_{p_i}^T \Sigma_{p_i}^{-1} + \mu_{g_k}^T \Sigma_{g_k}^{-1})(\Sigma_{p_i}^{-1} + \Sigma_{g_k}^{-1})^{-1},$$
$$\gamma_{ki} = (\Sigma_{p_i}^{-1} + \Sigma_{g_k}^{-1})^{-1}(\Sigma_{p_i}^{-1}\mu_{p_i} + \Sigma_{g_k}^{-1}\mu_{g_k}) = \beta_{ki}^T.$$

From (12), (13) (14),

$$2\frac{\partial Q}{\partial \sigma_{p_i}^{jl}} = \sum_{k=1}^{n} \psi_{ki}\left\{-(\Sigma_{p_i} + \Sigma_{g_k})^{-1} + \Sigma_{p_i}^{-1}(\mu_{p_i} - \gamma_{ki})(\mu_{p_i}^T - \beta_{ki})\Sigma_{p_i}^{-1}\right\} = 0.$$

Hence, we get

$$\Sigma_{p_i} = \frac{\sum_{k=1}^{n} \psi_{ki}\Lambda}{\Psi_i}. \tag{15}$$

Here

$$\Lambda_{ki} = -\Sigma_{g_k} + (\Sigma_{p_i} + \Sigma_{g_k})\Sigma_{p_i}^{-1}(\mu_{p_i} - \gamma_{ki})(\mu_{p_i}^T - \beta_{ki})\Sigma_{p_i}^{-1}(\Sigma_{p_i} + \Sigma_{g_k}).$$

## 2.4 Algorithm

We construct an EM-based clustering algorithm for uncertain data based on the above discussion.

---

**Algorithm 4.** EM-based Clustering Algorithm for Uncertain Data

---

    **EMU1** Set the initial parameters.
    **EMU2** Update the mixture ratio by (3).
    **EMU3** Update the average by (4) (one-dimension) or (10) (multi-dimension).
    **EMU4** Update the variance by (6) (one-dimension) or variance-covariance matrix by (15) (multi-dimension).
    **EMU5** If all parameters converge, finish. Otherwise, go back to **EMU1**.

---

# 3   Numerical Examples

We compare one-dimensional proposed and conventional EM-based clustering algorithms through numerical results for a real dataset. The real dataset is a set of nominal GDP in 12 countries in Table 1.

We classify the dataset into two clusters by those algorithms. The averages $\mu_{g_k}$ are the values in Table 1. The boldface in each table means the final cluster which the object belongs to.

**Table 1**  Nominal GDP in 12 Countries

| Country | GDP(US dollar) |
|---|---|
| France | 2608.7 |
| U.K. | 2440.51 |
| Brazil | 2395.97 |
| Russia | 2021.96 |
| Italy | 2014.08 |
| India | 1824.83 |
| Canada | 1819.08 |
| Australia | 1541.8 |
| Spain | 1352.06 |
| Mexico | 1177.12 |
| Corea | 1155.87 |
| Indonesia | 878.2 |

**Table 2**  Results by Proposed Algorithm

| Country | $\sigma_{g_k}^2$ | Belongingness | | |
|---|---|---|---|---|
| | | Class 1 | Class 2 | Class3 |
| France | 25 | 0 | **1.0** | 0 |
| U.K. | 25 | 0.000001 | **0.999998** | 0.000001 |
| Brazil | 25 | 0.000017 | **0.999980** | 0.000003 |
| Russia | 25 | **0.997436** | 0.000005 | 0.002559 |
| Italy | 25 | **0.997339** | 0.000003 | 0.002658 |
| India | 25 | **0.969030** | 0 | 0.030970 |
| Canada | 25 | **0.965129** | 0 | 0.034871 |
| Australia | 25 | 0.002736 | 0 | **0.997264** |
| Spain | 25 | 0 | 0 | **1.0** |
| Mexico | 25 | 0 | 0 | **1.0** |
| Corea | 25 | 0 | 0 | **1.0** |
| Indonesia | 25 | 0 | 0 | **1.0** |

**Table 3**  Results by Proposed Algorithm

| Country | $\sigma_{g_k}^2$ | Belongingness | | |
|---|---|---|---|---|
| | | Class 1 | Class 2 | Class3 |
| France | 40 | 0 | **1.0** | 0 |
| U.K. | 40 | 0.000001 | **0.999998** | 0.000001 |
| Brazil | 40 | 0.000017 | **0.999980** | 0.000003 |
| Russia | 40 | **0.997436** | 0.000005 | 0.002559 |
| Italy | 40 | **0.997339** | 0.000003 | 0.002658 |
| India | 40 | **0.969030** | 0 | 0.030970 |
| Canada | 40 | **0.965129** | 0 | 0.034871 |
| Australia | 40 | 0.002736 | 0 | **0.997264** |
| Spain | 40 | 0 | 0 | **1.0** |
| Mexico | 40 | 0 | 0 | **1.0** |
| Corea | 40 | 0 | 0 | **1.0** |
| Indonesia | 40 | 0 | 0 | **1.0** |

**Table 4** Results by Proposed Algorithm

| Country | $\sigma_{g_k}^2$ | Belongingness | | |
|---|---|---|---|---|
| | | Class 1 | Class 2 | Class3 |
| France | 70 | 0 | **1.0** | 0 |
| U.K. | 70 | 0.000001 | **0.999998** | 0.000001 |
| Brazil | 70 | 0.000017 | **0.999980** | 0.000003 |
| Russia | 70 | **0.997436** | 0.000005 | 0.002559 |
| Italy | 70 | **0.997339** | 0.000003 | 0.002658 |
| India | 70 | **0.969030** | 0 | 0.030970 |
| Canada | 70 | **0.965129** | 0 | 0.034871 |
| Australia | 70 | 0.002736 | 0 | **0.997264** |
| Spain | 70 | 0 | 0 | **1.0** |
| Mexico | 70 | 0 | 0 | **1.0** |
| Corea | 70 | 0 | 0 | **1.0** |
| Indonesia | 70 | 0 | 0 | **1.0** |

**Table 5** Results by Proposed Algorithm

| Country | $\sigma_{g_k}^2$ | Belongingness | | |
|---|---|---|---|---|
| | | Class 1 | Class 2 | Class3 |
| France | 90 | 0 | **1.0** | 0 |
| U.K. | 90 | 0.000001 | **0.999998** | 0.000001 |
| Brazil | 90 | 0.000017 | **0.999980** | 0.000003 |
| Russia | 90 | **0.997436** | 0.000005 | 0.002559 |
| Italy | 90 | **0.997339** | 0.000003 | 0.002658 |
| India | 90 | **0.969030** | 0 | 0.030970 |
| Canada | 90 | **0.965129** | 0 | 0.034871 |
| Australia | 90 | 0.002736 | 0 | **0.997264** |
| Spain | 90 | 0 | 0 | **1.0** |
| Mexico | 90 | 0 | 0 | **1.0** |
| Corea | 90 | 0 | 0 | **1.0** |
| Indonesia | 90 | 0 | 0 | **1.0** |

**Table 6** Results by Proposed Algorithm

| Country | $\sigma_{g_k}^2$ | Belongingness | | |
|---|---|---|---|---|
| | | Class 1 | Class 2 | Class3 |
| France | 65 | 0 | **1.0** | 0 |
| U.K. | 15 | 0.000001 | **0.999998** | 0.000001 |
| Brazil | 15 | 0.000017 | **0.999980** | 0.000003 |
| Russia | 15 | **0.997436** | 0.000005 | 0.002559 |
| Italy | 15 | **0.997339** | 0.000003 | 0.002658 |
| India | 15 | **0.969030** | 0 | 0.030970 |
| Canada | 15 | **0.965129** | 0 | 0.034871 |
| Australia | 15 | 0.002736 | 0 | **0.997264** |
| Spain | 15 | 0 | 0 | **1.0** |
| Mexico | 15 | 0 | 0 | **1.0** |
| Corea | 15 | 0 | 0 | **1.0** |
| Indonesia | 15 | 0 | 0 | **1.0** |

**Table 7** Results by Proposed Algorithm

| Country | $\sigma_{g_k}^2$ | Belongingness | | |
|---|---|---|---|---|
| | | Class 1 | Class 2 | Class3 |
| France | 65 | 0 | **1.0** | 0 |
| U.K. | 15 | 0.000001 | **0.999998** | 0.000001 |
| Brazil | 15 | 0.000017 | **0.999980** | 0.000003 |
| Russia | 15 | **0.997450** | 0.000005 | 0.002545 |
| Italy | 15 | **0.997353** | 0.000003 | 0.002644 |
| India | 15 | **0.969124** | 0 | 0.030876 |
| Canada | 15 | **0.965232** | 0 | 0.034768 |
| Australia | 15 | 0.002739 | 0 | **0.997261** |
| Spain | 15 | 0 | 0 | **1.0** |
| Mexico | 15 | 0 | 0 | **1.0** |
| Corea | 15 | 0 | 0 | **1.0** |
| Indonesia | 65 | 0 | 0 | **1.0** |

In comparison with Tables 2, 3, 4, 5, and Table 10, it is thought that the same variance dose not have effect on classification results. On the other hand, Tables 6 and 7 show that the results change by giving a part of data different variances except Table 6. Tables 8, 9, and 11 also show the same.

However, some calculations do not converge. Moreover, the dataset is classified into two clusters although we set cluster number as three. It can be thought that Newton's method to calculate $\sigma_{p_i}$ is unstable.

From these results, we have two future works. First, we have to consider more useful methods to calculate $\sigma_{p_i}$ than Newton's method. Second, we have to verify the feature for uncertain data of our proposed method through more numerical examples.

**Table 8** Results by Proposed Algorithm

| Country | $\sigma_{g_k}^2$ | Belongingness | |
|---|---|---|---|
| | | Class 1 | Class 2 |
| France | 65 | **1.0** | 0 |
| U.K. | 15 | **1.0** | 0 |
| Brazil | 15 | **1.0** | 0 |
| Russia | 15 | **0.999931** | 0.000069 |
| Italy | 15 | **0.999917** | 0.000083 |
| India | 15 | **0.995883** | 0.004117 |
| Canada | 15 | **0.995412** | 0.004588 |
| Australia | 15 | **0.700243** | 0.299757 |
| Spain | 15 | 0.190038 | **0.809962** |
| Mexico | 15 | 0.047836 | **0.952164** |
| Corea | 15 | 0.041471 | **0.958529** |
| Indonesia | 65 | 0.012863 | **0.987137** |

**Table 9** Results by Proposed Algorithm

| Country | $\sigma_{g_k}^2$ | Belongingness | |
|---|---|---|---|
| | | Class 1 | Class 2 |
| France | 65 | **1.0** | 0 |
| U.K. | 15 | **1.0** | 0 |
| Brazil | 15 | **1.0** | 0 |
| Russia | 15 | **0.999931** | 0.000069 |
| Italy | 15 | **0.999917** | 0.000083 |
| India | 15 | **0.995887** | 0.004113 |
| Canada | 15 | **0.995416** | 0.004584 |
| Australia | 25 | **0.700339** | 0.299661 |
| Spain | 15 | 0.190100 | **0.809900** |
| Mexico | 15 | 0.047846 | **0.952154** |
| Corea | 15 | 0.041480 | **0.958520** |
| Indonesia | 65 | 0.012868 | **0.987132** |

**Table 10** Results by Conventional Algorithm

| Country | Belongingness | | |
|---|---|---|---|
| | Class 1 | Class 2 | Class3 |
| France | 0 | **1.0** | 0 |
| U.K. | 0.000001 | **0.999998** | 0.000001 |
| Brazil | 0.000017 | **0.999980** | 0.000003 |
| Russia | **0.997436** | 0.000005 | 0.002559 |
| Italy | **0.997339** | 0.000003 | 0.002658 |
| India | **0.969030** | 0 | 0.030970 |
| Canada | **0.965129** | 0 | 0.034871 |
| Australia | 0.002736 | 0 | **0.997264** |
| Spain | 0 | 0 | **1.0** |
| Mexico | 0 | 0 | **1.0** |
| Corea | 0 | 0 | **1.0** |
| Indonesia | 0 | 0 | **1.0** |

**Table 11** Results by Conventional Algorithm

| Country | Belongingness | |
|---|---|---|
| | Class 1 | Class 2 |
| France | **1.0** | 0 |
| U.K. | **1.0** | 0 |
| Brazil | **1.0** | 0 |
| Russia | **0.999929** | 0.000071 |
| Italy | **0.999915** | 0.000085 |
| India | **0.995815** | 0.004185 |
| Canada | **0.995338** | 0.004662 |
| Australia | **0.698754** | 0.301246 |
| Spain | 0.189450 | **0.810550** |
| Mexico | 0.047705 | **0.952295** |
| Corea | 0.041356 | **0.958644** |
| Indonesia | 0.012774 | **0.987226** |

## 4   Conclusions

In this paper, we constructed an algorithm based on the EM-based clustering algorithm which handles uncertain data which are represented by the Gaussian distributions. First, we considered the expectation of log-likelihood for the uncertain data by the Gaussian distributions as an objective function. Next, we derived the optimal solutions to parameters which maximizes the objective function and construct new EM-based clustering algorithm for uncertain data by Gaussian distributions.

Moreover, effectiveness of the proposed algorithm was verified through numerical examples.

The concept of tolerance we proposed is very useful to handle uncertain data, however, the clustering method with tolerance can not handle the uncertainty by the Gaussian distribution. Considering that many uncertainty in the real world are represented by the Gaussian distribution, that is disadvantage of tolerance. From the viewpoint, the proposed EM-based clustering algorithm for uncertain data is expected to handle uncertain data usefully.

# References

1. MacQueen, J.: Some methods for classification and analysis of multivariate observations. In: Proceedings of the Fifth Berkeley Symposium on Mathematical Statistics and Probability, vol. 1, pp. 281–297 (1967)
2. Bezdek, J.C.: Pattern Recognition with Fuzzy Objective Function Algorithms. Plenum, New York (1981)
3. Dempster, A.P., Laird, N.M., Rubin, D.B.: Maximum Likelihood from Incomplete Data via the EM Algorithm. Journal of the Royal Statistical Society. Series B (Methodological) 39(1), 1–38 (1977)
4. Takata, O., Miyamoto, S.: Fuzzy clustering of Data with Interval Uncertainties. Journal of Japan Society for Fuzzy Theory and Systems 12(5), 686–695 (2000) (in Japanese)
5. Endo, Y., Horiuchi, K.: On Clustering Algorithm for Fuzzy Data. In: Proc. 1997 International Symposium on Nonlinear Theory and Its Applications, pp. 381–384 (November 1997)
6. Endo, Y.: Clustering Algorithm Using Covariance for Fuzzy Data. In: Proc. 1998 International Symposium on Nonlinear Theory and Its Applications, pp. 511–514 (September 1998)
7. Endo, Y., Murata, R., Haruyama, H., Miyamoto, S.: Fuzzy c-Means for Data with Tolerance. In: Proc. 2005 International Symposium on Nonlinear Theory and Its Applications, pp. 345–348 (2005)
8. Murata, R., Endo, Y., Haruyama, H., Miyamoto, S.: On Fuzzy c-Means for Data with Tolerance. Journal of Advanced Computational Intelligence and Intelligent Informatics 10(5), 673–681 (2006)

# An Algorithm for Fuzzy Clustering Based on Conformal Geometric Algebra

Minh Tuan Pham and Kanta Tachibana

**Abstract.** Geometric algebra(GA) is a generalization of complex numbers and quaternions. It is able to describe spatial objects and the geometric relations between them. Conformal GA(CGA) is a part of GA and it's vector is found that points, lines, planes, circles and spheres gain particularly natural and computationally amenable representations. So, CGA based hard clustering(hard conformal clustering(HCC)) is able to detect a cluster distributed over a sphere, plane, or their intersections such as a straight line or arc. However because HCC is a hard clustering, it is only divide data into crisp cluster. This paper applies fuzzy technique to HCC and proposes an algorithm of fuzzy conformal clustering(FCC). This paper shows that using the proposed algorithm, data was able to belong to more one cluster which is presented by a vector in CGA.

## 1 Introduction

Nowadays in world information explosion, data mining is very important. People will be difficult to manage or search for information without the aid of computers and the algorithms of data mining. Clustering is an important technique in data mining. Methods of clustering [1, 2] iteratively minimize an objective function based on a distance measure to label each individual data point. Manifold clusterings [3, 4] are able to discover the different manifolds using geodesic distance [5, 6, 7]. However when geometric structures are corrupted by noises or discontinuities, it is difficult to detect them from the clustering result. To take the geometric structure of

Minh Tuan Pham
The University of Danang, University of Science and Technology, Vietnam
e-mail: pmtuan@dut.udn.vn

Kanta Tachibana
Kogakuin University, Japan
e-mail: kanta@cc.kogakuin.ac.jp

V.-N. Huynh et al. (eds.), *Knowledge and Systems Engineering, Volume 2,*      83
Advances in Intelligent Systems and Computing 245,
DOI: 10.1007/978-3-319-02821-7_9, © Springer International Publishing Switzerland 2014

data into consideration, [8] adopted a distance measure based on straight lines, and [9, 10, 11] adopted a distance measure based on spherical shells. The straight line measure and the spherical shell measure are applicable to datasets which have clusters generated respectively on straight line or on spherical surface in $m$-dimensional space. But so far, a distance measure which detects a mixture of clusters generated on either straight line and on spherical surface has not been proposed. This paper proposes a clustering method which can detect clusters with either straight line and on spherical surface. Furthermore, the spherical shell measure cannot detect a cluster around a spherical surface with dimension less than $m$. For example, it cannot express a data distribution around a circle in three-dimensional space because it optimizes an objective function based on distance from a three-dimensional spherical shell. The conformal geometric algebra based clustering method(CGA clustering) [12] can also detect clusters around spheres and planes of any dimension less than $m$ because they are well expressed as elements of conformal geometric algebra.

The CGA clustering focus geometric algebra (GA), which can describe spatial vectors and the higher-order subspace relations between them [13, 14, 15]. There are already many successful examples of its use in image processing, signal processing, color image processing, and multi-dimensional time-series signal processing with complex numbers [16, 17, 18] or quaternions [19, 20, 21], which are low-dimensional GAs. In addition, GA-valued neural network learning methods for learning input–output relationships [22] and the feature extraction based on GA [23, 24] are well studied.

The CGA clustering uses the vectors in conformal GA (CGA) [25, 26] space, which is a part of GA to propose a approach of approximation and a clustering method for geometric data. A vector in CGA is called a conformal vector, and it can express both $m$-dimensional spheres and planes (infinite radius spheres). A 2-blade which is an exterior product of two conformal vectors, expresses the intersection of two $m$-dimensional elements, and thus it is able to express an $m - 1$-dimensional sphere or plane. Likewise, a $k$-blade in CGA expresses an $m - k$-dimensional element. A conformal vector can also express a point as a sphere with zero radius. A characteristic of the CGA clustering is that it is able to detect a cluster distributed over a sphere, plane, or their intersections such as a straight line or arc.

In this paper, we apply the fuzzy technique to the CGA clustering and propose an algorithm of fuzzy conformal clustering. Section 2 describes the related work of CGA clustering. Section 3 describes this paper's proposal to apply the fuzzy technique to the CGA clustering and shows an algorithm of the proposed clustering method. Section 4 demonstrates the effectiveness of the proposed method through three experiments. The clustering results demonstrate demonstrate that the kernel alignment [27] between the true labels and the labels detected by clustering using the fuzzy algorithm is better than those from using conventional algorithm.

## 2  Conformal Geometric Algebra Based Clustering

### 2.1  Conformal Geometric Algebra

CGA is a special part of GA, which is also called Clifford algebras. GA defines the signature of $p + q$ orthonormal basis vectors $\mathcal{O} = \{\mathbf{e}_1, \ldots, \mathbf{e}_p, \mathbf{e}_{p+1}, \ldots, \mathbf{e}_{p+q}\}$, $\{\mathbf{e}_i\}$, such that $\mathbf{e}_i^2 = +1, \forall i \in \{1, \ldots, p\}$ and $\mathbf{e}_i^2 = -1, \forall i \in \{p+1, \ldots, p+q\}$. This paper denotes the geometric algebra determined from $\mathcal{O}$ by $\mathscr{G}_{p,q}$. For example, the geometric algebra determined from the $m$-dimensional real Euclidean vector space $\mathbf{R}^m$ is denoted by $\mathscr{G}_{m,0}$.

A CGA space is extended from the real Euclidean vector space $\mathbf{R}^m$ by adding 2 orthonormal basis vectors. Thus, a CGA space is determined by $m + 2$ basis vectors $\{\mathbf{e}_1, \ldots, \mathbf{e}_m, \mathbf{e}_+, \mathbf{e}_-\}$, where $\mathbf{e}_+$ and $\mathbf{e}_-$ are defined so that

$$\mathbf{e}_+^2 = \mathbf{e}_+ \cdot \mathbf{e}_+ = 1,$$
$$\mathbf{e}_-^2 = \mathbf{e}_- \cdot \mathbf{e}_- = -1,$$
$$\mathbf{e}_+ \cdot \mathbf{e}_- = \mathbf{e}_+ \cdot \mathbf{e}_i = \mathbf{e}_- \cdot \mathbf{e}_i = 0, \forall i \in \{1, \ldots, m\}. \tag{1}$$

Therefore, a CGA space can be expressed by $\mathscr{G}_{m+1,1}$. This paper then uses the converted basis vectors $\mathbf{e}_0$ and $\mathbf{e}_\infty$ as

$$\mathbf{e}_0 = \frac{1}{2}(\mathbf{e}_- - \mathbf{e}_+), \mathbf{e}_\infty = (\mathbf{e}_- + \mathbf{e}_+). \tag{2}$$

From Eq. (1) and Eq. (2), we get

$$\mathbf{e}_0 \cdot \mathbf{e}_0 = \mathbf{e}_\infty \cdot \mathbf{e}_\infty = 0,$$
$$\mathbf{e}_0 \cdot \mathbf{e}_\infty = \mathbf{e}_\infty \cdot \mathbf{e}_0 = -1,$$
$$\mathbf{e}_0 \cdot \mathbf{e}_i = \mathbf{e}_\infty \cdot \mathbf{e}_i = 0, \forall i \in \{1, \ldots, m\}. \tag{3}$$

We also make use of a transform based upon that proposed by Hestenes [26]. A real Euclidean vector $\mathbf{x} = \sum_i^m x_i \mathbf{e}_i \in \mathbf{R}^m$ is extended to a point $P \in \mathscr{G}_{m+1,1}$ according to the equation

$$P = \mathbf{x} + \frac{1}{2}\|\mathbf{x}\|^2 \mathbf{e}_\infty + \mathbf{e}_0. \tag{4}$$

Note that the inner product of $P$ with itself, $P \cdot P$, is 0. Then, a sphere is represented as a conformal vector

$$S = P - \frac{1}{2}r^2 \mathbf{e}_\infty$$
$$= \mathbf{x} + \frac{1}{2}\{\|\mathbf{x}\|^2 - r^2\}\mathbf{e}_\infty + \mathbf{e}_0, \tag{5}$$

where the sphere has center $\mathbf{x}$ and radius $r$ in real Euclidean space $\mathbf{R}^m$. The inner product $S \cdot Q$ is 0 for any point $Q$ on the surface of the sphere. From Eqs. 4 and 5, we can see that a point is naturally expressed as a sphere with radius $r = 0$ in CGA space $\mathscr{G}_{m+1,1}$.

CGA also expresses a plane as a conformal vector:

$$L = \mathbf{n} + d\mathbf{e}_\infty, \tag{6}$$

where $\mathbf{n} \in \mathbf{R}^m$ is the normal vector, and $d$ is a scalar coefficient of the plane $\mathbf{n} \cdot \mathbf{x} - d = 0$ in real space $\mathbf{R}^m$. The inner product $L \cdot Q$ is 0 for any point $Q$ in the plane.

A conformal vector $S$ in $\mathscr{G}_{m+1,1}$ is generally written in the following form:

$$S = \mathbf{s} + s_\infty \mathbf{e}_\infty + s_0 \mathbf{e}_0. \tag{7}$$

$\mathbf{s} = \sum_i^m s_i \mathbf{e}_i$ is a vector in the real Euclidean space $\mathbf{R}^m$. $s_\infty$ and $s_0$ are the scalar coefficients of the basis vectors $\mathbf{e}_\infty$ and $\mathbf{e}_0$.

## 2.2 Estimation of Hyper-spheres (-planes) from Geometric Data Sets

The inner product between a point $P$ and a conformal vector $S$ in CGA space is proportional to the distance between them. From Eqs. (3), (4), and (7), the inner product of $P$ and $S$ is

$$
\begin{aligned}
d(P,S) &\propto P \cdot S \\
&= \left( \mathbf{x} + \frac{1}{2}\|\mathbf{x}\|^2 \mathbf{e}_\infty + \mathbf{e}_0 \right) \cdot (\mathbf{s} + s_\infty \mathbf{e}_\infty + s_0 \mathbf{e}_0) \\
&= \mathbf{x} \cdot \mathbf{s} - s_\infty - \frac{1}{2}\|\mathbf{x}\|^2 s_0.
\end{aligned}
\tag{8}
$$

When $d(P,S) = 0$, we get

$$
\mathbf{x} \cdot \mathbf{s} - s_\infty - \frac{1}{2}\|\mathbf{x}\|^2 s_0 = 0
$$

$$
\Leftrightarrow
\begin{cases}
\mathbf{x} \cdot \mathbf{s} - s_\infty = 0 & (s_0 = 0), \\
\left\| \mathbf{x} - \frac{1}{s_0}\mathbf{s} \right\|^2 = \dfrac{\|\mathbf{s}\|^2 - 2s_\infty s_0}{s_0^2} & (s_0 \neq 0).
\end{cases}
\tag{9}
$$

Eq. (9) shows that when $d(P,S) = 0$, the point $P$ lies on the hyper-plane ($s_0 = 0$) or hyper-sphere surface ($s_0 \neq 0$) which is expressed as conformal vector $S$ in CGA space.

Given a set of vectors $\mathscr{X} = \{ \mathbf{x}_k = \sum_i^m x_{k,i}\mathbf{e}_i \in \mathbf{R}^m, k = 1, \ldots, n \}$, we transform each $m$-dimensional vector to a point in CGA space using Eq. (4) to yield the set $\mathscr{P} = \{ P_k = \mathbf{x}_k + \frac{1}{2}\|\mathbf{x}_k\|^2 \mathbf{e}_\infty + \mathbf{e}_0, k = 1, \ldots, n \}$ in $\mathscr{G}_{m+1,1}$. Then, we approximate the set of points $\mathscr{P}$ using a vector $S = \mathbf{s} + s_\infty \mathbf{e}_\infty + s_0 \mathbf{e}_0$ in $\mathscr{G}_{m+1,1}$, which is estimated using least squares. The error function $E$ is defined as

$$E = \sum_{k=1}^{n} d^2 (P_k, S)$$
$$= \sum_{k=1}^{n} \left( \mathbf{x}_k \cdot \mathbf{s} - s_\infty - \frac{1}{2} \|\mathbf{x}_k\|^2 s_0 \right)^2. \tag{10}$$

From Eq. (9), we can see that the hyper-plane or hyper-sphere surface does not change when we multiply both sides by a scalar value. It means that when we minimize the error function $E$, $\mathbf{s}$ can be restricted to $\|\mathbf{s}\|^2 = 1$. In this case, the optimization problem becomes

$$\min \sum_{k=1}^{n} \left( \mathbf{x}_k \cdot \mathbf{s} - s_\infty - \frac{1}{2} \|\mathbf{x}_k\|^2 s_0 \right)^2 \tag{11}$$

subject to

$$\|\mathbf{s}\|^2 = 1. \tag{12}$$

Therefore, we might be tempted to express the previous problem by means of a non-negative Lagrange multiplier $\lambda$ as the minimization of

$$L = \frac{1}{n} \sum_{k=1}^{n} \left( \mathbf{x}_k \cdot \mathbf{s} - s_\infty - \frac{1}{2} \|\mathbf{x}_k\|^2 s_0 \right)^2$$
$$- \lambda \left( \|\mathbf{s}\|^2 - 1 \right). \tag{13}$$

The critical values of $L$ occur when its gradient is zero. The partial derivatives are

$$\frac{\partial L}{\partial \mathbf{s}} = \frac{2}{n} \sum_{k=1}^{n} \left( \mathbf{x}_k \cdot \mathbf{s} - s_\infty - \frac{1}{2} \|\mathbf{x}_k\|^2 s_0 \right) \mathbf{x}_k - 2\lambda \mathbf{s}$$
$$= 0, \tag{14}$$

$$\frac{\partial L}{\partial s_\infty} = -\frac{2}{n} \sum_{k=1}^{n} \left( \mathbf{x}_k \cdot \mathbf{s} - s_\infty - \frac{1}{2} \|\mathbf{x}_k\|^2 s_0 \right) = 0, \tag{15}$$

$$\frac{\partial L}{\partial s_0} = -\frac{1}{n} \sum_{k=1}^{n} \left( \mathbf{x}_k \cdot \mathbf{s} - s_\infty - \frac{1}{2} \|\mathbf{x}_k\|^2 s_0 \right) \|\mathbf{x}_k\|^2$$
$$= 0. \tag{16}$$

From Eqs. (15) and (16),

$$s_\infty = \mathbf{f}_\infty \cdot \mathbf{s}, \tag{17}$$
$$\frac{1}{2} s_0 = \mathbf{f}_0 \cdot \mathbf{s}, \tag{18}$$

where

$$\mathbf{f}_\infty = \frac{-\sum_{k=1}^{n} \|\mathbf{x}_k\|^4 \sum_{k=1}^{n} \mathbf{x}_k + \sum_{k=1}^{n} \|\mathbf{x}_k\|^2 \sum_{k=1}^{n} \|\mathbf{x}_k\|^2 \mathbf{x}_k}{\sum_{k=1}^{n} \|\mathbf{x}_k\|^2 \sum_{k=1}^{n} \|\mathbf{x}_k\|^2 - \sum_{k=1}^{n} \|\mathbf{x}_k\|^4 \sum_{k=1}^{n} 1}, \tag{19}$$

$$\mathbf{f}_0 = \frac{\sum_{k=1}^{n} \|\mathbf{x}_k\|^2 \sum_{k=1}^{n} \mathbf{x}_k - \sum_{k=1}^{n} 1 \sum_{k=1}^{n} \|\mathbf{x}_k\|^2 \mathbf{x}_k}{\sum_{k=1}^{n} \|\mathbf{x}_k\|^2 \sum_{k=1}^{n} \|\mathbf{x}_k\|^2 - \sum_{k=1}^{n} \|\mathbf{x}_k\|^4 \sum_{k=1}^{n} 1}. \tag{20}$$

More simply, we define the equation

$$\mathbf{f}(\mathbf{x}) = \mathbf{x} - \mathbf{f}_\infty - \|\mathbf{x}\|^2 \mathbf{f}_0 \in \mathbf{R}^m. \tag{21}$$

Then, Eq. (13) can be rewritten as

$$L = \mathbf{s}^T \mathbf{A}\mathbf{s} - \lambda \left( \|\mathbf{s}\|^2 - 1 \right), \tag{22}$$

where

$$\mathbf{A} = \frac{1}{n} \sum_{k=1}^{n} \mathbf{f}(\mathbf{x}_k) \mathbf{f}^T (\mathbf{x}_k). \tag{23}$$

Therefore, the optimization problem is solved by the eigen decomposition

$$\mathbf{A}\mathbf{s} = \lambda \mathbf{s}. \tag{24}$$

An eigen vector $\mathbf{s}$ is an eigen vector defining the hyper-sphere (-plane) $S = \mathbf{s} + s_\infty \mathbf{e}_\infty + s_0 \mathbf{e}_0$, and the eigen-value $\lambda$ is the variance, i.e., the sum of the squared distances between $P_k$ and $S$. The estimated hyper-sphere or -plane will exactly fit the data set if the eigen-value $\lambda$ is 0. Because there are $m$ eigen-values when the original data space is $m$-dimensional, we can find $m$ solutions of hyper-spheres (or planes). Thus, we can also detect the $m - k$-dimensional hyper-spheres (or planes) using the $k$ eigen vectors corresponding to the smallest $k$ eigen-values. For example, in the case of three-dimensional data, we can detect a circle which is a intersection between two spheres having the smallest eigen-values. Furthermore, we can detect a cluster of data with an arc, i.e., part of a circle. So, this estimation method is expected to improve clustering performance.

## 2.3   Conformal Geometric Algebra Based Clustering Method

For each cluster $\mathscr{X}_c = \{\mathbf{x}_k = \sum_i^m x_{k;i}\mathbf{e}_i \in \mathbf{R}^m \mid y_k = c, k \in \{1,\dots,n\}\}$, the conformal geometric algebra based clustering method (CGA Clustering) uses Eqs. (17), (18), and (23) to estimate the $m$ pairs of hyper-spheres (-planes) $S_{c,l} = \mathbf{s}_{c,l} + s_{\infty c,l}\mathbf{e}_\infty + s_{0c,l}\mathbf{e}_0, l \in \{1,\dots,m\}$ and eigen-values $\lambda_{c,l}$, where the eigen-values are the variances based on the distance $d\left(P, S_{c,l}\right)$ between a point $P = \mathbf{x} + \frac{1}{2}\|\mathbf{x}\|^2\mathbf{e}_\infty + \mathbf{e}_0, \mathbf{x} \in \mathscr{X}_c$ and the hyper-sphere (-plane) $S_{c,l}$. Herein, it assumes that the variance based on $d\left(P, S_{c,l}\right)$ follows a Gaussian distribution. Then the density function is

$$p\left(\mathbf{x} \mid \lambda_{c,l}\right) = \frac{1}{\sqrt{2\pi\lambda_{c,l}}}\exp\left(-\frac{d^2\left(P, S_{c,l}\right)}{2\lambda_{c,l}}\right). \tag{25}$$

The posterior probability density function for $\mathbf{x}$ given the cluster $y = c$ can be defined as

$$p\left(\mathbf{x} \mid c\right) = \prod_l^m p\left(\mathbf{x} \mid \lambda_{c,l}\right). \tag{26}$$

The CGA clustering minimizes the following objective function:

$$\min_{u,\mathbf{c}} \sum_{k=1}^n \sum_{c=1}^C d^2\left(P_k, S_c\right), \tag{27}$$

and it algorithm is following:

Step1   Choose the number of clusters, $C$.
Step2   Randomly assign label $y_k \in \{1,\dots,C\}$ for each data point $k$.
Step3   Estimate the posterior probability density function $p\left(\mathbf{x} \mid c\right)$ for all clusters $\mathscr{X}_c = \{\mathbf{x}_k \mid y_k = c\}$.
Step4   Update the data label by $y_k = \arg\max_c p\left(\mathbf{x} \mid c\right)P\left(c\right)$, where $P\left(c\right) = \frac{|\mathscr{X}_c|}{n}$ is the prior distribution of cluster $\mathscr{X}_c$, and $|\mathscr{X}_c|$ is the number of data in cluster $\mathscr{X}_c$.
Step5   Repeat the two previous steps until the algorithm has converged.

## 3   Algorithm for Conformal Geometric Algebra Based Fuzzy Clustering

In this paper, we apply the fuzzy technique for conformal geometric algebra based clustering method. We minimize the following objective function:

$$\min_{u,\mathbf{c}} \sum_{k=1}^n \sum_{c=1}^C \left(u_{kc}\right)^w d^2\left(P_k, S_c\right), \tag{28}$$

$$\text{s.t. } \sum_{c=1}^C u_{kc} = 1, u_{kc} \in [0,1].$$

where, $u_{kc}$ is the grade of membership of point $P_k$ in cluster $S_c$, and $w \in [1, \infty)$ is the weighting exponent on each fuzzy membership. Same with the case of hard CGA clustering, we are easy to minimize (Eq. 28) by solving the eigen decomposition (Eq. 24), where

$$\mathbf{f}_{\infty,c} = \frac{-\sum_{k=1}^{n} u_{kc} \|\mathbf{x}_k\|^4 \sum_{k=1}^{n} u_{kc} \mathbf{x}_k + \sum_{k=1}^{n} u_{kc} \|\mathbf{x}_k\|^2 \sum_{k=1}^{n} u_{kc} \|\mathbf{x}_k\|^2 \mathbf{x}_k}{\sum_{k=1}^{n} u_{kc} \|\mathbf{x}_k\|^2 \sum_{k=1}^{n} u_{kc} \|\mathbf{x}_k\|^2 - \sum_{k=1}^{n} u_{kc} \|\mathbf{x}_k\|^4 \sum_{k=1}^{n} u_{kc}}, \tag{29}$$

$$\mathbf{f}_{0,c} = \frac{\sum_{k=1}^{n} u_{kc} \|\mathbf{x}_k\|^2 \sum_{k=1}^{n} u_{kc} \mathbf{x}_k - \sum_{k=1}^{n} u_{kc} \sum_{k=1}^{n} u_{kc} \|\mathbf{x}_k\|^2 \mathbf{x}_k}{\sum_{k=1}^{n} u_{kc} \|\mathbf{x}_k\|^2 \sum_{k=1}^{n} u_{kc} \|\mathbf{x}_k\|^2 - \sum_{k=1}^{n} u_{kc} \|\mathbf{x}_k\|^4 \sum_{k=1}^{n} u_{kc}}, \tag{30}$$

and

$$\mathbf{A_c} = \sum_{k=1}^{n} u_{kc} \mathbf{f}(\mathbf{x}_k) \mathbf{f}^T(\mathbf{x}_k). \tag{31}$$

The memberships is updated as following:

$$u_{kc} = \frac{\prod_{l}^{m} p(\mathbf{x}_k \mid \lambda_{c,l})}{\sum_{c'=1}^{C} \prod_{l}^{m} p(\mathbf{x}_k \mid \lambda_{c',l})}, \tag{32}$$

where,

$$p(\mathbf{x} \mid \lambda_{c,l}) = \frac{1}{\sqrt{2\pi\beta\lambda_{c,l}}} \exp\left(-\frac{d^2(P, S_{c,l})}{2\beta\lambda_{c,l}}\right). \tag{33}$$

Here, the fuzziness of the clusters can be controlled by $\beta$, and the algorithm of fuzzy CGA clustering is following:

Step1    Choose the number of clusters $C$ and fuzzy controller $\beta$.
Step2    Randomly set memberships $u_{kc} \in [0, 1]$ for each data point $k$.
Step3    Estimate $\mathbf{f}_{\infty,c}$ and $\mathbf{f}_{0,c}$ for all clusters.
Step4    Update the memberships $u_{kc}$.
Step5    Repeat the two previous steps until the algorithm has converged.

## 4  Experiments and Discussion

This experiment uses two kinds of two-dimensional geometric data to compare hard algorithm and fuzzy algorithm of CGA clustering. Fig. 1 shows an example of geometric data which was used in this experiment. This data maybe changed by noises when we make them randomly. Fig. 1(a) shows data from two arcs. Fig. 1(b) shows data from one line and one circle. In this paper, we choose true cluster number $C = 2$ for all data sets.

Because the results depend on the randomly assigned initial labels in most clustering methods, this experiment is performed using 100 different random sets of initial labels for each clustering method. The following kernel alignment function is used to evaluate the performance of the clustering results:

$$A(Y_t, Y_e) = \frac{\sum\limits_{k,l} y_{kl;t} y_{kl;e}}{\sqrt{\sum\limits_{k,l} y_{kl;t}^2} \sqrt{\sum\limits_{k,l} y_{kl;e}^2}} \in [0,1], \tag{34}$$

where $Y_t = [y_{kl;t}]$ and $Y_e = [y_{kl;e}]$ are symmetric matrices with $n$ rows and $n$ columns corresponding respectively to the generating data labels and the estimated data labels. $y_{kl}$ indicates whether labels are different between instance $k$ and instance $l$, and is defined by

$$y_{kl} = \begin{cases} 1 & (y_k = y_l), \\ 0 & (y_k \neq y_l). \end{cases} \tag{35}$$

Table 1 and Table 2 shows the results of clustering in terms of average and standard deviation of the kernel alignments according to a variety of fuzzy control parameter values $\beta$. As shown in Table 1 and Table 2, the kernel alignment when using the proposed fuzzy algorithm was better than when using hard algorithm.

**Table 1** Average of the kernel alignments

|  | Hard algorithm | Fuzzy algorithm (proposed method) | | | | | | | | | |
| --- | --- | --- | --- | --- | --- | --- | --- | --- | --- | --- | --- |
|  |  | $\beta = 0.5$ | 0.6 | 0.7 | 0.8 | 0.9 | 1.0 | 1.1 | 1.2 | 1.3 | 1.4 |
| 2 arcs | 0.94 | 0.74 | 0.88 | 0.88 | 0.88 | 0.88 | **0.97** | **0.97** | **0.97** | **0.97** | **0.97** |
| ten mark | 0.81 | 0.77 | 0.74 | 0.90 | 0.90 | **1.00** | **1.00** | **1.00** | **1.00** | **1.00** | **1.00** |

**Table 2** Standard deviation of the kernel alignments

|  | Hard algorithm | Fuzzy algorithm (proposed method) | | | | | | | | | |
| --- | --- | --- | --- | --- | --- | --- | --- | --- | --- | --- | --- |
|  |  | $\beta = 0.5$ | 0.6 | 0.7 | 0.8 | 0.9 | 1.0 | 1.1 | 1.2 | 1.3 | 1.4 |
| 2 arcs | 0.11 | 0.19 | 0.18 | 0.18 | 0.18 | 0.18 | **0.00** | **0.00** | **0.00** | **0.00** | **0.00** |
| ten mark | 0.21 | 0.19 | 0.21 | 0.19 | 0.19 | **0.00** | **0.00** | **0.00** | **0.00** | **0.00** | **0.00** |

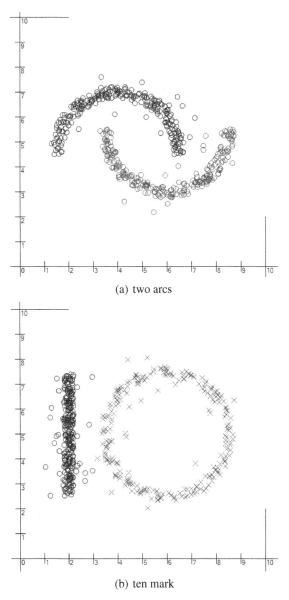

(a) two arcs

(b) ten mark

**Fig. 1** Geometric data for clustering

## 5 Conclusion

The conformal geometric algebra based clustering is able to detect the geometric objects from data set. This paper proposed the fuzzy algorithm for the conformal geometric algebra based clustering. The experiment uses two kinds of two-dimensional geometric data to compare hard algorithm and fuzzy algorithm of conformal geometric algebra based clustering. In this experiment, we used the kernel alignment to compare the performance of the clustering results. The results of the experiment showed that the fuzzy algorithm was better than the non-fuzzy algorithm of conformal geometric algebra based clustering.

# References

1. MacQueen, J.B.: Some Methods for classification and Analysis of Multivariate Observations. In: Proceedings of 5th Berkeley Symposium on Mathematical Statistics and Probability, vol. 1, pp. 281–297 (1967)
2. Bezdek, J.C., Ehrlich, R., Full, W.: FCM: the fuzzyc-means clustering algorithm. Computers and Geosciences 10, 191–203 (1984)
3. Souvenir, R., Pless, R.: Manifold clustering. In: IEEE International Conference on Computer Vision, vol. I, pp. 648–653 (2005)
4. Goh, A.: Riemannian manifold clustering and dimensionality reduction for vision-based analysis. In: Machine Learning for Vision-Based Motion Analysis: Theory and Techniques, pp. 27–53. Springer (2011)
5. Feil, B., Abonyi, J.: Geodesic Distance Based Fuzzy Clustering. In: Saad, A., Dahal, K., Sarfraz, M., Roy, R. (eds.) Soft Computing in Industrial Applications. ASC, vol. 39, pp. 50–59. Springer, Heidelberg (2007)
6. Kim, J., Shim, K.H., Choi, S.: Soft Geodesic Kernel K-means. In: Proc. ICASSP 2007, vol. 2, pp. 429–432 (2007)
7. Asgharbeygi, N., Maleki, A.: Geodesic K-means Clustering. In: Proc. ICPR 2008, pp. 1–4 (2008)
8. Bezdek, J.C., Coray, C., Gunderson, R., Watson, J.: Detection and characterization of cluster substructure I. Linear structure fuzzy c-lines. SIAM Jour. of Appl. Math. 40(2), 339–357 (1981)
9. Krishnapuram, R., Nasraoui, O., Frigui, H.: The fuzzy c spherical shells algorithms: a new approach. IEEE Trans. on Neural Networks 3(5), 663–671 (1992)
10. Cruz, B., Barrón, R., Sossa, H.: A new unsupervised learning for clustering using geometric associative memories. In: Bayro-Corrochano, E., Eklundh, J.-O. (eds.) CIARP 2009. LNCS, vol. 5856, pp. 239–246. Springer, Heidelberg (2009)
11. Li, M., Guan, J.: Possibilistic C-Spherical Shell Clustering Algorithm Based on Conformal Geometric Algebra. In: Proceedings of 2010 IEEE 10th International Conference on Signal Processing (ICSP), pp. 1347–1350 (2010)
12. Pham, M.T., Tachibana, K., Yoshikawa, T., Furuhashi, T.: A clustering method for geometric data based on approximation using conformal geometric algebra. In: 2011 IEEE International Conference on Fuzzy Systems, pp. 2540–2545 (2011)
13. Doran, C., Lasenby, A.: Geometric algebra for physicists. Cambridge University Press (2003)
14. Hestenes, D.: New foundations for classical mechanics, Dordrecht (1986)
15. Dorst, L., Fontijne, D., Mann, S.: Geometric Algebra for Computer Science: An Object-oriented Approach to Geometry. Morgan Kaufmann Series in Computer Graphics (2007)
16. Sekita, I., Kurita, T., Otsu, N.: Complex Autoregressive Model for Shape Recognition. IEEE Trans. on Pattern Analysis and Machine Intelligence 14(4) (1992)
17. Hirose, A.: Complex-Valued Neural Networks: Theories and Applications. Series on Innovative Intelligence, vol. 5 (2006)

18. Nitta, T.: An Extension of the Back-Propagation Algorithm to Complex Numbers. Neural Networks 10(8), 1391–1415(25) (1997)
19. Matsui, N., Isokawa, T., Kusamichi, H., Peper, F., Nishimura, H.: Quaternion neural network with geometrical operators. Journal of Intelligent and Fuzzy Systems 15(3-4), 149–164 (2004)
20. Buchholz, S., Le Bihan, N.: Optimal separation of polarized signals by quaternionic neural networks. In: 14th European Signal Processing Conference, EUSIPCO 2006, Florence, Italy, September 4-8 (2006)
21. Hitzer, E.: Quaternion Fourier Transform on Quaternion Fields and Generalizations. Advances in Applied Clifford Algebras 17(3), 497–517 (2007)
22. Sommer, G.: Geometric Computing with Clifford Algebras. Springer (2001)
23. Pham, M.T., Tachibana, K., Hitzer, E.M.S., Yoshikawa, T., Furuhashi, T.: Classification and Clustering of Spatial Patterns with Geometric Algebra. In: AGACSE 2008, Leipzig (2008)
24. Pham, M.T., Tachibana, K., Hitzer, E.M.S., Buchholz, S., Yoshikawa, T., Furuhashi, T.: Feature Extractions with Geometric Algebra for Classification of Objects. In: IEEE World Congress on Computational Intelligence / International Joint Conference on Neural Networks, Hong Kong (2008)
25. Hildenbrand, D., Hitzer, E.: Analysis of point clouds using conformal geometric algebra. In: 3rd International Conference on Computer Graphics Theory and Applications, Funchal, Madeira, Portugal (2008)
26. Hestenes, D., Sobczyk, G.: Clifford Algebra to Geometric Calculus: A unified language for mathematics and physics, Reidel (1984)
27. Cristianini, N., Kandola, J., Elisseeff, A., Shawe-Taylor, J.: On kernel target alignment. Journal of Machine Learning Research (2002)

# MOSS: A Formalism for Ontologies Including Multilingual Features

Jean-Paul A. Barthès and Claude Moulin

**Abstract.** The dominant approach for developing ontologies is currently based on description logics. We offer here a different approach, MOSS, based on semantic nets. The simple frame-based formalism that we are using allows specifying multilingual concepts easily and includes features like virtual classes and properties equivalent to rules in other approaches. Reasoning is done by traversing the graph of classes and instances in an efficient manner thanks to indexes. When restricted, the approach can be compiled into the OWL-DL formalism augmented with rules compatible with the JENA framework or using SPARQL queries. MOSS has been used in several applications including a multi-agent platform.

## 1 Introduction

A number of formalisms for representing knowledge and in particular ontologies have been proposed since the early seventies. There has been basically two approaches: the first one favoring logics at the expense of reducing the representation power to have a precise control over reasoning; the second one favoring representational power at the expense of less strict reasoning. Among the first category, one finds the early proposals like KL-ONE from Brachman and Schmolze [2] that through a number of intermediate proposals among which CLASSIC [12], can be considered an ancestor of OWL-DL[1]. The second category started with a number of proposals focusing on semantic nets, was developed in commercial systems like ART, KEE or KC in the mid eighties [6, 7], led to CYC, developed under the supervision of Lenat at MCC [8] which became a commercial product[2]. There have been

Jean-Paul A. Barthès · Claude Moulin
Heudiasyc, JRU CNRS 7253, Université de Technologie de Compiègne,
60200 Compiègne, France
e-mail: {jean-paul.barthes,claude.moulin}@utc.fr

[1] http://www.w3.org/TR/owl-guide/
[2] http://www.cyc.com/

V.-N. Huynh et al. (eds.), *Knowledge and Systems Engineering, Volume 2,*
Advances in Intelligent Systems and Computing 245,
DOI: 10.1007/978-3-319-02821-7_10, © Springer International Publishing Switzerland 2014

many discussions among the researchers which were tenants of one approach vs. the other. For example Doyle and Patil [4] maintained that a logical approach was not expressive enough to represent medical knowledge. Currently however, OWL-DL seems to be the main approach to operationalize ontologies, coupled with the Protégé[3] tool. Our approach was mainly developed for modeling environments that are progressively discovered by a robot and needed features like default values and classless objects. The resulting environment, MOSS, is thus more of the second type rather than belonging to the logical approaches. Recently, we developed the formalism to represent ontologies and knowledge bases adding virtual classes and properties rather than multiplying the number of subsumed classes. We also introduced multilinguism. CYC can be difficult to use (see Mili and Pachet [9]). MOSS is less ambitious but easier to use.

MOSS offers a compact ontology formalism. Reasoning is implemented through a query mechanism. MOSS can be restricted in order to model ontologies that can be translated partly into an OWL-DL representation and partly into additional rules.

This paper first introduces briefly the MOSS formalism showing how it can be used directly. In particular we explain virtual concepts and properties, the reasoning process, and multilingual features. Afterwards, we show how MOSS expressions can be translated into OWL, augmented by rules expressed in a JENA or SPARQL formalism. We then mention two examples of applications in which MOSS has been used.

## 2   The MOSS Formalism

The MOSS formalism is a simple reflexive frame-based formalism in which concepts are expressed as classes with attributes (OWL would use the term datatype property) and relations (OWL would use the term object property). Attributes and relations are first class objects and can exist independently of concepts. The approach is based on the notion of *typicality* rather than prescription, meaning that a concept is the structural abstraction of a number of individuals each of which can exhibit additional properties not included in the concept. In addition, default values are allowed at different levels. A few examples will illustrate some of the points.

### 2.1   Concepts and Properties

Let us consider the concept of PERSON that can be defined as follows:

```
(defconcept "Person"
    (:att "name" (:min 1)(:default "John")(:index))
    (:att "first name" (:max 5))
    (:att "birth date" (:unique))
    (:rel "employer" (:to "organization"))
    )
```

---

[3] http://protege.stanford.edu/

Executing the expression in the MOSS environment produces the creation of generic properties has-name, has-first-name, has-birth-date, and has-employer with specific local properties has-person-name, has-person-first-name, has-person-birth-date and has-person-employer.

There are constraints on some properties: name must have at least one value, first name must have at most five values, birth date must have exactly one value. There are no constraints on the relation employer. It can have zero or more values. Such constraints are recorded at the level of the local properties. Note that the attribute name has a default value "John" which will be shadowed by actual values, and is declared as an index (which will be used to accelerate access). Inverse relations are also maintained automatically by the system. However, they do not have a specific semantics, and are only used to traverse the graph using a relation in the reverse direction.

We declare subconcepts as follows:

```
(defconcept "Member"
  (:is-a "Person")
  (:rel "group" (:to "Group"))
  )
```

As usual MEMBER is subsumed by PERSON and inherits all the properties and associated constraints from PERSON and has the additional relation has-group linking a member to a group.

## 2.2 Virtual Classes and Properties

Traditionally, new concepts are specified when some constraints are added to a concept. For example we could define the concept of ADULT as a subconcept of PERSON. However, if we do that and create adult individuals, we are going to spread individuals among a set of concepts like ADULT, STUDENT, PROFESSIONAL, GOLFER, etc. To avoid this situation we proposed the concept of virtual concept, which defines a concept by introducing a constraint to a previously defined concept. An adult would be defined as follows:

```
(defvirtualconcept
  (:name  "Adult")
  (:is-a "person")
  (:def
    (?* "age" :ge 18))
  (:doc "Adults are persons over 18 years."))
```

A virtual concept does not span individuals, thus individuals are persons and adults are obtained by filtering persons using the age constraint.

The same approach can be applied to properties, as shown here:

```
(defvirtualattribute  "age"
  (:class "person")
  (:def
```

```
   (?* "birth date" ?y)
   (:fcn "getYears" ?y))
(:type :int) (:unique)
(:doc "the property age is computed from the birth date of a person
      by applying the function getYears to the birthdate."))
```

This amounts to inserting a user-defined function when retrieving the value of the age.

In the same fashion one can do composition of relations:

```
(defvirtualrelation "uncle"
  (:class "person")
  (:to "person")
  (:compose (:or ("mother" "brother") ("father" "brother")))
  (:doc "an uncle is a person who is the brother of the mother or
        the brother of the father.")
  )
```

Virtual concepts or properties are used when querying the knowledge base either directly with the MOSS formalism or translated into JENA or SPARQL rules when translating the ontology into an OWL format.

## 3  Reasoning Using Queries

Reasoning is done through the querying of the database. The format is a tree structure including operators, e.g.

```
- ("person" ("name" :is "Smith"))
- ("adult" ("employer" ("organization" ("name" :is "UTC"))))
- ("organization" (">employer" (> 10) ("person")))
```

The first query is a straightforward extraction of all the persons named Smith. The second query is an extraction of all the persons older than 18 and working for UTC. In that case subsumption and expansion of the constraint are done automatically. The last query looks for organizations that employ more than 10 persons. Note that the employer relation is traversed on the reverse direction (expressed by the "¿" prefix).

Execution of the queries in the MOSS context is done using indexes (e.g. SMITH or UTC) and traversing the graph modeling the ontology and knowledge base. When we can use indexes, the answer time is independent from the number of individuals (persons) in the knowledge base. This must be compared with the approach consisting of exploding the ontology and knowledge base into a set of triples and then using a RETE algorithm to try to optimize the filtering.

The syntax of a MOSS query is rather complex, allowing disjunction, negation and the use of variables. Some operators are predefined, but user-defined operators are allowed through the use of virtual entities. The exact syntax is summarized in the following table, where the term **entry-point** corresponds to the value of an index:

```
<general-query> ::= <entry-point> | <query>
<query>            ::= (<class> <clause>*)
<clause>           ::= (<simple-clause>) | (<or-clause>)
                                          | (<or-constrained-sub-query>)
<simple-clause> ::= <attribute-clause> | (<sub-query>)
<attribute-clause> ::= (<attribute> <attribute-operator> <value>)
                       | (<attribute> <attribute-operator> <variable>)
<or-clause>        ::= (OR <simple-clause>+)
<mixed-clause>  ::= (<sub-query>) | (<simple-clause> . <mixed-clause>*)
                                  | (<sub-query> . <mixed-clause>)
<sub-query>     ::= (<relationship> <cardinality-condition>  <query>)
<relationship>  ::= <relation> | <inverse-property>
<or-constrained-sub-query> ::= (OR <cardinality-condition> <sub-query>+)
<cardinality-condition> ::= (<numeric-operator> <arg>+)
<numeric operator> ::= < | <= | = | >= | > | <> | BETWEEN | OUTSIDE
<attribute-operator> ::= < | <= | = | >= | > | ? | BETWEEN | OUTSIDE
                       | IS | IS-NOT | IN | ALL-IN
                       | CARD= | CARD< | CARD<= | CARD> | CARD>=
<variable>         ::= symbol starting with ?
<entry-point>      ::= symbol | string

where * means 0 or more, and + one or more
```

Note that we can express disjunction, structural constraints and some kind of negation.

The algorithm for executing a query is the following:

When the query reduces to a single entry point then the answer is obtained immediately. Otherwise, we use the following approach, in which the goal node is the name of the class starting the query.

1. First expand all virtual items into the query, replacing them by the actual ones and inserting constraints as clauses of the query.
2. Estimate a list of candidates: Use the query entry points in positive branches to compute a subset of the class of the goal node. The result may be *none* meaning no answer, a list of candidates, or nil, meaning that we must do a class access (i.e. load all the objects of the class).
3. Build a *constrained graph*, i.e. a graph in which each node corresponds to a class of the query and has the set of constraints (clauses) attached to this node.
4. Collect objects
   For each object in the candidate list obtained in Step 1, navigate through the constrained graph and apply all filters to determine if the object must be kept in the final solution.

   Thus, we start with a list of candidates, and a global test on the subset of objects to be checked. Possible test operators are ALL (for checking all objects), $<, <=, =>, >=, <>$, BETWEEN, OUTSIDE, or user-defined operators coming from the virtual items. If the test is verified, then we return a list of the accepted candidates; otherwise we return nil.

   For each node of the constrained graph, corresponding to a particular set of objects, we operate as follows:

a. First, we split the query into simple clauses from which we dynamically build and compile a local filter, OR-clauses (disjunctive clauses) and sub-queries.

b. At each cycle, we have a list of candidates waiting to be examined, and a *good-list* of objects which were candidates and which passed the tests.

   i. Applying the global test to the good-list:
      A. If we have an immediate success, then we return the set of objects in the good-list.
      B. If we have a failure, we return nil.
      C. If we cannot decide, then we examine the current candidate.

   ii. Examining the current candidate:
       We first apply the local filter. If the test fails, then we proceed with the next candidate in the candidate list.

   iii. Then, we must check the set of all clauses (conjunction of OR-clauses and sub-queries):
        A. if there are no more clauses to check, we have a success and we proceed with the next candidate
        B. if the clause is a sub-query, we proceed recursively.
        C. if the clause is an OR-clause, we check each sub-clause in turn until we find one that is verified.

When accessing objects at each node, subsumption and defaults are taken into account. The result of the query process is a list of objects ids that satisfy the query. The algorithm is efficient due to the "pre-joined" nature of the constrained graph.

## 4   Multilingual Features

Multilingualism becomes unavoidable in the context of World Wide Web [11]. Most current approaches to multilingual ontologies usually take the perspective of aligning several monolingual ontologies (Trojhan et al. [13]), eventually using a meta-model [10]. Others use translation to rebuild the ontology in a different language [5]. Recently Cardeosa et al. [3] recommended using the Universal Networking Language (UNL) from the United Nations University to develop concepts around universal words then add the various language tags.

MOSS does not advocate any particular method for developing multilingual ontologies but provides a simple formalism that can be used as follows:

```
(defconcept
  (:name :en "Revenue" :fr "Rmunration" :it "Reddito" :pl "Dochd")
  (:att (:en "amount" :fr "montant" :pl "Ilo" :it "importo")
        (:type :float))
  (:doc :en "A Revenue is an amount of money earned by a person."
        :fr "Prix et avantages divers que l'on accorde  une personne
             en retour d'un travail ou de services rendus."
        :pl "Dochd to kwota pienidzy, ktr zarabia dana osoba."
        :it "Il Reddito  la somma di denaro guadagnata da una persona."))
```

Here REVENUE concept, amount attribute and documentation are defined in English, French, Italian and Polish by using the special language tags :en, :fr, :it and :pl. The concept is stored as a single object with different languages and can be read using any of the specified language. The same applies to virtual concepts or properties.

MOSS has been implemented as a full environment including message passing, methods and versioning. It can be downloaded from the following address: http://www.utc.fr/~barthes/MOSS/. A detailed manual giving full information can be found at this address.

# 5  OWL and Rule Translation

We found out that a subset of MOSS can be used as a simple format for OWL-DL. It then can be compiled into OWL. Virtual concepts or virtual properties implement constraints on a particular item. They cannot be translated solely in terms of OWL-DL concepts and produce additional rules to be used during the reasoning process.

## 5.1  Translating Concepts into OWL-DL

Let us consider the concept of CITY expressed in English, French and Italian with synonyms:

```
(defconcept
    (:name :en "City; town" :fr "ville; cit" :it "Citta")
    (:rel (:en "country" :fr "pays")(:to "stato"))
    (:is-a "territorio")
    (:doc :en "A large and densely populated urban area; may include several ~
               independent administrative districts."
          :fr "Une ville est une zone urbaine peuple de faon dense pouvant inclure ~
               des districts administratifs indpendants."))
```

Assuming that COUNTRY has already been translated, CITY will translate in turn into a concept and an object type property:

```
<!-- +++++++++++++++++++++++++++++++++ Classes ++++++++++++++++++++++++++++++++++ -->

<owl:Class rdf:ID="Z-City">
  <rdfs:label xml:lang="en">City; town</rdfs:label>
  <rdfs:label xml:lang="fr">ville; cit</rdfs:label>
  <rdfs:label xml:lang="it">Citta</rdfs:label>
  <rdfs:subClassOf rdf:resource="#Z-Territory"/>
  <rdfs:comment xml:lang="en">A large and densely populated urban area; may include several
independent administrative districts.</rdfs:comment>
  <rdfs:comment xml:lang="fr">Une ville est une zone urbaine peuple de faon dense pouvant
 inclure des districts administratifs indpendants.</rdfs:comment>
</owl:Class>

<!-- +++++++++++++++++++++++++++++++++ End Classes ++++++++++++++++++++++++++++++++++ -->
```

```
<!-- ++++++++++++++++++++++++++ Object Properties ++++++++++++++++++++++++++ -->

  <owl:ObjectProperty rdf:ID="hasCountry">
    <rdfs:label xml:lang="en">country</rdfs:label>
    <rdfs:label xml:lang="fr">pays</rdfs:label>
    <rdfs:domain rdf:resource="#Z-City"/>
    <rdfs:range rdf:resource="#Z-Country"/>
    <owl:inverseOf>
      <owl:ObjectProperty rdf:about="#isCountryOf"/>
    </owl:inverseOf>
  </owl:ObjectProperty>

  <owl:ObjectProperty rdf:ID="isCountryOf">
    <rdfs:label xml:lang="en">country of</rdfs:label>
    <rdfs:label xml:lang="fr">pays de</rdfs:label>
    <rdfs:domain rdf:resource="#Z-Country"/>
    <rdfs:range rdf:resource="#Z-City"/>
    <owl:inverseOf>
      <owl:ObjectProperty rdf:about="#hasCountry"/>
    </owl:inverseOf>
  </owl:ObjectProperty>

<!-- ++++++++++++++++++++++++++ End Object Properties ++++++++++++++++++++++++++ -->
```

Note that two object properties have been produced: a link between CITY and COUNTRY and a reverse link from COUNTRY to CITY. The reverse link however is not an inverse property.

## 5.2 Translating Virtual Concepts

When MOSS virtual items are defined, one must include in OWL the corresponding classes to be able to do reasoning. Thus, we first extend OWL with additional classes with a declaration of virtual class, datatype property and object property:

```
<!-- +++++++++++++++++++++ Declaration for Virtual Categories +++++++++++++++++++++ -->

  <rdfs:Class rdf:ID="VirtualClass">
    <rdfs:comment xml:lang="en">Class of all the virtual classes</rdfs:comment>
    <rdfs:label xml:lang="fr">Classe Virtuelle</rdfs:label>
    <rdfs:subClassOf rdf:resource="http://www.w3.org/2002/07/owl#Class"/>
    <rdfs:label xml:lang="en">Virtual Class</rdfs:label>
  </rdfs:Class>

  <rdfs:Class rdf:ID="VirtualObjectProperty">
    <rdfs:label xml:lang="fr">Proprit Objet Virtuelle</rdfs:label>
    <rdfs:comment xml:lang="en">The class of the virtual object properties</rdfs:comment>
    <rdfs:subClassOf rdf:resource="http://www.w3.org/2002/07/owl#ObjectProperty"/>
    <rdfs:label xml:lang="en">Virtual Object Property</rdfs:label>
  </rdfs:Class>

  <rdfs:Class rdf:ID="VirtualDatatypeProperty">
    <rdfs:subClassOf rdf:resource="http://www.w3.org/2002/07/owl#DatatypeProperty"/>
    <rdfs:comment xml:lang="en">The class of virtual datatype properties</rdfs:comment>
```

```
        <rdfs:label xml:lang="fr">Proprit datatype virtuelle</rdfs:label>
        <rdfs:label xml:lang="en">Datatype Virtual Property</rdfs:label>
    </rdfs:Class>
```

```
<!-- +++++++++++++++++++ End Declaration for Virtual Categories +++++++++++++++++++ -->
```

Now, the translation for Adult yields a virtual class:

```
<VirtualClass rdf:ID="Z-Adult">
    <rdfs:label xml:lang="en">adult</rdfs:label>
    <rdfs:label xml:lang="fr">Adulte</rdfs:label>
    <rdfs:subClassOf rdf:resource="#Z-Person"/>
    <rdfs:comment xml:lang="en">Adults are persons over 18 years.</rdfs:comment>
    <rdfs:comment xml:lang="fr">Personne ayant 18 ans ou plus.</rdfs:comment>
</VirtualClass>
```

Note that the language uses the rdfs:label property with an xml:lang language tag.

The MOSS virtual elements must be translated into something that can be used for reasoning in addition to the corresponding classes. The solution was to produce rules implementing the corresponding constraints to be added to the reasoner. When reasoning with JENA, we produce rules compatible with the JENA framework[4]. When using SPARQL, we produce SPARQL CONSTRUCT queries[5]. Rules are output into different files.

When necessary user-defined operators have to be provided and added to the reasoning system to be fired during the reasoning process.

The output compatible with the JENA framework is the following:

```
######-----------------------------------------------------------######
###### Jena

###### Operator List
# operator=net.eupm.terregov.categorization.jena.getYears

###### Prefix List
@prefix tgv: <http://www.terregov.eupm.net/virtual/tg-virtual.owl#>.
@include <RDFS>.

###### Include List
@include <OWLMicro>.

###### Virtual Entities

### the property age is computed from the birthdate of a person by applying
### the function getYears to the birthdate.

[Role_Aage:
    (?SELF tgv:hasAage ?RESULT)
<-
    (?SELF rdf:type tgv:Z-Person)
    (?SELF tgv:hasBirthDate ?Y)
    getYears(?RESULT, ?Y)
]
######-----------------------------------------------------------######
```

---

[4] http://jena.apache.org
[5] http://www3.org/TR/rdf-sparql-query

The output producing SPARQL CONSTRUCT is the following:

```
###----------------------------------------------------------------###
###--- SPARQL ---###

###---Rule: Aage ---###
PREFIX ref: <http://www.w3.org/1999/02/22-rdf-syntax-ns#>
PREFIX tgv: <http://www.terregov.eupm.net/virtual/tg-virtual.owl#>.
PREFIX tgfn: <java:net.eupm.terregov.categorization.fn.>
CONSTRUCT{
    ?SELF tgv:hasAge ?RESULT .
}
WHERE{
    ?SELF rdf:type tgv:Z-Person .
    ?SELF tgv:hasBirthDate ?Y .
    ?Y tgfn:getYears ?RESULT .
}
###----------------------------------------------------------------###
```

## 5.3   Taking Indexing into Account

Indexing is an interesting and efficient feature. We have shown in a previous article
[1] how the indexing mechanism can be added to the OWL language using annotation properties. In MOSS, an attribute (datatype property), that can be used for
indexing, is simply declared with the (:index) feature. In the following example,
the attribute acronym is intended to be used for indexing the instances of the PRO-
GRAM concept. If the attribute **acronym** were defined in another concept it could
be used for indexing or not, according to the presence of (:index) in its declaration.

```
(defconcept
    (:name :en "Program" :fr "Programme")
    (:att (:en "acronym" :fr "acronyme") (:unique) (:index))
    (:att (:en "documentation" :fr "documentation")) ...)
```

In the OWL syntax, an annotation property is first created. It is also a functional
datatype property whose range is **xsd:boolean**. It is considered as a restriction on a
datatype property for some concepts and its domain is the class **owl:Restriction**:

```
<owl:FunctionalProperty rdf:ID="indexing">
  <rdf:type rdf:resource="http://www.w3.org/.../owl#AnnotationProperty"/>
  <rdf:type rdf:resource="http://www.w3.org/.../owl#DatatypeProperty"/>
  <rdfs:range rdf:resource="http://www.w3.../XMLSchema#boolean"/>
  <rdfs:domain rdf:resource="http://www.w3.org/.../owl#Restriction"/>
</owl:FunctionalProperty>
```

A concept having a datatype property intended to be used for indexing is declared
as a subclass of an **owl:Restriction**. This restriction is annotated by the indexing
property with true as the Boolean value. For example, for the following OWL code
indicates that each instance of the class **Z-Program** must be indexed by the value
of the property **hasAcronym**.

```
<owl:Class rdf:about="#Z-Program">
  <rdfs:subClassOf>
    <owl:Restriction>
```

```
        <owl:onProperty rdf:resource="#hasAcronym"/>
        <owl:cardinality
         rdf:datatype="&xsd;nonNegativeInteger">1</owl:cardinality>
        <indexing
         rdf:datatype="http://www.w3.../XMLSchema#boolean">true</indexing>
      </owl:Restriction>
    </rdfs:subClassOf>
</owl:Class>
```

However, although indexes can be translated into OWL, the reasoners have to take
advantage of the mechanism to optimize accesses.

## 6  Examples

Several applications were developed using MOSS ontologies. We selected two: The
first one, NEWS is an international multi-agent application in which each agent has
a MOSS ontology and knowledge base; the second was a European project, TER-
REGOV, in which multilingual ontologies were meant to be translated into OWL.

NEWS is an international multi-agent application developed between Brazil,
France, Japan, Mexico and UK. It is meant to share information. Local agents (personal assistant agents) use the local language having ontologies in the national language. A central multilingual PUBLISHER agent gathers pieces of information and
composes periodically a journal distributed to subscribers. Each local subscriber can
ask for various types of information in her own language or subscribe to categories
of events, keywords or authors. In this application MOSS is integrated seamlessly
with the multi-agent platform.

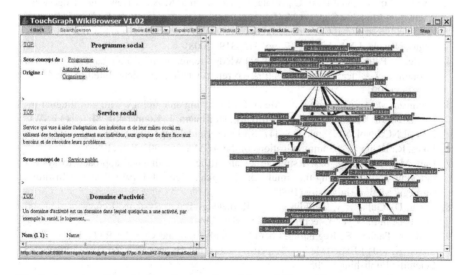

**Fig. 1** An extract from the French extraction of the multilingual Terregov ontology

TERREGOV[6] was an FP6 European project (IST-507749), the goal of which was to develop e-government applications related to social services. Tests were selected to be conducted in France, Italy, Poland, and UK. The main ontology contained several hundreds of concepts, and several thousands of properties in four different languages. A compiler was used to extract national sub-ontologies and convert the MOSS formalism into OWL with an adjunction of rules. An extract of the ontology in French is shown Fig. 1.

## 7   Conclusion

Wa have presented in this paper an approach to developing ontologies different from the common descriptive logics approach. The formalism allows specifying defaults, virtual entities (corresponding to rule constraints), multilinguism. Reasoning is implemented by a query mechanism. The formalism can be used to express OWL-DL entities, translating into OWL-DL structures, complemented by rules to be used by reasoners. More information can be found in the mentioned web sites from which the environment can be downloaded.

## References

1. Bettahar, F., Moulin, C., Barths, J.-P.A.: Adding an index mechanism to an ontology. In: Wegrzyn-Wolska, K.M., Szczepaniak, P.S. (eds.) Advances in Intelligent Web Mastering. ASC, vol. 43, pp. 56–61. Springer, Heidelberg (2007)
2. Brachman, R.J., Schmolze, J.: An Overview of the KL-ONE Knowledge Representation System. Cognitive Sci. 9(2), 171–216 (1985)
3. Jess Cardeosa, J., Gallardo, C., Iraola, L., De la Villa, M.A.: A New Knowledge Representation Model to Support Multilingual Ontologies. A case Study. In: Proceedings of the 2008 International Conference on Semantic Web & Web Services, SWWS 2008, Las Vegas, Nevada, USA, July 14-17, pp. 313–319 (2008)
4. Doyle, J., Patil, R.S.: Two theses of knowledge representation: Language restrictions, taxonomic classification, and the utilisty o representation. Artificial Intelligence 48(3), 261–297 (1991)
5. Espinoza, M., Gómez-Pérez, A., Mena, E.: Enriching an ontology with multilingual information. In: Bechhofer, S., Hauswirth, M., Hoffmann, J., Koubarakis, M. (eds.) ESWC 2008. LNCS, vol. 5021, pp. 333–347. Springer, Heidelberg (2008)
6. Fikes, R., Kehler, T.: Commiunications of the ACM 9(28), 905–920 (1985)
7. Fox, M.: Knowledge based simulation: an artificial intelligence approach to system modeling and automating the simulation life cycle, Memo 1-1-1988 Robotics Institute, Carnegie Mellon University (1988)
8. Lenat, D., Guha, R.V.: Building Large Knowledge-Based Systems: Representation and Inference in the Cyc Project. Addison-Wesley (1990)
9. Mili, H., Pachet, F.: Regularity, document generation and Cyc. In: Rada, R., Tochtermann, K. (eds.) ExpertMedia - Expert Systems and Hypermedia, pp. 171–206. World Scientific Publishing (1995)

---

[6] http://www.terregov.eupm.net/

10. Montiel-Ponsoda, E., Aguado de Cea, G., Gómez-Prez, A., Peters, W.: Enriching ontologies with multilingual information. Natural Language Engineering 1(1), 1–27
11. Monteil-Ponsoda, H.: Multilingualism in Ontologies: Building Patterns and Representation Models. LAP LAMBERT Academic Publishing (2011)
12. Patel-Schneider, P.F., McGuinness, D.L., Brachman, R.J., Resnick, L.A.: The CLASSIC knowledge representation system: guiding principles and implementation rationale. SIGART Bull. 2(3), 108–113 (1991)
13. Trojhan, C., Quaresma, P., Vieira, R.: A framework for multilingual ontology mapping. In: European Language Resources Association (ELRA), Proceedings of the 6th Edition of the Language Resources and Evaluation Conference, Marrakech, Morocco (May 2008)

# Integrating Social Network Data for Empowering Collaborative Systems

Xuan Truong Vu, Marie-Hélène Abel, and Pierre Morizet-Mahoudeaux

**Abstract.** Over the past years, online social networks with websites such as Facebook, Twitter or LinkedIn, have become a very important part of our everyday life. These websites are increasingly used for creating, publishing and sharing information by users. This creates a huge amount of information a part of which may match the interests of a given group. However the distributed and protected nature of these information make it difficult for retrieving. In this paper, we present a user-centered approach for aggregating social data of members of a group to promote the collaboration and the sharing of knowledge inside collaborative systems. The members will be able to delegate the proposed system to aggregate their different social profiles and to make available the relevant part of information to other members of the group.

## 1 Introduction

Online social networks, or more commonly known as social networks, have become very popular with websites such as Facebook, Twitter and LinkedIn. These websites provide constantly open and evolving socio-technical ecosystems which promote the interactions and the inter-relationships between distributed users - both individuals and organisations, and facilitate the creation and the sharing of cross-domain information and knowledge.

A huge amount of information is therefore available and increasingly grows every day. A part of these information may match the scope of interest of a given group which could be a professional group, a community of interest or any other community. It is interesting to gather and share such information within the group.

Xuan Truong Vu · Marie-Hélène Abel · Pierre Morizet-Mahoudeaux
UMR CNRS 7253 Heudiasyc, Université de Technologie de Compiègne, France
e-mail: {xuan.vu,marie-helene.abel,
         pierre.morizet-mahoudeaux}@utc.fr

V.-N. Huynh et al. (eds.), *Knowledge and Systems Engineering, Volume 2,*  109
Advances in Intelligent Systems and Computing 245,
DOI: 10.1007/978-3-319-02821-7_11, © Springer International Publishing Switzerland 2014

Commercial solutions such as GNIP[1] allow an access to real-time social media data streams from dozens of social websites via one single API. However, they are addressed to the enterprises for divers business processes (i.e. Business Intelligence, Community Management, etc.) not for the collaboration and the sharing of knowledge.

Today social networks had not been initially designed to be interoperable together. This creates disconnected websites and subsequently isolated data silos. Moreover, most of social networks require the provision of authentications to access to their user data. Such disconnected and protected nature of social networks make it difficult and time-consuming if one want to manually browse one by one in order to retrieve relevant information.

We present, in this paper, the idea of using collaborative peer-sourcing to integrate data available on social networks into collaborative systems with the objective to enhance the sharing of knowledge. Information to gather is not from all over the social networks but had been published by the members of the group. Moreover, it is intended that only information relevant to his/her respective owner' private preferences and the group' common interests will be capitalized and available to other members.

The paper is organized as follows. In the next section, we present some of the main beneficial features of such a social and collaborative approach. Then we introduce an extensible system architecture designed for aggregating the users' social data and for filtering collected data. An illustrative case is also provided in the fourth section to show the applicability of our proposed system. In the fifth section, we discuss about some related works and our originality. Finally, we conclude and present our future works.

## 2  Motivation

Users publish and exchange different types of information on social networks such as user profile information ranging from demographic information to personal interests, relationships between users, *user-created contents* including photos, videos, statuses, bookmarks, blogs, etc. [8]. All of them form so-called *social data*. These social data are therefore very rich and frequently updated.

It is interesting to consider social data as a very useful source of cross-domain information. A part of them could be relevant to a group which is driven by some common interests. If each of its members agrees to share their social data with other members, then more knowledge can be reachable and even new knowledge may emerge. Let consider the following possibilities :

1. *Members' additional interests/expertise* : when joining a group, the person only declares a part of his/her profile including his/her interests and/or expertise. However, these information are evolving and changing over time. Social data can be an alternative source for updating and enriching one's profile by allowing

---
[1] http://gnip.com/

to uncover his/her last interests and/or expertise [12, 6]. Therefore, the group will learn more about each of its members.

2. *New web resources* : social networks are intensively used for publishing and spreading news and Web resources. For example, a significant part of tweets, *i.e.* short messages published by Twitter users, can be considered as *information sharing* [11]. Most of them contain *URLs* referring to web pages, thus allowing to discover new resources matching the interests of the group.

3. *Emerging topics* : by watching recently captured members' additional interests and new Web resources, emerging topics could be identified.

4. *Possible sub-groups* : Each member can be connected to some other members on one or several social networks. These relationships and their interaction degree [16] will provide extra indicators beside the similarity indicators [7] for efficiently locating some possible sub-groups.

5. *Extended membership* : some external users, not actually belong to the group, might be considered as extended memberships when they are in the lists of contacts of several members of the group. These people could be invited to join the group or given a certain truth if other members would like to reach them for information.

These identified benefits, among others, provide the motivation for the members of a given group to collaboratively share their social data to other members. However, it is obviously impossible to ask to each of them to duplicate information considered pertinent that his/her has already published in other social networks. Thus, to help users avoid making extra manual efforts, an automated or semi-automated process is needed for aggregating their social data and subsequently for filtering the relevant part of gathered information.

As people are the key components of our approach, we must keep in mind that one of the main tasks is to provide them with the entire control over the information that they will possibly share with others. Therefore, we have adopted a user-centered approach which allows each member to delegate the system by suitable permissions to aggregate his/her different social profiles. He/she is also able to set a variety of personal private settings which will be taken in consideration during the information filtering process.

## 3   The Proposed System

In this section, we describe a general and extensive system architecture for aggregating the social data of the members of a given group and for filtering out the relevant part of information with respect to the personal private setting of each member as well as to the common interests of the group. Our proposed system is made up of three main modules (Figure 1) : (i) a user data integration module, (ii) an information filtering module, and (iii) a collaborative platform. The first module replies on different social networks to gather members' social data and then output preprocessed data that the second module should filter out by the means of a number

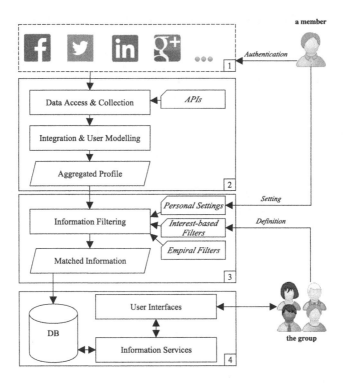

**Fig. 1** System architecture for integrating users' aggregated profiles into collaborative systems : (1) social networks (2) user-centered social network aggregation (3) information filtering (4) collaborative knowledge management system

of techniques. Finally, the collaborative platform provides the group with a digital support so that they can access to and share capitalized knowledge.

## 3.1 Social Networks

There are recently a lot of social websites[2]. Amongst others, Facebook[3], Twitter[4] and LinkedIn[5] are the three most popular websites in terms of the number of users, the volume of daily generated data and the traffic.

Each of them have been trying to provide users with different services and user experiences. A public social networking service like Facebook allows anyone aged 13 and over to register. They come and use Facebook as a communication platform

---

[2] http://en.wikipedia.org/wiki/
List_of_social_networking_websites
[3] https://www.facebook.com
[4] https://www.twitter.com
[5] https://www.linkedin.com

to connect with others and to join common-interest user groups [4]. LinkedIn is also a social networking website but since it is mainly devoted to people in professional occupations. Unlike Facebook and LinkedIn, Twitter is rather a microblogging service that enables users to send messages, known as "tweets", about any topic within the 140-character limit and to follow others for receiving their tweets. Twitter has therefore become a new and powerful medium of information sharing [9].

Social networks all allow users to set up visible profiles and to link to other individuals' profiles. User profile is a unique page where one can type oneself and display an articulated list of friends. This page might also include frames, where different kinds of information may appear such as user-created contents (i.e. posts, statuses, tags, messages, etc.).

By investigating the coverage of user profiles handled by these three services and those from Google+[6], OpenSocial[7] [17], we have identified the most frequent user data and categorized into six dimensions as follows :

- *Personal Characteristics* includes the user's basic information such as name, city, gender, and so forth;
- *Friends* includes the user' social connections;
- *Interests* contains user interests, preferences, and expertise;
- *Groups* contains the user's memberships:
- *Studies* and *Works* describe the user's schools and the user's workplaces respectively.
- *User-created contents* includes contents, social activities produced by the user.

These dimensions also form our general social user model.

## 3.2   User Data Integration

The *user data integration* is the process dealing fist with accessing and collecting each member's social data then with merging various data and modelling them into a unified profile.

### 3.2.1   Data Access and Collection

Most of actual social networks make it possible for their users to grant selected third-party applications an access to user data via their own application programming interface (API). With respect to this policy, the system always asks users for permission to access their profiles on subscribed social networks.

Since each social network support its proprietary API (e.g. Graph API for Facebook, REST API 1.1 for Twitter), different *aggregators* are then needed. Each of them is dedicated to a specific social network and recover information corresponding to the previous user dimensions if available.

---

[6] https://plus.google.com/
[7] http://opensocial.org/

**Fig. 2** User aggregated profile representation

Moreover, the aggregators are programmed to regularly (i.e. daily) crawl members' profiles so that their last published information on social networks will be early taken into account for filtering and possibly sharing with others.

### 3.2.2 Integration and User Modelling

Once members' social data are gathered, they should be merged together. For that purpose, a target common representation is needed. For this purpose, we have used the Friend Of A Friend (FOAF) vocabulary[8].

To map each gather information to a specific FOAF concept we simply base on a hand-crafted set of rules [17]. For example, *friends* will be represented as *foaf:knows* while *interests* can be described by *foaf:topic_interest*.

Each member is identified by an identifier, for example a unique email. His various social accounts will be represented as separated entities and linked to the member by the *owl:sameAs* concept (Figure 2). Each entity therefore has its own attributes and corresponding values. Such a representation allows to preserve in an explicit way the provenance of data (i.e. source) which is an important feature in our system cause it eases users' control over their data and the information sharing.

To improve the members' plain-text profiles, we have followed the *semantic web* approach introduced in [12, 2] matching the concepts contained in profiles to DBpedia[9] resources. Such approach allows to semantically enrich user information and to possibly refer to other linked resources. For example, the resource found at http://dbpedia.org/resource/Compigne stands for the Compi'egne city and is linked to the http://dbpedia.org/resource/Picardy resource which is the region where the city is located.

DBpedia Spotlight[10] and DBpedia keyword search API[11] are thus used for extracting concepts including named entities from texts end finding related DBpedia resources.

---

[8] http://xmlns.com/foaf/spec/

[9] http://dbpedia.org

[10] https://github.com/dbpedia-spotlight/dbpedia-spotlight

[11] https://github.com/dbpedia/lookup

Regarding user interests, the process is less straightforward. We should extract the user' interests either from a list of things for which users have explicitly claimed their interest or from user-created contents, in particular those containing *URLs*. In the second case, the referred web-pages will be explored to extract the titles and the keyword tags.

## 3.3   Information Filtering

The *Information Filtering* step is very important. Because, it should decide which information will be ignored or kept according to the member' private settings and the group's common interests. To this end, we have developed different techniques including manual as well as automated methods following :

- *User Private Settings* : the users will be free to set and modify the following settings :

  - The social account(s) the information of which could be aggregated and filtered,
  - The dimension(s) the information of which could be shared,
  - An extra verification might be needed before sharing. Either users can verify which information is shareable one by one, or all matched information will be available to other members.

- *Hashtag Method* : a hashtag is a word or a phrase prefixed with the symbol #, for example, #UTC could stand for the University of Technology of Compiegne. Hashtags have become very popular and efficient means for grouping and retrieving messages related to a given topic on social websites. The users of our system will be encouraged to use their commonly defined hashtags across social networks. Gathered contents including such hashtags will be considered pertinent for the group and directly accessible.

- *Keyword-based Method* : keywords and their synonyms can be also used for matching information. There is a need however for an analysis before constructing the keyword list due to a lot of ambiguities and noises.

- *Ontology-based Method* : in comparison to keywords, ontology gives more powerful performance for matching information. Firstly, there would be much less ambiguities. Secondly, it would not be necessary to list all concepts, named entities in particular, which belong to certain categories. Let's consider an example from DBpedia, in which the *Social Networking Services* category[12] is the subject (i.e. *dcterms:subject*) of a lot of networking services such as *dbpedia:Facebook, dbpedia:Twitter, dbpedia:Myspace, dbpedia:Instagram, dbpedia:FOAF_(software)*, etc. In such a case, only the category will be needed in order to match information related to one of its members.

- *Empirical Methods,* additional empiric methods can be also used to reduce the number of information extracted from user-created contents. They actually tend

---

[12] dbpedia.org/resource/Category:Social_networking_services

to filter out personal messages which are self-describing or addressed to a particular person. For this purpose, they rely on some simple detection patterns such as containing emoticons (e.g. "tired and upset :(") or including other usernames (e.g. "take a look at these photos http://bit.ly/Ywg7p6 @truongci5" ).

## 3.4 Collaborative Platform

The collaborative platform could be any collaborative system available on the market. It should provide among others, two essential functions such as the storage of data and the user interfaces for collaborating (e.g. defining interest-based filters) and accessing to capitalized knowledge.

Since, we have oriented to Semantic Web technologies in our approach, we prefer collaborative systems which reply on or support Semantic Web too. One of the identified candidates is the Memorae[13] platform which has been developed at the Université de Technologies de Compiègne. This platform is an ontology-based collaborative environment easing organizational learning and knowledge capitalization and is recently improved to integrate and index resources from social networks[5].

## 4   An Illustrative Scenario

In this section, we present a use case example of our system previously proposed. The example is taken for a group composed of several professors and Ph.D. students from the University of Technology of Compiègne.

Being recently interested by the topics like *knowledge management* and *digital ecosystem*, they have been collaborating within the Memorae platform with the objective to share interesting resources related to the two topics. Thought, a certain amount of knowledge have been therefore capitalized thanks to the manual input of each member, they would like to enrich and expand further this common knowledge base.

On the other hand, they are all using online social networks but for divers purposes. They did not necessarily join to the same types of social networks neither put the same efforts for publishing and sharing information. For example, the professors use essentially professional networks such as LinkedIn while the students are much more active on large-scale social networking services such as Facebook and Twitter.

Each of them have been publishing various information and content on his/her profiles handled by different social networks. There is a part of these information matching the two interested topics such as specialised web resources, related scientific events or researchers in the field. Unfortunately, they are not all reachable by all members of the group. Thus, the group wants to integrate these social data into the shared knowledge base.

To this end, the members of the group use our previously proposed aggregation and filtering extension which is supposed to be already operational on the platform

---

[13] http://www.hds.utc.fr/memorae/

Memorae. A hashtag #KE for *knowledge ecosystem* and several keywords such as *knowledge engineering, digital ecosystem, knowledge ecosystem* and *collaborative platform* have been then defined by the group for filtering relevant information. We suppose moreover that in *private settings*, they all agree to share the entire aggregated data including interests, contacts, published contents, etc.

Now, let consider three specific cases Pierre, Etienne and Xuan. Pierre is a professor while two others are both students. Each of them grants specific permissions to the system for accessing their social profiles and for collecting data. For example, Pierre has given access to his LinkedIn profile and Etienne has authorized access to his Facebook and LinkedIn profiles while Xuan has given access to his LinkedIn and Twitter profiles. During a couple of weeks, they continue normally using their subscribed social networks when the collaborative platform starts to detect some relevant information from the collected data. Here follows several examples.

Xuan has seen a tweet from one of his Twitter contacts concerning *the fifth international conference on knowledge and systems engineering* (KSE2013), then he re-posts this tweet by adding the #KE hashtag. By containing the predefined hashtag, this message will be straightforward selected to share.

Etienne has found out an interesting document about *knowledge ecosystem* available at[14]. He shares this link on his Facebook profile. The system first detects the url and then investigates the referred webpage. Since the page contains *Knowledge ecosystem - Management - Part 1* as its title, it will be retrieved.

The professor Pierre has published on LinkedIn his latest research paper. Since its title includes *digital ecosystem*, this publication will also be considered relevant to the group. Besides, Pierre has some LinkedIn contacts who have been self-qualified as expert in *knowledge engineering*. Such information allows other members to follow these people on social networks so that more interesting resources can be discovered. They could be moreover invited to join the group.

Thus, the shared knowledge base of the group is increased.


## 5  Discussion

Social networks, by providing rich user data and social interactions, have recently received a wide interest in many areas of Computer Science literature. In particular, many authors have focused on the identification of trust groups [3], *recommender systems* running on multiple social networks [10] and the detection of real-world event using crowd-sourcing [13].

There are some works related to the integration of user profiles in social networks, especially for constructing users' profiles of interests[1, 12], preferences[14] or expertise [16]. The users' output profiles are generally used in the context of recommender systems or social search engines.

The originality of our work is a new use of users' aggregated and enriched profiles in a collaborative way. The members of a given group, supported by our system,

---

[14] http://www.slideshare.net/Presentationsat24point0/
knowledge-ecosystem-powerpoint-slide

agree to put together their own unified profiles. Once aggregated and shared, more information is accessible and new knowledge can even emerge. This therefore allows to feed and to enhance the collective intelligence within the same group of users around mutual interests.

In other words, our work consists in contribution to a collaborative knowledge ecosystem by the use of other well-known digital ecosystems such as social networks. To our knowledge, this is the first time that such an social and collaborative approach has been proposed.

Another interesting work that we could take into consideration has been introduced in [15]. The authors have defined a set of requirements, sketched a security model and presented a framework of cryptographic protocols for securing friendship requests and user-generated content within groups formed outside of the social network, around secret, sensitive or private topics.

## 6  Conclusion and Future Work

In this paper, we have presented a new approach for using social network data to empower collaborative systems. We have introduced some interesting benefits that can motivate each member of a given group to collaborate in sharing their social data. A user-centered approach is recommended for enabling the access to the members' cross-social-network data and for allowing them to have a better control on their information to share with other member of the group. Therefore, we have described a general and extensive system architecture for aggregating and integrating member' social profiles into collaborative systems. A variety of methods for matching and filtering information have been also developed.

On-going work will mainly focuses on the implementation of the social extension on a concrete collaborative system, namely Memorae. It will then be possible to investigate the willingness of users, the potential of social data and to subsequently to evaluate actual benefits of our proposed approach for real groups of end-users whose shared divers interests.

**Acknowledgments.** Part of this work has been developed in cooperation with the 50A Company[15] who is funding this work.

## References

1. Abel, F., Gao, Q., Houben, G.-J., Tao, K.: Analyzing Temporal Dynamics in Twitter Profiles for Personalized Recommendations in the Social Web. In: Proceedings of ACM WebSci 2011, 3rd International Conference on Web Science, Koblenz, Germany. ACM (June 2011)
2. Abel, F., Henze, N., Herder, E., Krause, D.: Linkage, aggregation, alignment and enrichment of public user profiles with mypes. In: Proceedings of the 6th International Conference on Semantic Systems, I-SEMANTICS 2010, pp. 11:1–11:8. ACM, New York (2010)

---

[15] http://www.50a.fr

3. Caverlee, J., Liu, L., Webb, S.: The socialtrust framework for trusted social information management: Architecture and algorithms. Inf. Sci. 180(1), 95–112 (2010)

4. Chiu, P.-Y., Cheung, C.M.K., Lee, M.K.O.: Online social networks: Why do we use facebook? In: Lytras, M.D., Carroll, J.M., Damiani, E., Tennyson, R.D., Avison, D., Vossen, G., De Pablos, P.O. (eds.) WSKS 2008. CCIS, vol. 19, pp. 67–74. Springer, Heidelberg (2008)

5. Deparis, E., Abel, M.-H., Mattioli, J.: Modeling a social collaborative platform with standard ontologies. In: 2012 Eighth International Conference on Signal Image Technology and Internet Based Systems, pp. 167–173 (2011)

6. Gao, Q., Abel, F., Houben, G.-J.: Genius: generic user modeling library for the social semantic web. In: Pan, J.Z., Chen, H., Kim, H.-G., Li, J., Wu, Z., Horrocks, I., Mizoguchi, R., Wu, Z. (eds.) JIST 2011. LNCS, vol. 7185, pp. 160–175. Springer, Heidelberg (2012)

7. Horowitz, D., Kamvar, S.D.: The anatomy of a large-scale social search engine. In: Proceedings of the 19th International Conference on World Wide Web, WWW 2010, pp. 431–440. ACM, New York (2010)

8. Kim, W., Jeong, O.-R., Lee, S.-W.: On social web sites. Inf. Syst. 35(2), 215–236 (2010)

9. Kwak, H., Lee, C., Park, H., Moon, S.: What is twitter, a social network or a news media? In: Proceedings of the 19th International Conference on World Wide Web, WWW 2010, pp. 591–600. ACM, New York (2010)

10. De Meo, P., Nocera, A., Terracina, G., Ursino, D.: Recommendation of similar users, resources and social networks in a social internetworking scenario. Information Sciences 181(7), 1285–1305 (2011)

11. Naaman, M., Boase, J., Lai, C.-H.: Is it really about me?: message content in social awareness streams. In: Proceedings of the 2010 ACM Conference on Computer Supported Cooperative Work, CSCW 2010, pp. 189–192. ACM, New York (2010)

12. Orlandi, F., Breslin, J., Passant, A.: Aggregated, interoperable and multi-domain user profiles for the social web. In: Proceedings of the 8th International Conference on Semantic Systems, I-SEMANTICS 2012, pp. 41–48. ACM, New York (2012)

13. Sakaki, T., Okazaki, M., Matsuo, Y.: Earthquake shakes twitter users: real-time event detection by social sensors. In: Proceedings of the 19th International Conference on World Wide Web, WWW 2010, pp. 851–860. ACM, New York (2010)

14. Shapira, B., Rokach, L., Freilikhman, S.: Facebook single and cross domain data for recommendation systems. User Modeling and User-Adapted Interaction 23(2-3), 211–247 (2013)

15. Sorniotti, A., Molva, R.: Secret interest groups (sigs) in social networks with an implementation on facebook. In: Proceedings of the 2010 ACM Symposium on Applied Computing, SAC 2010, pp. 621–628. ACM, New York (2010)

16. Vu, T., Baid, A.: Ask, don't search: A social help engine for online social network mobile users. In: 2012 35th IEEE Sarnoff Symposium (SARNOFF), pp. 1–5 (2012)

17. Vu, X.T., Morizet-Mahoudeaux, P., Abel, M.-H.: User-centered social network profiles integration. In: Proceedings of the 9th International Conference on Web Information Systems and Technologies, Aachen, Germany (2013)

# Recommendation of a Cloud Service Item Based on Service Utilization Patterns in Jyaguchi

Shree Krishna Shrestha, Yasuo Kudo, Bishnu Prasad Gautam, and Dipesh Shrestha

**Abstract.** One of the most determining factors for mining the sequences in terms of service mining in cloud services is time sequence. However, this factor is found to be often ignored and recommendation of services in cloud system is done based on the item mining approach. The problem that we discussed in this paper is addressed by applying the concept of time weight factor in the collection of sequence from which we achieved better result of recommendation from relational sequences. In this paper, we describe a recommendation method of service to user based on his service usage pattern in the system. The recommendation algorithm is based on the mining result of TWSMA algorithm which adopts an innovative approach based on sequences of service usage pattern and then characterizes each set of sequences using multidimensional properties based on user id, time series, and usage frequencies. We take advantage of implementing recommendation in Jyaguchi cloud system in which the user are recommended the services according to the log of service used by the users.

## 1 Introduction

The emergence of new network media, applications and services in Internet increases the data flow as well as the services that can be utilized by end users. This is positive trend in terms of selecting services from given options, however, at the

Shree Krishna Shrestha · Yasuo Kudo
Muroran Institute of Technology
e-mail: 12054071@mmm.muroran-it.ac.jp, kudo@csse.muroran-it.ac.jp

Bishnu Prasad Gautam
Wakkanai Hokusei Gakuen University
e-mail: gautam@wakhok.ac.jp

Dipesh Shrestha
DynaSystem Co., Ltd., Sapporo, Hokkaido, Japan

V.-N. Huynh et al. (eds.), *Knowledge and Systems Engineering, Volume 2,*                 121
Advances in Intelligent Systems and Computing 245,
DOI: 10.1007/978-3-319-02821-7_12, © Springer International Publishing Switzerland 2014

same time it has increased the level of complexities while selecting services from tremendous numbers of items. These complexities are most of the time related with the way of filtering out targeted services from the service pool. In order to sort out this problem, filtering functionalities provided in search engines may support users to pinpoint needful information and service acquisition. Furthermore, the objectives of these tools are to rank information according to partly revealed preferences, which are encapsulated in previously undertaken surveys done by marking users' provided information and properties itself [1]. In this research we have highlighted the concept of sequential information, sequential pattern mining and multidimensional concept. This concept is further enhanced with a new parameter of time sequence which was not applied until the inception of our research in service mining. Hence, the mining of service usage pattern on the basis of sequential patterns extracted through multidimensional information results into achieving very reliable recommendation. This recommendation helps the end users to select services that they are looking for. In this paper, we have proposed a recommendation system which is based on the service usage of user on same system, to apply the pattern of service search on the basis of time factor at which the time element will considered and propose the users the needful services as per their consumption history. Generally, the recommendation algorithm proposed in this paper is based on a mining algorithm Time Weight Sequence Mining Algorithm (TWSMA) [2], which we will explain in later section of this paper. This algorithm recommends the service to the users from the pool of mined data from TWSMA according to his current service usage log. The algorithm is capable to recommend services for not logged in users, just logged in users and users who have already used services after login in the system. This algorithm is suitable to recommend the service to new users and existing user with enough log data and very less log data.

## 2 Background and Related Works

### 2.1 *Jyaguchi Cloud System and Its Overview*

In order to perform our experiment, we utilize the Jyaguchi cloud system [3, 4, 5]. The term Jyaguchi was introduced by the author [5] and was derived from the Japanese language, in which the term means "an outlet portion of a tube or tap, which has opening and closing valves to regulate the rate of water flow." Accordingly, such a behavior of regulating resources is incorporated in the field of service usage, which was introduced in the Jyaguchi architecture. Jyaguchi proposed a hybrid architectural model because no single architectural model sufficiently provides a solution that is capable of regulating services on a pay per use basis, thereby providing features of SaaS. Furthermore, Jyaguchi is an architectural model for the development of distributed applications that can be extended to architecture for cloud services and demonstrates how this style can be used to enhance the architectural design of a next-generation service cloud.

We choose to use Jyaguchi in our experiment because of a rapid development of cloud computing technology as well as diverse methods of obtaining data. Consequently, the requirement for data mining in this infrastructural environment has been dramatically increased. At present, however, there are few data analysis tools to process large-scale data that float in the cloud service environment. Also, data mining technology is gradually emerging in the backdrop of such circumstances. In fact, massive data mining over cloud services could be a very important guide to scientific research and business decision making. In order to propose a prototype data mining technique in cloud services utilizing sequential pattern mining, we have use the Jyaguchi architecture, in which the authors have substantial experiences.

## 2.2 TWSMA Algorithm

The recommendation algorithm we proposed in this paper is based on the result of the sequential pattern mining algorithm named TWSMA algorithm proposed by the authors [2]. TWSMA is a sequential pattern mining algorithm for mining multidimensional sequences considering service usage time as a service weight parameter. The whole algorithm has two parts: the first one covers creation of multidimensional service weight sequences and the second one covers mining multidimensional sequences. In creation of multidimensional service-weight sequence, whole log database is read and relative service weight is calculated for each service in each position to generate sequence pairs of services and relative service weights. Mining a multidimensional sequence [6] is done in three steps: 1. Mining Sequential Pattern, 2. Forming Projected Database, and 3. Mining MD-patterns. We modified prefixSpan [7] algorithm to support service weight for becoming factor of frequent service while mining sequential pattern. The mean weight of services in a subset is used to check if the subset is frequent.

Here, we are going to explain the algorithm with the help of an example taking log database of 27 rows with 3 users (user id: 10, 15 and 16) and 7 services (service id: 1, 2, 3, 4, 123, 234, 456). The multidimensional table is created from the log database as shown in the Table 1, whose entry consists of the sequence id, user id and a set of pairs of service and corresponding service usage time. The services used in a single session are considered as a single sequence. From this table, we calculate the relative service weight of each service in each sequence and position, after which we create a multidimensional service weight sequential database.

We take a case of user 10 and service 2 for following calculations:

Time of usage of service 2 at position 1 and sequence $1 = 6$ min
Total time of usage of service $i$ for user $j$ $(ST_{i,j})$ = sum of time of use of service $i$ by user $j$.
Total time of usage of service 2 for user 10 $(ST_{2,10}) = (6 + 33 + 21 + 20 + 22 + 21)$ min $= 123$ min
Total service usage time for user $j$ $(T_j)$ = sum of all time of user $j$
Total service usage time for user 10 $(T_{10}) = 227$ min
Then, weight of service i for user $j$ $(ASW_{i,j}) = ST_{i,j}/T_j$

**Table 1** Multi-dimensional Sequence with Service Usage Time

| Sequence id | User id | Sequence |
|---|---|---|
| 1 | 10 | (2,6), (123,16), (456,31), (2,33), (456,35) |
| 2 | 10 | (2,21), (2,20), (2,22), (1,22), (2,21) |
| 3 | 16 | (2,1), (123,9), (456,1), (123,1), (456,15) |
| 4 | 15 | (456,19), (456,24), (234,24), (456,43) |
| 5 | 15 | (234,20), (234,11), (234,30), (456,38) |
| 6 | 16 | (456,19), (123,39), (456,30), (234,30) |

Weight of service 2 for user 10 $(ASW_{1,10}) = 123/227 = 0.542$

Let, unit time $(ut) = 5$ min. Then service usage count for service $i$ at position $k$ and sequence $l$ $(SC_{i,k,l}) = (T_{i,k,l})/ut$

Service usage count for service 2 at position 1 and sequence 1 $(SC_{2,1,1}) = 6/5 = 1.2$

Now, relative service weight of service $i$ at sequence $k$ and position $l$ $(RSW_{i,k,l}) = SC_{i,k,l} \times ASW_{i,j}$

Relative service weight of service 2 at sequence 1 and position $l$ $(RSW_{2,1,1}) = 1.2 \times 0.542 = 0.650$

Similarly, relative weight of service 2 at sequence 1 and position 4 is 3.577. From this result we can know that, for same service and same user also service usage time makes difference in the weight of service.

After calculating relative service weight of each service in each position with the above method, we can create multi-dimensional service weight sequential database as shown in Table 2.

**Table 2** Multi-dimensional Sequence with Relative Service Weight

| Seq. id | User id | Sequence |
|---|---|---|
| 1 | 10 | (2,0.650), (123,0.224), (456,1.804), (2,3.577), (456,2.037) |
| 2 | 10 | (2,2.276), (2,2.168), (2,2.385), (1,0.427), (2,2.276) |
| 3 | 16 | (2,0.0014), (123,0.608), (456,0.089), (123,0.068), (456,1.344) |
| 4 | 15 | (456,2.253), (456,2.846), (234,1.954), (456,5.1) |
| 5 | 15 | (234,1.628), (234,0.895), (234,2.442), (456,4.507) |
| 6 | 16 | (456,1.702), (123,2.636), (456,2.688), (234,1.242) |

The total weight of sequence database $SWD = \sum\limits_{j=1,q=1,r=1}^{n,x,y} RSW_{s_j,o,p}$, where $n = $ total no. of service used by user $i$, $z = $ total no. of sequences, $y = $ total no. of positions in sequence

$SDW = 49.83$

For minimum support $m\%$, relative minimum weight $(W_m) = SDW \times m\%$

For minimum support 5%, $W_m = 49.83 \times 5\% = 2.4915$

For first scanning of sequence, if the condition "total weight of service in whole database $\geq W_m$" holds, we treat the service as a frequent service

The total weight of service 2 in whole database

$= 0.650 + 3.577 + 2.276 + 2.168 + 2.385 + 2.276 + 0.001 = 13.333$

Since total weight of service 2 is greater than minimum weight, service 2 is frequent service.

This will make the 3 projected databases with service 2 as $(123, 456, 2, 456)$, $(2, 2, 1, 2)$ and $(123, 456, 123, 456)$. By scanning the $< 2 >$ −projected database once, its locally frequent services are generated checking sum of weight of service is higher than $W_m$ and all the length-2 sequential patterns prefixed with $< 2 >$ will be found. The repetition of process for all frequent services will give the frequent pattern from the frequent service and user_id dimension.

## 2.3   Related Works

After proposal of first recommendation algorithm on mid 1990's many research were done in this field proposing different algorithms and techniques. These researches not only proposed new algorithm but also improved the existing algorithms. Research in the recommendation algorithm seems to be influenced and geared up by the Netfix prize [8]. In the years 2006-2009, the Netfix DVD rental company held a competition on improving their recommender with a reward of USD1,000,000.

In the recommendation system mainly 3 type of approaches are being used to recommend: Collaborative filtering [9], Content based filtering [10], and Hybrid forms [11]. Based on those technique different researchers has purposed different algorithm that uses sequential pattern mining algorithm also. Zhou et al. [12] have proposed a sequential access-based web recommender system which is based on CSB-mine (Conditional Sequence Base mining algorithm) which algorithm is based directly on the conditional sequence bases of each frequent event. Han et al. [13] and Zhou et al. [14] have also proposed a recommendation algorithm based on the sequential pattern mining. Khonsha et al. [15] have proposed a recommendation algorithm for mining the user's behavior pattern and user's browsing pattern and recommended accordance with result of pattern mining calling hybrid recommendation system.

In evaluation of recommendation system different approaches are being proposed by different researchers and used also. There are mainly 2 methods for evaluating the recommendation algorithm which are widely used, off-line evaluation where no actual users are involved and an existing data set is used and on-line evaluation where users interact with a running recommender system and actually receive recommendation.

## 3   Algorithms

A lot of algorithms are being proposed for sequential pattern mining but very less to recommend items or services from the frequent pattern. In this paper we propose a step wise recommendation system. In this system for any user sequence of current service from log is used to recommend the service. In this paper we assume if the user has long sequence of used service for a single session, the service in beginning of sequence has less significance. According to which the beginning services will be dropped while finding recommended services. The whole recommendation algorithm is divided for 3 cases: 1. Recommendation to Anonymous User Group, 2. Registered User without Previous Usage of Services, and 3. Registered Users with Previous History of Service Usage (has service use log file). Different algorithm has implemented for each case which will be defined following:

**Recommendation to Anonymous User Group**

For these users, sequence which has highest support will be found from all frequent patterns and recommend service from that sequence.

| **Algorithm of recommendation for user who has not logged in** |
| --- |
| 1: Initialize minimum Support |
| 2: Select output frequent sequence for active user |
| 3: For every frequent sequence of given user and all user, check if support is greater or equal to minimum support |
| 4: Find out the sequence which have largest support from all users' frequent pattern |
| 5: Recommend Service from sequence with highest support from all users' frequent pattern and exit |

**Registered User without Previous Usage of Services**

For these users, sequence which has highest support will be found from current user's frequent pattern and service will be recommended from that sequence.

| **Algorithm of recommendation for user who has logged in but do not have service used log** |
| --- |
| 1: Initialize minimum Support |
| 2: Select output frequent sequence for active user |
| 3: For every frequent sequence of given user and all user, check if support is greater or equal to minimum support |
| 4: Find out the sequence which have largest support from current user's frequent pattern |
| 5: Recommend Service from sequence with highest support and exit |

**Registered Users with Previous History of Service Usage**

Firstly this recommendation algorithm will try to recommend the service by matching all current service sequence with frequent patterns. If match could not be found, the first service in current service usage sequence, i.e., first service used after logged in, will be dropped and try to match remaining sequence with frequent patterns. This process will follow until match is found or until current service sequence finishes. If current service sequence finished without finding matched pattern a service which has highest support for that user will be recommended.

---

**Algorithm of recommendation for user who has logged in and have service used log**

1: Initialize minimum support
2: Select output frequent sequence for active user
3: For every frequent sequence of given user and all user, check if support is greater or equal to minimum support
4: Find the sequence of current service uses from log file
5: Find the sequence from frequent pattern from current user's frequent pattern in which user has same sequence
if (Pattern found)
    6: Recommend the next service and exit
else {
    // Exclude the first service and search the user's frequent pattern again
    for (service is not last service of current service sequence) {
        6: Set sequence by excluding first service
        7: Search the new sequence from current user's frequent pattern
        if (Pattern found)
            8: Recommend next service that has highest support from current
                user's frequent pattern and exit
    } // End for loop

    // Recommendation based on all users' frequent patterns
    for (service is not last service of current service sequence) {
        8: Search the new sequence from all users' frequent pattern
        if (Pattern found)
            9: Recommend next service that has highest support from all user's
                frequent pattern and exit
        else
            9: Set sequence by excluding first service
    } // End for loop

    // Recommendation by current user's frequent pattern with highest support
    10: Recommend a service from the current user's frequent pattern with the
        highest support and exit
}

Here, we explain the recommendation system with an example. Table 3 is the frequent patterns obtained from TWSMA algorithm which patterns are used to recommend services for each user. The value inside [ ] is user id and a sequence of values inside {} is frequent pattern for that user.

**Table 3** Frequent patterns

| Pattern | Support | Pattern | Support | Pattern | Support |
|---|---|---|---|---|---|
| [*]{2} | 3 | [*]{2,123} | 2 | [*]{2,456,456} | 2 |
| [10]{123} | 2 | [*]{2,456} | 2 | [*]{2,123,456} | 2 |
| [*]{123} | 3 | [10]{2,2} | 2 | [*]{123,456,456} | 2 |
| [16]{123} | 2 | [*]{123,456} | 3 | [*]{456,456,234} | 2 |
| [*]{456} | 5 | [16]{123,456} | 2 | [16]{456,123,4} | 2 |
| [16]{456} | 2 | [*]{456,456} | 4 | [16]{456,123,3} | 1 |
| [15]{456} | 2 | [*]{456,123} | 5 |  | 2 |
|  |  |  |  | [*]{2,123,456,456} |  |
| [*]{234} | 3 | [*]{456,234} | 2 |  |  |
|  |  | [15]{234,456} | 2 |  |  |

For the user 16 who have used service 456 and 123 in a row, this algorithm will search for same sequence of service used by user 16. In the above frequent pattern, there is a pattern with sequence 456,123. Hence recommend next service, i.e., service 4. For same user 16, if the sequence of his service uses is 2,456,123, there are not exact matching frequent pattern for user 16. So the algorithm will leave very first service, i.e., service 2 and search for the sequence 456,123 and recommend service 4. Let us take another example for user 15 who have used service 2, 123 in sequence. Our algorithm will search for the sequence in frequent pattern but it could not found same sequence for user 15 so it will search from frequent pattern of all users. In this case algorithm will find the frequent pattern in frequent pattern of all users and recommend next service, i.e., 456 for user 15. Table 4 includes some examples of users' service used sequence and relative recommended service using our recommendation algorithm.

## 4  Experiments, Results and Evaluation

In this research we accomplished user based evaluation. Two types of experiment are conducted to evaluate and verify the efficiency and usability of the recommendation algorithm for cloud users in Jyaguchi System. In first type of experiment, we asked the Jyaguchi user to use the implemented recommendation system and evaluate according to the data of recommendation system used by the users. In the second type of experiment, Jyaguchi user is asked to answer 5 recommendation algorithm related questions and gathered the feedback of the users.

**Table 4** Recommended services

| User id | Service id | Recommended |
|---------|------------|-------------|
| 16 | 456, 123 | 4 |
| 16 | 2, 456, 123 | 4 |
| 15 | 456 | 123 |
| 15 | 234 | 234 |
| 15 | 2 | 123 |
| 15 | 2, 123 | 456 |
| 15 | 456, 2, 123 | 456 |

The evaluation process is divided into two phases, training phase and recommendation phase. In training phase, the service usage pattern of users is taken by asking users to use the Jyaguchi system without implementation of recommendation system. Then that usage pattern is used in mining and data from mining is used in recommending services to the users. In recommendation phases, firstly, the recommendation algorithm is implemented in the Jyaguchi System. Then users of Jyaguchi used the recommendation in the real environment. The data of user using the recommendation is saved in database and used for evaluation purpose. We took the log of each user and save the log in database to know either they are using the services recommended for them or not. The experiment was done in the system on the set of 11 services and 10 users.

Figure 1 is the experiment result of the recommended service used percentage by the users of Jyaguchi system. From the experiment result about 37% of service used within the experiment time was used from the recommended service. Among them

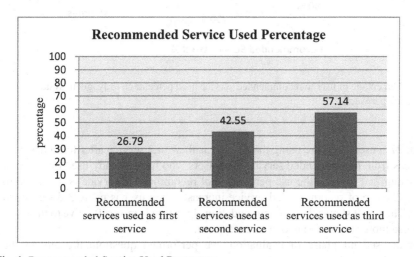

**Fig. 1** Recommended Service Used Percentage

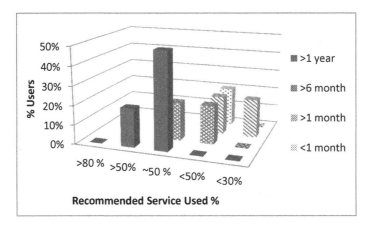

**Fig. 2** Percentage of user to recommended service used percentage to Jyaguchi system usage time

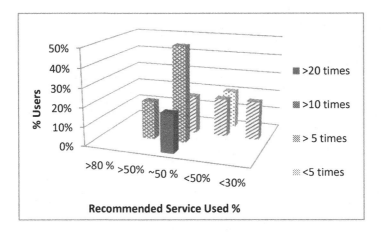

**Fig. 3** Percentage of user to recommended service used percentage to Jyaguchi system usage frequency par week

about 27% users who are using the service as their first service after login has used services from recommended service. The users who have already used one service and using 2nd service, the use percentage of recommended service is increased to 43%. The use of recommended service increase when user is using 3rd service to 57%. The result shows that recommendation algorithm is more effective to the users who use more services than one in the system.

In the second phase of evaluation, we performed questionnaire among the Jyaguchi users. Users are asked five questions about the implementation of recommendation algorithm. Then the evaluation is done using the feedback data we got from the users. Following figure shows the result of our survey.

Figures 2 and 3 are the results of questionnaire and from whose result we can conclude that for the users who are using the system for last more than 1 year and using the system more frequently per week, the recommendation system is much helpful and has used the recommended services more often than other users who are relatively new to the system and use system less frequently.

# 5    Future Works and Conclusion

## 5.1    Future Works

One major problem in this method comes from the fact that during the construction of a multidimensional sequence pattern, we need to formulate a tree structure in order to reduce searching and constructing time for the sets of sequences. However, we have just utilized a prefix span algorithm based approach during our search of the sequence. This approach is well suited while there are few dimensions; however, a hierarchical tree structure or graph algorithm needs to be applied in order to formulate and effectively construct our multidimensional sequence pattern. This would improve performance during the search and construction of multidimensional sets, but it would be costly to set up. Nevertheless, it is recommended that this trade-off needs to be investigated further to find the optimal type of searching algorithm. An additional performance gain could be achieved through utilizing parallelized processing in the database of multidimensional sequence sets. Furthermore, we have identified service category, user category as other dimensions to increase number of dimensions as future task. We will also be focused on calculating appropriate unit time and distributed behavior mining in our research.

## 5.2    Conclusion

We found that there are several researches going on in regards to multi-dimensional sequence pattern of data mining approach that has been successfully applied to data search in social networks and e-commerce sites. However the works related to service search as ours and recommendation to such services still lacks in the literature. This work takes into account the possibility to mine complex patterns of services, in which sequences may occur along different patterns and dimensions. Furthermore, these sequences are also characterized by time factor which were not sufficiently discussed in previous studies. In our work, we have proposed a complete framework and an inductive object oriented programming algorithm to tackle this problem. Our algorithm was implemented in Java supported with Jyaguchi service cloud as its infrastructure. The result achieved from our experiment is a dedicated Jyaguchi system in which multi-dimensional sequence patterns were analyzed. We utilized the user data stored in Jyaguchi cloud service at which we categorized the service users as below:

- Recommendation to Anonymous User Group.
- Registered User without Previous Usage of Services
- Registered Users with Previous History of Service Usage

By categorizing as above we were able to recommend the service item to the end users. However, these experiments were done in academic environment and we conclude that in order to execute our system into the business communities, sufficient real world experiment in the future are recommended.

## References

1. Agrawal, R., Srikant, R.: Mining Sequential Patterns. In: Proc. of the Eleventh International Conference on Data Engineering, pp. 3–14 (1995)
2. Shrestha, S.K., Kudo, Y., Gautam, B.P., Shrestha, D.: Multidimensional Service Weight Sequence Mining based on Cloud Service Utilization in Jyaguchi. In: Proc. of the International Multi Conference of Engineers and Computer Scientists, IMECS 2013, vol. I, pp. 301–306 (2013)
3. Gautam, B.P.: An Architectural Model for Time Based Resource Utilization and Optimized Resource Allocation in a Jini Based Service Cloud. Master Thesis, Shinshu University, Nagano, Japan (2009)
4. Gautam, B.P., Shrestha, D.: A Model for the Development of Universal Browser for Proper Utilization of Computer Resources Available in Service Cloud over Secured Environment. In: Proc. of the International Multi Conference of Engineers and Computer Scientists, IMECS 2010, vol. I (2010)
5. Gautam, B.P., Shrestha, S.K., Paudel, D.R.: Utilization of Jyaguchi Architecture for development of Jini Based Service Cloud. Wakkanai Hokusei Gakuen University Journal (11), 7–21 (2011)
6. Pinto, H., Han, J., Pei, J., Wang, K., Chen, Q., Dayal, U.: Multidimensional Sequential Pattern Mining. In: Proc. of the Tenth International Conference on Information and Knowledge Management (CIKM 2001), pp. 81–88 (2001)
7. Pei, J., Han, J., Mortazavi-Asl, B., Wang, J., Pinto, H., Chen, Q., Dayal, U., Hsu, M.C.: Mining Sequential Patterns by Pattern-Growth: The PrefixSpan Approach. IEEE Transactions on Knowledge and Data Engineering 16(10), 1424–1440 (2004)
8. http://en.wikipedia.org/wiki/NetfixPrize
9. Oku, K., Tung, T.S., Hattori, F.: Collaborative Filtering for Predicting Users Potential Preferences. In: König, A., Dengel, A., Hinkelmann, K., Kise, K., Howlett, R.J., Jain, L.C. (eds.) KES 2011, Part IV. LNCS (LNAI), vol. 6884, pp. 44–52. Springer, Heidelberg (2011)
10. Pazzani, M.J., Billsus, D.: Content-Based Recommendation Systems. In: Brusilovsky, P., Kobsa, A., Nejdl, W. (eds.) The Adaptive Web. LNCS, vol. 4321, pp. 325–341. Springer, Heidelberg (2007)
11. Burke, R.: Hybrid Web Recommender Systems. In: Brusilovsky, P., Kobsa, A., Nejdl, W. (eds.) The Adaptive Web. LNCS, vol. 4321, pp. 377–408. Springer, Heidelberg (2007)
12. Zhou, B., Hui, S.C., Fong, A.C.M.: Efficient Sequential Access Pattern Mining for Web Recommendations. International Journal of Knowledge-based and Intelligent Engineering Systems 10(2), 155–168 (2006)
13. Han, M., Wang, Z., Yuan, J.: Mining Constraint Based Sequential Patterns and Rules on Restaurant Recommendation System. Journal of Computational Information Systems 9(10), 3901–3908 (2013)

14. Zhou, B., Hui, S.C., Chang, K.: An intelligent recommender system using sequential web access patterns. In: Proc. of the 2004 IEEE Conference on Cybernetics and Intelligent Systems, vol. 1, pp. 393–398 (2004)
15. Khonsha, S., Sadreddini, M.H.: New hybrid web personalization framework. In: Proc. of IEEE 3rd International Conference on Communication Software and Networks, ICCSN, pp. 86–92 (2011)

# Heyting-Brouwer Rough Set Logic

Seiki Akama, Tetsuya Murai, and Yasuo Kudo

**Abstract.** A rough set logic based on Heyting-Brouwer algebras *HBRSL* is proposed as a basis for reasoning about rough information. It is an extension of Düntsch's logic with intuitionistic implication, and is seen as a variant of Heyting-Brouwer logic. A Kripke semantics and natural deduction for the logic are presented and the completeness theorem is proved.

**Keywords:** rough set logic, regular double Stone algebra, Heyting-Brouwer logic, Kripke semantics, natural deduction.

## 1 Introduction

In 1982, Pawlak proposed a *rough set* to represent coarse (rough) information; see Pawlak [5, 6]. The formalization of rough information has been the subject of investigation in rough set theory which is closely related to other areas. In formal logic, it is very important to develop a logic for rough sets.

Initial work in this direction has been done in Orlowska [3, 4]. The most significant is probably due to Düntsch [2] who proposed a propositional logic for rough sets with an algebraic semantics based on *regular double Stone algebras*.

It is a famous fact that the collection of all subsets of a set constitutes a Boolean algebra and that its logic is exactly the classical propositional logic. J. Pomykala

Seiki Akama
C-Republic, 1-20-1, Higashi-Yurigaoka, Asao-ku, Kawasaki 215-0012, Japan
e-mail: akama@jcom.home.ne.jp

Tetsuya Murai
Hokkaido University, Kita 13, Nishi 8, Kita-ku, Sapporo 080-8628, Japan
e-mail: murahiko@main.ist.hokudai.ac.jp

Yasuo Kudo
Muroran Institute of Technology, 27-1, Muroran, 050-8585, Japan
e-mail: kudo@csse.muroran-it.ac.jp

V.-N. Huynh et al. (eds.), *Knowledge and Systems Engineering, Volume 2*,
Advances in Intelligent Systems and Computing 245,
DOI: 10.1007/978-3-319-02821-7_13, © Springer International Publishing Switzerland 2014

and J.A. Pomykala [7] showed that the collection of rough sets of an approximation space forms a *regular double Stone algebra*. Based on their results, Düntsch succeed in developing a logic for rough sets.

There are, however, two problems with Düntsch's logic. The first problem is that he did not give a Kripke-type relational semantics. This means that we cannot intuitively understand his logic. The second problem is the lack of proof theory. Indeed in his presentation the Hilbert system is implicit, but it is not adequate for practical inferences.

The purpose of this paper is to develop another rough set logic which extends Düntsch's logic with intuitionistic implication. Our approach starts with *Heyting-Brouwer logic*, also known as bi-intuitionistic logic, which was proposed by Rauszer [8], and the idea leads interesting proof-theoretic and semantical characterization of the new rough set logic.

The structure of this paper is as follows. In section 2, we briefly review rough sets. In section 3, we present an exposition of Düntsch's logic for rough sets. In section 4, we introduce a new rough set logic called *Heyting-Brouwer rough set logic* with Kripke semantics and natural deduction. We also prove the completeness theorem based on a canonical model. The final section makes some conclusions with the discussion on future work.

## 2   Rough Set

The concept of *rough set* was proposed by Pawlak [5]; also see Pawlak [6]. A rough set can be seen as an approximation of a set denoted by a pair of sets, called the lower and upper approximation of the set to deal with reasoning from imprecise data.

We here sketch the background of rough sets. Let $U$ be a non-empty finite set, called the *universe* of objects in question. Then, any subset $X \subseteq U$ is called a *concept* in $U$ and any family of concepts in $U$ is called *knowledge* about $U$. If $R$ be the equivalence relation on $U$, then $U/R$ denotes the family of all equivalence classes of $R$ (or *classification* about $U$), called *categories* or *concepts* of $R$. We write $[x]_R$ for a category in $R$ containing an element $x \in U$. If $P \subseteq R$ and $P \neq \emptyset$, then $\cap P$ is also an equivalence relation called *indiscernibility relation* on $P$, designated as $IND(P)$.

An *approximation space* is a pair $\langle U, R \rangle$. Then, for each subset $X \subseteq U$ and equivalence relation $R$, we associate two subsets, i.e.,

$$\underline{R}X = \{x \in U : [x]_R \subseteq X\}, \quad \overline{R}X = \{x \in U : [x]_R \cap X \neq \emptyset\}.$$

Here, $\underline{R}X$ is called the *lower approximation* of $X$, and $\overline{R}X$ is called the *upper approximation* of $X$, respectively. A *rough set* is designated as the pair $\langle \underline{R}X, \overline{R}X \rangle$. Intuitively, $\underline{R}X$ is the set of all elements of $U$ which can be certainly classified as elements of $X$ in the knowledge $R$, and $\overline{R}X$ is the set of elements which can be possibly classified as elements of $X$ in the knowledge $R$. Then, we can define three types of sets, i.e.,

$POS_R(X) = \underline{R}X$ ($R$-positive region of $X$),

$NEG_R(X) = U - \overline{R}X$ ($R$-negative region of $X$),

$BN_R(X) = \overline{R}X - \underline{R}X$ ($R$-boundary region of $X$).

These sets enable us to classify our knowledge. Pawlak [6] contains comprehensive account of rough sets.

## 3   Rough Set Logic

Düntsch [2] developed a propositional logic for rough sets inspired by the topological construction of rough sets using Boolean algebras. His work is based on the fact that the collection of all subsets of a set forms a Boolean algebra under the set-theoretic operation, and that the collection of rough sets of an approximation space is a regular double Stone algebra. Thus, we can assume that regular double Stone algebras give a semantics for a logic for rough sets.

Here, we need to survey Düntsch's work. To understand his logic, we need some concepts. A *double Stone algebra DSA* is denoted by $\langle L, +, \cdot, *.+, 0, 1 \rangle$ with the type $\langle 2, 2, 1, 1, 0, 0 \rangle$ satisfying the following conditions:

(1)  $\langle L, +, \cdot, 0, 1 \rangle$ is a bounded distributed lattice.

(2)  $x^*$ is the pseudocomplement of $x$, i.e.,

$$y \leq x^* \Leftrightarrow y \cdot x = 0.$$

(3)  $x^+$ is the dual pseudocomplement of $x$, i.e.,

$$y \geq x^+ \Leftrightarrow y + x = 1.$$

(4)  $x^* + x^{**} = 1, x^+ \cdot x^{++} = 0$

*DSA* is called *regular* if it satisfies the additional condition: $x \cdot x^+ \leq x + x^*$. Let $B$ be a Boolean algebra, $F$ be a filter on $B$, and

$$\langle B, F \rangle = \{\langle a, b \rangle \mid a, b \in B, a \leq b, a + (-b) \in F\}$$

We define the following operations on $\langle B, F \rangle$ as follows:

$\langle a, b \rangle + \langle c, d \rangle = \langle a + c, b + d \rangle$,

$\langle a, b \rangle \cdot \langle c, d \rangle = \langle a \cdot c, b \cdot d \rangle$,

$\langle a, b \rangle^* = \langle -b, -b \rangle$,

$\langle a, b \rangle^+ = \langle -a, -a \rangle$.

If $\langle U, R \rangle$ is an approximation space, the classes o $R$ can be viewed as a complete subalgebra of the Boolean algebra $B(U)$. Conversely, any atomic complete subalgebra $B$ of $B(U)$ yields an equivalence relation $R$ on $U$ by the relation:

$xRy \Leftrightarrow x$ and $y$ are contained in the same atom of $B$,

and this correspondence is bijective. If $\{a\} \in B$, then for every $X \subseteq U$ we have:

If $a \in \underline{R}X$, then $a \in X$,

and the rough sets of the corresponding approximation space are the elements of the
regular double Stone algebra $\langle B, F \rangle$, where $F$ is the filter of $B$ which is generated by
the union of the singleton elements of $B$.

Based on the construction of regular double Stone algebras, Düntsch proposed a
propositional rough set logic *RSL*. The language $\mathscr{L}$ of *RSL* has two binary connectives $\wedge$ (conjunction), $\vee$ (disjunction), two unary connectives $^*, ^+$ for two types of
negation, and the logical constant $\top$ for truth.

Let $P$ be a non-empty set of propositional variables. Then, the set **Fml** of formulas with the logical operators constitutes an absolutely free algebra with a type
$\langle 2, 2, 1, 1, 0 \rangle$.

Let $W$ be a set and $B(W)$ be a Boolean algebra based on $W$. Then, a *model M* of
$L$ is seen as a pair $(W, v)$, where $v : P \to B(W) \times B(W)$ is the *valuation function* for
all $p \in P$ satisfying:

if $v(p) = \langle A, B \rangle$, then $A \subseteq B$.

Here, $v(p) = \langle A, B \rangle$ states that $p$ holds at all states of $A$ and does not hold at any
state outside $B$.

Düntsch relates the valuation to Lukasiewicz's three-valued logic by the follow-
ing construction. For each $p \in P$, let $v_p : W \to \mathbf{3} = \{0, \frac{1}{2}, 1\}$. $v : P \to B(W) \times B(W)$
is defined as follows:

$v(p) = \langle \{w \in W : v_p(w) = 1\}, \{w \in W : v_p(w) \neq 0\} \rangle$.

In addition, Düntsch connected the valuation and rough sets as follows:

$v_p(w) = 1$ if $w \in A$,
$v_p(w) = \frac{1}{2}$ if $w \in B \setminus A$,
$v_p(w) = 0$ otherwise.

Given a model $\mathtt{M} = (W, v)$, the *meaning function* $\mathtt{mng} : \mathbf{Fml} \to \mathrm{B}(W) \times \mathrm{B}(W)$ is defined to give a valuation of arbitrary formulas in the following way:

$\mathtt{mng}(\top) = \langle W, W \rangle$,
$\mathtt{mng}(p) = v(p)$ for $p \in P$.

If $\mathtt{mng}(\phi) = \langle A, B \rangle$ and $\mathtt{mng}(\psi) = \langle C, D \rangle$, then

$\mathtt{mng}(\phi \wedge \psi) = \langle A \cap C, B \cap D \rangle$,
$\mathtt{mng}(\phi \vee \psi) = \langle A \cup C, B \cup D \rangle$,
$\mathtt{mng}(\phi^*) = \langle -B, -B \rangle$,
$\mathtt{mng}(\phi^+) = \langle -A, -A \rangle$.

Here, $-A$ denotes the complement of $A$ in $B(W)$. We can understand that the meaning function assigns the meaning to formulas.

A formula $A$ *holds* in a model $M = \langle W, v \rangle$, written $M \models A$, if $\mathtt{mng}(A) = \langle W, W \rangle$.
A set $\Gamma$ of sentences *entails* a formula $A$, written $\Gamma \vdash A$, if every model of $\Gamma$ is
a model of $A$. Düntsch proved that *RSL* is sound and complete with respect to the
above semantics, where he seemed to assume a Hilbert system as a proof theory.

## 4   Heyting-Brouwer Rough Set Logic

As noted in section 1, Düntsch did not provide a Kripke semantics for his logic. In addition, his Hilbert system seems to be abstractly presented. To overcome these difficulties, we introduce a new logic, i.e., Heyting-Brouwer rough logic denoted *HBRSL* whose language is the one of *RSL* with intuitionistic implication $\to$. The addition of $\to$ is essential in that we can construct a rough set logic as a variant of Heyting-Brouwer logic.

Heyting-Brouwer logic is the system founded on Heyting algebras and Brouwerian algebras. *Heyting algebra* $\langle L, \vee, \wedge, \to, 0, 1 \rangle$ is a lattice with the bottom 0, the top 1, and the binary operation called *implication* $\to$ satisfying $a \wedge b \leq c \Leftrightarrow a \leq b \to c$. $\neg a = a \to 0$ is called the *pseudocomplement* of $a$. Heyting algebra is an algebraic model for intuitionistic logic in which Heyting (intuitionistic) negation is denoted by $\neg$.

The dual of Heyting algebra is called *Brouwerian algebra* $\langle L, \vee, \wedge, -<, 0, 1 \rangle$ is a lattice with 0 and 1, and the binary operation called *dual implication* $-<$ satisfying $x -< y \leq z \Leftrightarrow x \leq y \vee z$. $-a = 1 -< a$ is called the *dual pseudocomplement* of $a$. Brouwerian algebra is an algebraic model for dual intuitionistic logic in which Brouwerian (dual intuitionistic) negation is denoted by $-$.

Heyting-Brouwer logic *HBL* is an extension of positive intuitionistic logic with implication and dual implication (and intuitionistic negation and dual intuitionistic negation, if needed). Rauszer [7] extensively studied proof and model theory for Heyting-Brouwer logic.

We are now ready to turn to an exposition of a new rough set logic called *Heyting-Brouwer rough set logic* denoted by *HBRSL*. The language of *HBRSL* is that of *RSL* with intuitionistic implication $\to$, truth $\top$ and falsity $\bot$. We write atomic formula by $p, q, r, \ldots$ and arbitrary formula by $A, B, C, \ldots$, respectively.

Note that $*$ and $+$ denote intuitionistic-like negation and dual intuitionistic-like negation, respectively. We here say intuitionistic-like and dual intuitionistic-like negation, because they do not correspond to intuitionistic and dual intuitionistic negation in the sense of Rauszer. The addition of $\to$ is essential in that it enables us to work out an elegant theoretical foundation for *HBRSL*. Of course, one could also add $-<$ to *HBRSL*, but its addition may not be important for practical purposes.

We start with a *Kripke semantics* for *HBRSL*, which is a modification of that for *HBL* in Rauszer [7]. A *Kripke model* for *HBRSL* is a tuple $\mathscr{M} = \langle W, R, V \rangle$. Here, $W$ is a non-empty set of *worlds*. $R$ is a binary relation on $W$, which is reflexive and transitive, and directed, i.e., $\exists v \forall w (wRv)$, dual directed, i.e., $\exists v \forall w (vRw)$, and bridged, i.e., $\forall w \forall v (wRv \Rightarrow w = v$ or $\forall u (wRu \Rightarrow w = u))$ for $w, v, u \in W$. $V$ is a valuation function from $W \times At$ to $\{0, 1\}$, where $At$ is a set of atomic formulas, satisfying that $V(w, \top) = 1$ and $V(w, \bot) = 0$ for any $w \in W$.

Then, we define the truth relation $\models$ such that $V(w, p) = 1 \Leftrightarrow w \models p$ and $V(w, p) = 0 \Leftrightarrow w \not\models p$. Here, $w \models p$ reads "$p$ is true at $w$" and $w \not\models p$ reads "$p$ is not true at $w$", respectively. The truth relation $\models$ is then defined for any formula $A, B$ as follows.

$w \models A \wedge B \ \Leftrightarrow w \models A$ and $w \models B$

$w \models A \vee B \ \Leftrightarrow w \models A$ or $w \models B$

$w \models A \rightarrow B \Leftrightarrow \forall v(wRv$ and $v \models A \ \Rightarrow \ v \models B)$

$w \models *A \quad \Leftrightarrow \forall v(wRv \Rightarrow \ v \not\models A)$

$w \models +A \quad \Leftrightarrow \exists v(vRw$ and $v \not\models A)$

Although *HBRSL* has no dual implication $-\!\!<$, it can be added to *HBRSL* and interpreted as follows:

$w \models A -\!\!< B \ \Leftrightarrow \ \exists v(vRw$ and $v \models A$ and $v \not\models B)$

In the Kripke model, both persistency (P) and dual-persistency (DP) with respect to $\models$ and $\not\models$ hold:

(P) $\forall w \forall v(w \models p$ and $wRv \Rightarrow v \models p)$

(DP) $\forall w \forall v(w \not\models p$ and $vRw \Rightarrow v \not\models p)$

for any atomic $p$. We write $w \models \Gamma$ to mean that for all formulas in $\Gamma$ are true at $w$. We say that a formula $A$ is *valid*, written $\models A$, if it is true for all worlds for all models.

**Lemma 1.** *For any formula A, both* (P) *and* (DP) *hold:*

$\forall w \forall v(w \models A$ and $wRv \Rightarrow v \models A)$,

$\forall w \forall v(w \not\models A$ and $vRw \Rightarrow v \not\models A)$.

*Proof.* By induction $A$.

Next, we describe a proof theory of *HBRSL* denoted *NHBRSL* using *natural deduction* in a sequential form. *NHBRSL* is formalized by *axiom* and *rule*. Let $\Gamma, \Delta$ be sets of formulas, $A, B, C, D$ be formulas. An expression of the form $\Gamma \vdash A$ is called a *sequent*. If $\Gamma = \{A_1, ..., A_n\}$, then $\Gamma \vdash B$ iff $(A_1 \wedge ... \wedge A_n) \rightarrow B$. When $\Gamma$ is empty, $\Gamma \vdash A$ is written as $\vdash A$. Then, a rule is of the form:

$$\frac{\Gamma_1 \vdash A_1 .... \Gamma_i \vdash A_i}{\Delta \vdash B}$$

which says that if $\Gamma_1 \vdash A_1, ... \Gamma_i \vdash A_n$ (premises) holds then $\Delta \vdash B$ (consequent) holds. An axiom can be regarded as the rule without premises.

There are two types of rules, i.e., *introduction rule* and *elimination rule*. An introduction rule introduces a logical symbol in the consequent, and an elimination rule eliminates a logical symbol in the consequent. We denote, for example, the introduction rule for $\wedge$ by $(\wedge I)$ and the elimination rule for $\wedge$ by $(\wedge E)$, respectively. Additionally, we use some special rules. A *proof* is constructed as a tree in which all leaves are axioms, and in this case the formula in the root is a formula to be proved. We write $\vdash_{NHBRSL} A$ when $A$ is provable in *NHBRSL*.

Below are axioms and rules for *NHBRSL*. $\Gamma$ is a (possibly empty) set of formulas and $A, B, C, D$ are formulas, respectively.

## Natural Deduction System $NHBRSL$

**Axioms**

$(A1)$ $\Gamma, A \vdash A$ $\qquad$ $(A2)$ $\Gamma, \perp \vdash A$

$(A3)$ $\Gamma \vdash \top$ $\qquad$ $(A4)$ $\vdash *A \vee **A$

$(A5)$ $+A \wedge ++A \vdash$ $\qquad$ $(A6)$ $\Gamma, A \wedge +A \vdash A \vee *A$

$(A7)$ $*+A \vdash A$

**Rules**

$$\frac{\Gamma \vdash A \quad \Gamma \vdash B}{\Gamma \vdash A \wedge B}(\wedge I) \qquad \frac{\Gamma \vdash A \wedge B \quad \Gamma \vdash A \wedge B}{\Gamma \vdash A \qquad \Gamma \vdash B}(\wedge E)$$

$$\frac{\Gamma \vdash A \quad \Gamma \vdash B}{\Gamma \vdash A \vee B \quad \Gamma \vdash A \vee B}(\vee I) \qquad \frac{\Gamma \vdash A \vee B \quad \Gamma, A \vdash C \quad \Gamma, B \vdash C}{\Gamma \vdash C}(\vee E)$$

$$\frac{\Gamma, A \vdash B}{\Gamma \vdash A \rightarrow B}(\rightarrow I) \qquad \frac{\Gamma \vdash A \quad \Gamma \vdash A \rightarrow B}{\Gamma \vdash B}(\rightarrow E)$$

$$\frac{\Gamma, A \vdash}{\Gamma \vdash *A}(*I) \qquad \frac{\Gamma \vdash A \quad \Gamma \vdash *A}{\Gamma \vdash \perp}(*E)$$

$$\frac{D \vdash \top \quad A \vdash C}{D \vdash +A}(+I) \qquad \frac{\Gamma \vdash +A \quad \Gamma, \top \vdash A}{\Gamma \vdash B}(+E)$$

Here, we can dispense with rules for $*$, since $*A$ is defined as $A \rightarrow \perp$. Observe that the condition in $(+I)$, namely $D$ in the premise $D \vdash \top$ and in the consequent $D \vdash +A$ and $A$ in the premise $A \vdash C$ must be a single formula, not a set of formulas, is crucial to our formalization. One could also describe the rules for dual implication $-<$ as follows:

$$\frac{D \vdash A \quad B \vdash C}{D \vdash A -< B}(-< I) \qquad \frac{\Gamma \vdash A -< B \quad \Gamma, A \vdash B}{\Gamma \vdash C}(-< E)$$

The natural deduction system with axioms $(A1) - (A3)$ and rules for $\wedge, \vee, *$ is for intuitionistic propositional logic *Int*, in which $*$ can be identified with intuitionistic negation $\neg$.

If we delete the axiom $(A6), (A7)$ and rules for $(\rightarrow)$, the natural deduction system $NHBRSL_0$ for the logic based on double Stone algebras is available. A natural deduction system $NHBRSL_1$ is obtainable from $NHBRSL_0$ by adding $(A6)$.

**Lemma 2.** *The following formulas are provable in NHBRSL.*

*(i)* $\vdash_{NHBRSL} A \rightarrow \top$

*(ii)* $\vdash_{NHBRSL} \perp \rightarrow A$

*(iii)* $\vdash_{NHBRSL} *(+A \wedge ++A)$

*(iv)* $\vdash_{NHBRSL} (A \wedge +A) \rightarrow (A \vee *A)$

*(v)* $\vdash_{NHBRSL} A \leftrightarrow *+A$

*Here, $A \leftrightarrow B$ abbreviates $(A \rightarrow B) \wedge (B \rightarrow A)$.*

Next, we present the soundness result of *HBRSL*. As noted above, $*$ is intuitionistic-like negation and $+$ is dual intuitionistic-like negation. The fact is technically important here. To validate $(A4)$, we need the condition of directedness. Logics stronger than intuitionistic logic but weaker than classical logic are called the *intermediate logics* or *superintuitionistic logics*.

Intuitionistic logic with the axiom called *the weak law of excluded middle*: $\neg A \vee \neg\neg A$ is the intermediate logic often denoted by $LQ$; e.g., see Akama [1]. Similarly, the condition of dual directedness is needed to validate $(A5)$. The intermediate extensions of dual intuitionistic logic did not seem to be fully studied in the literature. The condition of bridge is added for the validity of $(A6)$.

**Theorem 1 (soundness).** $\vdash_{NHBRSL} A \Rightarrow \models_{NHBRSL} A$

*Proof.* It can be proved by checking that all axioms are valid and all rules preserve validity. Most cases are immediate from the soundness proof for intuitionistic logic. Checking of rules is trivial and omitted. For axioms, the validity of $(A1), (A2), (A3)$, and $(A7)$ are obvious. Thus, we here only consider $(A4), (A5)$ and $(A6)$.

$(A4)$: Suppose $(A4)$ is not valid. Then, there is a Kripke model satisfying that $w \not\models *A \vee **A$ for some $w$. From the truth definition of $*$, we have that $w \not\models *A$ and $w \not\models **A$ which is equivalent to the following:

$$\exists v(wRv \text{ and } v \models A) \text{ and } \exists v(wRv \text{ and } \forall u(vRu \Rightarrow u \not\models A)).$$

From the first conjunct, $v \models A$ holds. Since $R$ is directed, $\exists u \forall v(vRu)$. By persistency (P), we have $u \models A$. The second conjunct says that $u \not\models A$, which contradicts $u \models A$ from the first conjunct. Consequently, $(A4)$ is shown to be sound.

$(A5)$: Suppose $(A5)$ is not valid. Then, we have a Kripke model satisfying that $w \models +A \wedge ++A$ for some $w$. From the truth definition of $+$, we have:

$$\exists v(wRv \text{ and } v \not\models A) \text{ and } \exists v(wRw \text{ and } \forall u(uRv \Rightarrow u \models A)).$$

From the first conjunct, $v \not\models A$ holds. By the dual directedness of $R$, i.e., $\exists u \forall v(uRv)$, together with the dual persistency (DP), $u \not\models A$ is derived. But it contradicts with the second conjunct $u \models A$. Consequently, $(A5)$ is a sound rule.

$(A6)$: It suffices to see the validity of $A \wedge +A \vdash A \vee *A$. Suppose $A \wedge +A \vdash A \vee *A$ is not valid. Then, there is a Kripke model satisfying that $w \models A \wedge +A$ but $w \not\models A \vee *A$ for some $w$, which is equivalent to the following:

$$w \models A \text{ and } \exists v(vRw \text{ and } v \not\models A) \text{ and } w \not\models A \text{ and } \exists u(wRu \text{ and } u \models A).$$

Since $R$ is bridged, $\forall w \forall v(wRv \Rightarrow w = v \text{ or } \forall u(wRu \Rightarrow w = u))$ holds. We must consider two cases. First, if the condition $(wRv \Rightarrow w = v)$ in the bridge condition holds, then the first and second conjuncts give contradiction. Second, if the condition $(wRu \Rightarrow w = u)$ in the bridge condition holds, then the third and fourth conjuncts give contradiction. From these considerations, we obtain the fact that $(A6)$ is sound.

Next, we prove the completeness of *HBRSL* by means of *canonical model*. Our method for proving completeness is a suitable modification of the one used in Kripke semantics for intuitionistic logic. We need some preliminary definitions. If $\Gamma = \{A_1, ..., A_n\}, \Delta = \{B_1, ..., B_m\}$, then we set $\bigwedge \Gamma = A_1 \wedge ... \wedge A_n, \bigvee \Delta = B_1 \vee ... \vee B_m$. The pair $(\Gamma, \Delta)$ is *consistent* iff there are no finite subsets $\Gamma_0 \subset \Gamma$ and $\Delta_0 \subset \Delta$ such that $\vdash_{NHBRSL} \bigwedge \Gamma_0 \to \bigvee \Delta_0$, where $\bigwedge \emptyset = \top, \bigvee \emptyset = \bot$. $\Gamma$ is *consistent* iff $(\Gamma, \emptyset)$ is consistent. A pair $(\Gamma', \Delta')$ is an *extension* of a pair $(\Gamma, \Delta)$ iff $\Gamma \subseteq \Gamma'$ and $\Delta \subseteq \Delta'$.

A set $\Gamma$ of formulas is *saturated* if the following conditions hold: (i) $\Gamma$ is consistent, (ii) $\Gamma \vdash_{NHBRSL} A \Rightarrow A \in \Gamma$, (iii) $\Gamma \vdash_{NHBRSL} A \vee B \Rightarrow \Gamma \vdash_{NHBRSL} A$ or $\Gamma \vdash_{NHBRSL}$. If $\Gamma \nvdash_{NHBRSL} A$, then $\Gamma$ can be extended to saturated $\Gamma' \supset \Gamma$ such that $\Gamma' \nvdash_{NHBRSL} A$ by standard construction.

For our setting, we generalize the notion of saturated set for the pair of sets of formulas defined above. Let $T, S$ be sets of formulas of the language of $HBRSL$. The pair $(T, S)$ is *saturated* iff the following hold:

$T \cap S = \emptyset$
$A \wedge B \in T \Rightarrow A \in T$ and $B \in T$
$A \vee B \in T \Rightarrow A \in T$ or $B \in T$
$A \to B \in T \Rightarrow A \in S$ or $B \in T$
$*A \in T \Rightarrow A \in S$ or $\bot \in T$
$+A \in T \Rightarrow \top \in T$ and $A \in S$
$A \wedge B \in S \Rightarrow A \in S$ or $B \in S$
$A \vee B \in S \Rightarrow A \in S$ and $B \in S$
$A \to B \in S \Rightarrow A \in T$ and $B \in S$
$*A \in S \Rightarrow A \in T$ and $\bot \in S$
$+A \in S \Rightarrow \top \in S$ or $A \in T$

We are now ready to define a canonical model $\mathcal{M}^c = \langle W^c, R^c, V^c \rangle$. Here, $W^c$ is a set of all sets $\Gamma = T \cup S$ where $(T, S)$ is a saturated pair. $R^c$ is $\subseteq$ satisfying $\exists \Delta \forall \Gamma (\Gamma \subseteq \Delta), \exists \Sigma \forall \Gamma (\Sigma \subseteq \Gamma)$, and $\forall \Gamma \forall \Delta (\Gamma \subseteq \Delta \Rightarrow \Gamma = \Delta$ or $\forall \Sigma (\Gamma \subseteq \Sigma \Rightarrow \Gamma = \Sigma))$. $V^c(\Gamma, p) \Leftrightarrow p \in \Gamma$ for atomic $p$. We can then define $\models^c$ for any formula as described before.

Lemma 3 is a key lemma to prove completeness.

**Lemma 3.** *For any $\Gamma \in W^c$ and any formula: we have:*

$$\Gamma \models^c A \Leftrightarrow A \in \Gamma$$

*Proof.* The cases in which $A$ is of the form $B \wedge C, B \vee C, B \to C$ or $*B$ are proved as in intuitionistic logic. It thus suffices to only consider the case in which $A = +B$.

$$\Gamma \models^c +B \Leftrightarrow \exists \Delta (\Delta \subseteq \Gamma \text{ and } \Delta \nvDash^c B)$$
$$\Leftrightarrow \exists \Delta (\Delta \subseteq \Gamma \text{ and } B \notin \Delta)$$
$$\Leftrightarrow +B \in \Gamma$$

Here, we must prove that $\exists \Delta (\Delta \subseteq \Gamma \text{ and } B \notin \Delta) \Leftrightarrow +B \in \Gamma$. For ($\Leftarrow$), suppose $+B \in \Gamma$. Then, by the definition of saturated pair, we have that $\top \in T$ and $B \in S$. In $\Gamma = T \cup S$ set $T = \Delta$, then $\Delta \subseteq \Gamma$ follows. Since $T \cap S = \emptyset$, from $B \in S$ we have $B \notin \Delta = T$. Then, $\Delta \subseteq \Gamma$ and $B \notin \Delta$ hold.

For ($\Rightarrow$), suppose $\exists \Delta (\Delta \subseteq \Gamma \text{ and } B \notin \Delta)$. Set $\Gamma = \{D\}$, then $D \vdash_{NHBRSL} \top$ by the axiom (A3). As $B \notin \{D\}$, $B \vdash_{NHBSRL} \bot$ by saturatedness. By applying $(I \to)$ to the axiom (A2) $B, \bot \vdash_{NHBRSL} C$, we have that $B \vdash_{NHBRSL} \bot \to C$. By $(\to E)$, $B \vdash C$ follows. Using the rule $(+I)$ enables us to obtain $D \vdash +B$. Thus, $+B \in \Gamma$ holds.

Then, we can conclude the completeness of $HBRSL$:

**Theorem 2 (completeness).** $\Gamma \vdash_{NHBRSL} A \Leftrightarrow \Gamma \models A$.

*Proof.* For soundness ($\Rightarrow$), $\Gamma \vdash_{NHBRSL} A$ iff $\vdash_{NHBRSL} \Gamma \to A$. Then, applying theorem 1 to it leads the soundness.

For completeness ($\Leftarrow$), we use contrapositive argument. Assume that $\Gamma \not\vdash_{NHBRSL} A$. Then, there is a saturated pair $\Gamma' = (T, S)$. Thus, the completeness follows by Lemma 3.

We can similarly establish completeness results of $HBRSL_0$ and $HBRSL_1$ by considering the corresponding conditions of Kripke models. Our result also implies the Kripke completeness of Düntsch's logic. If we add dual intuitionistic implication to $HBRSL$, we can show the completeness proof for Heyting-Brouwer logic and some of its extensions.

## 5 Conclusion

We proposed a rough set logic called the Heyting-Brouwer rough set logic $HBRSL$, which extends Düntsch's rough set logic with intuitionistic implication. A model theory was supplied by a Kripke model to give an intuitive semantics, and proof theory based on natural deduction is presented. We established a completeness result by means of a canonical model. Thus, an alternative foundation for rough set logics was outlined in this paper. We believe that our logic can serve as a logical framework for reasoning about rough information.

There are some interesting research topics related to our logic. Although we use a natural deduction system as a proof theory, other proof methods like sequent calculus and tableau calculus can be explored. In particular, cut-free sequent formulation seems important to advance a practical proof method.

It would be also possible to investigate other types of logics based on double Stone algebras with Kripke or algebraic semantics. For instance, introducing different types of implications is one of the important problems.

Another line of work in this research will be the modal and three-valued characterizations of rough set logics. A modal approach would be promising because many connections of intuitionistic (and intermediate) logic and modal logic are known. A three-valued approach is also interesting. For example, a model theory based on three-valued Lukasiewicz algebra appears to provide some extensions of rough set logics.

**Acknowledgments.** Thanks are due to an anonymous referee for useful comments and suggestions.

## References

1. Akama, S.: The Gentzen-Kripke construction of the intermediate logic $LQ$. Notre Dame Journal of Formal Logic 33, 148–153 (1992)
2. Düntsch, I.: A logic for rough sets. Theoretical Computer Science 179, 427–436 (1997)

3. Orlowska, E.: Modal logics in the theory of information systems. Zeitschrift für Mathematische Logik und Grundlagen der Mathematik 30, 213–222 (1988)
4. Orlowska, E.: Logic for reasoning about knowledge. Zeitschrift für Mathematische Logik und Grundlagen der Mathematik 35, 559–572 (1989)
5. Pawlak, Z.: Rough sets. International Journal of Computer and Information Sciences 11, 341–356 (1982)
6. Pawlak, Z.: Rough Sets: Theoretical Aspects of Reasoning about Data. Kluwer, Dordrecht (1990)
7. Pomykala, J., Pomykala, J.A.: The stone algebra of rough sets. Bulletin of Polish Academy of Science, Mathematics 36, 495–508 (1988)
8. Rauszer, C.: Semi-Boolean algebras and their applications to intuitionistic logic with dual operations. Fundamenta Mathematicae 83, 219–249 (1974)

# Bicluster-Network Method and Its Application to Movie Recommendation

Tatsuya Saito and Yoshifumi Okada

**Abstract.** Biclustering can find sub-matrices (biclusters), namely a subset of rows that exhibit similar behavior across a subset of columns, by simultaneous clustering of both rows and columns in a data matrix. Since biclustering usually outputs a huge number of biclusters, it is not easy for user to find only necessary biclsters. In this study, we propose a bicluster-network method that groups biclusters overlapping with each other by representing them as a connected graph. Moreover, a new item recommender system is presented as an application of bicluster-network method. As experiments, we apply this method to a movie rating dataset and evaluate the performance by comparing the recommendation accuracy with that of an existing bicluster-based recommendation method.

## 1 Introduction

Recent online shopping sites or movie-sharing sites have enabled us to obtain a wide variety of products and information. However, it is not easy to search only for necessary information or products (hereinafter "items") that match interests or preference of users. Recommender system is one of the most promising technologies to solve such issue. Currently, the most mainstream form of recommendation technology is a method known as collaborative filtering [1]. This method predicts items suitable for user's tastes on the basis of a large number of profiles of other users, such as purchase history and rating values.

Tatsuya Saito
Department of Information and Electronic Engineering, Muroran Institue of Technology, 27-1, Mizumoto-cho, Muroran 050-8585, Japan
e-mail: saito@cbrl.csse.muroran-it.ac.jp

Yoshifumi Okada
College of Information and Systems, Muroran Institute of Technology, 27-1, Mizumoto-cho, Muroran 050-8585, Japan
e-mail: okada@csse.muroran-it.ac.jp

V.-N. Huynh et al. (eds.), *Knowledge and Systems Engineering, Volume 2*,     147
Advances in Intelligent Systems and Computing 245,
DOI: 10.1007/978-3-319-02821-7_14, © Springer International Publishing Switzerland 2014

Recently, Symeonidis et al. [2] proposed a method of collaborative filtering known as Nearest-Biclusters Collaborative Filtering (NBCF). This method first identifies groups of items that have been rated highly by multiple users. Here a tuple of the users, items and the rating values is called a bicluster. Next, this method identifies biclusters that are similar to rating values of a user requesting recommendation (hereinafter "active user"). Finally, items liked by users belonging to those biclusters are recommended to the active user. It was reported that NBCF showed higher recommendation accuracy than user-based method and item-based method [3], which are the most widely-used recommendation methods. However, as described above, this method focuses only on biclusters that are similar to rating values of an active user. Thus, it can cause a problem that only trivial or known items for the active user are recommended because of the limited search space.

In this paper, we propose a recommendation method based on bicluster network that utilizes not only biclusters being similar to active user's rating but also biclusters overlapping with them (hereinafter "overlapping biclusters"). By this approach, we expect that it enables more wider and adequate search for items matching active user's tastes. The aim of this study is to clarify the effectiveness of using bicluster network. For this purpose, the bicluster being most similar to active user's rating (hereinafter "nearest bicluster") and the overlapping biclusters are employed for evaluating the recommendation accuracy. The recommendation accuracy of the proposed method can be affected by the overlapping degree between the nearest bicluster and the overlapping biclusters. In the evaluation experiments, we examine the recommendation accuracy with the variation of the overlapping degree using a movie rating dataset, and compare it to that of a method of using only the nearest bicluster.

## 2 Method

### 2.1 Bicluster Extraction

Figure 1 outlines the bicluster extraction. In this study, we use BiModule that can exhaustively extract maximal biclusters on the basis of closed itemset mining [4]. BiModule achieves efficient and fast bicluster extraction using a high-speed closed itemset mining algorithm called LCM [5]. Closed itemset mining is performed on a transaction database that is composed of a set of transactions (i.e. "user" in this study) including one or more items. A closed itemset is defined as a maximal set of items which is shared by a subset of transactions under a specified minimum support value. A pair of a closed itemset and its supporting transactions corresponds to a maximal bicluster.

### 2.2 Generation of Bicluster Network

Suppose that an active user input one or more favourite items as a query. Then, on the basis of Symeonidis's manner [2], we calculate the degree of similarity between

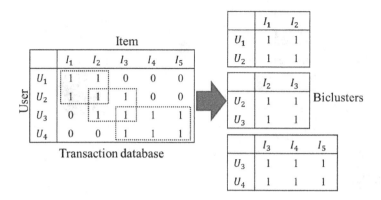

**Fig. 1** Bicluster generation

the query and each bicluster enumerated in the previous section, and extract the bicluster with highest degree of similarity as the nearest bicluster. Next, we identify overlapping biclusters that share most of elements with the nearest bicluster. The overlapping degree between the nearest bicluster and the other biclusters is calculated according to the following formula:

$$overlapping\,(nb,b_i) = \frac{2*usersim\,(nb,b_i)*itemsim\,(nb,b_i)}{(usersim\,(nb,b_i)+itemsim\,(nb,b_i))},\qquad(1)$$

where $nb$ is the nearest bicluster and $b_i$ is the other bicluster. The functions $usersim$ and $itemsim$ are defined as the following expressions,

$$usersim\,(nb,b_i) = \frac{\left|U_{nb} \cap U_{b_i}\right|}{\sqrt{|U_{nb}|}\sqrt{|U_{b_i}|}},\qquad(2)$$

$$itemsim\,(nb,b_i) = \frac{\left|I_{nb} \cap I_{b_i}\right|}{\sqrt{|I_{nb}|}\sqrt{|I_{b_i}|}},\qquad(3)$$

where $U_x$ and $I_x$ are the set of users and a set of items in a bicluster $x$, respectively. Namely, $usersim$ and $itemsim$ exhibit the similarities of the users or the items between $nb$ and $b_i$, respectively. Overlapping degree is an index that takes the balance between the $usersim$ and $itemsim$. Subsequently, Biclusters with overlapping degrees more than a specified threshold are extracted as the overlapping biclusters. The nearest bicluster and the overlapping biclusters can be represented by a connected graph as shown in Fig.2. We call such connected graph "bicluster network". Fig.2(a) exhibits the nearest bicluster and the overlapping biclusters. Fig.2(b) shows the bicluster network that abstracts the relation between the nearest biclusters and the overlapping biclusters. In Fig.2(b), a black node and white node stand for the nearest bicluster and an overlapping bicluster, respectively. An edge indicates an overlap

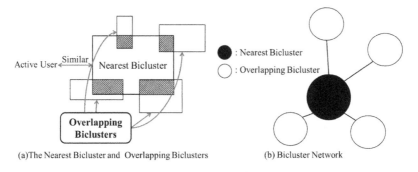

(a)The Nearest Bicluster and Overlapping Biclusters                    (b) Bicluster Network

**Fig. 2** Bicluster network

between two biclusters, and the length of each edge corresponds to the overlapping degree. Finally, the items included in the bicluster network are recommended for the active user.

## 3 Evaluation Experiment

### 3.1 Dataset

In this study, we use a movie rating dataset that was published by GroupLens [6] . The dataset contains rating data from 943 users for 1,682 movies. Movies are ranked from one to five in ascending order of rating. Each user rates a minimum of 10 movies and there are a total of 100,000 ratings.

### 3.2 Pre-processing

The rating data are binary-processed and the transaction database is constructed in the following way. Ratings of four or five (highly-rated movies) are converted to one and all other ratings (movies with lower ratings) are converted to zero. The transaction database contains matrix data: rows are users, columns are items (movies), and the elements are binary rating values (1, 0).

### 3.3 Cross Validation

Figure 3 outlines the evaluation experiment. In this experiment, the recommendation accuracy of the bicluster network method is evaluated on the basis of three-fold cross-validation. First, the transaction database is divided into three sets with regard to users. The first two sets are training datasets, and the third is a test dataset. The training datasets are used to generate biclusters. The test dataset is divided into movies for test queries and for evaluation of recommendation accuracy. At this time, a fixed number of movies (determined in advance) are used as movies for test queries, and the remaining movies are used as the evaluation. The recommendation accuracy is evaluated in terms of whether the movies for evaluation were comprehensively and correctly recommended in relation to the test queries. With the above operation, the evaluation experiment is conducted on a total of three patterns, substituting the training dataset for the test dataset.

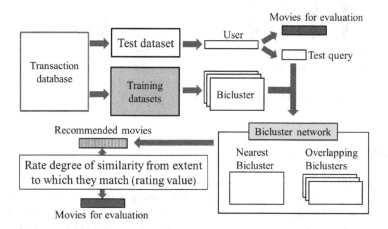

**Fig. 3** Evaluation experiment

## 3.4 Evaluation Index

Recommendation accuracy is evaluated using *precision*, *recall*, and *F − measure*. Here *precision* and *recall* represent the correctness and completeness of the recommendation, respectively. *F − measure* is an index that takes the balance between the *precision* and *recall* into consideration. The *precision*, *recall*, and *F − measure* are calculated according to the following formulae:

$$Precision = \frac{|A \cap B|}{|B|}, \tag{4}$$

$$Recall = \frac{|A \cap B|}{|A|}, \tag{5}$$

$$F - measure = \frac{2 * Recall * Precision}{(Recall + Precision)}, \tag{6}$$

where $A$ is the set of movies rated highly by the user and $B$ is the set of recommended movies. In this experiment, the minimum number of users is set to 20, the minimum support is set to 10, and the number of movies for test queries is set to five.

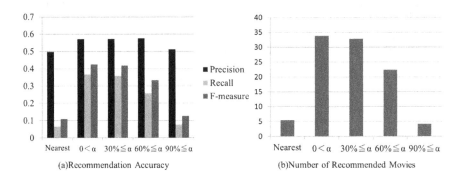

**Fig. 4** Results of experiment

## 4    Results and Discussion

Figure 4(a) shows the recommendation accuracies with the variation of the over-lapping degree . In this figure, we show the averages of *precision*, *recall* and $F-measure$ which are calculated by three-fold cross-validation. Fig.4(b) shows the average of the numbers of movies recommended by the cross-validation in the Fig.4(a). In these figures, the results using only the nearest bicluster (indicated with "Nearest"), i.e. the results in the case of not using bicluster network, are also shown as a comparative method.

As seen in Fig.4(a), the bicluster network method shows higher scores in all over-lapping degrees than "Nearest". In particular, both the *recalls* and the $F-measures$ are drastically improved. In addition, from Fig.4(b), we can see that the numbers of the recommended movies also increase compared to "Nearest". These results show that the bicluster network method can recommend more items matching user's query than "Nearest". The main reason for it lies in using overlapping biclusters as local solutions in addition to the nearest bicluster as an optimal solution.

From the above results, we found that the bicluster network method can widely and adequately search for items that are suitable for user's tastes.

## 5    Conclusion

In this study, we proposed an item recommendation method based on the bicluster network. As an experiment, we applied this method to a movie rating dataset and evaluated the recommendation accuracy as well as the number of recommended movies according to increasing overlap degree through comparison to an existing method. As a result, we showed that our method can drastically improve the *recall* and $theF-measure$ and also can recommend more movies. These results mean that the bicluster network method can widely and adequately search for items that are suitable for user's tastes.

In future, we will perform evaluation experiments with different minimum support value in addition to using other datasets.

## References

1. Su, X., Khoshgoftaar, T.M.: A Survey of Collaborative Filtering Techniques. Advances in Artificial Intelligence (2009)
2. Symeonidis, P., Nanopoulos, A., Papadopoulos, A., Manolopoulos, Y.: Nearest Biclusters Collaborative Filtering. In: Nasraoui, O., Spiliopoulou, M., Srivastava, J., Mobasher, B., Masand, B. (eds.) WebKDD 2006. LNCS (LNAI), vol. 4811, pp. 36–55. Springer, Heidelberg (2007)
3. Sarwar, B., Karypis, G., Konstan, J., Riedl, J.: Item-Based Collaborative Filtering Recommendation Algorithms. In: Proceedings of the 10th International Conference on World Wide Web, pp. 285–295 (2001)
4. Okada, Y., Fujibuchi, W., Horton, P.: A biclustering method for gene expression module discovery using closed itemset enumeration algorithm. IPSJ Transactions on Bioinformatics 48, 39–48 (2007)
5. Uno, T., Asai, T., Uchida, Y., Arimura, H.: An Efficient Algorithm for Enumerating Closed Patterns in Transaction Databases. In: Suzuki, E., Arikawa, S. (eds.) DS 2004. LNCS (LNAI), vol. 3245, pp. 16–31. Springer, Heidelberg (2004)
6. GroupLens, http://www.grouplens.org

# Item Recommendation by Query-Based Biclustering Method

Naoya Yokoyama and Yoshihumi Okada

**Abstract.** We propose a new recommender system that explores useful items by a biclustering method based on user's query. The advantage of our method is that the computational time can be reduced because the search space of biclusters is restricted to the transactions (users) which rate items within a query. In this study, the performance of our method is compared to that of a previous method that executes biclustering for entire transaction database. As a result, it is shown that our method enables item recommendation with higher accuracy at a considerably lower computational cost than the previous method.

## 1 Introduction

In online shopping sites or video-sharing sites, recommendation technology has been utilized as the useful tool for predicting products or information (hereafter "items") matching user's taste/purpose. To date, various recommendation methods such as user-based correlation method and item-based correlation method have been proposed and have been put into practical use at many commercial systems [1].

Recently, biclustering-based approaches were proposed [2, 3], and it was reported that these methods presented higher recommendation accuracy than existing popular algorithms such as user- or item-based correlation method. Biclustering-based approach first finds groups of items that are rated commonly by multiple users. Here, a tuple of users, items and rating values is called a bicluster. Next, this approach

Naoya Yokoyama
Department of Information and Electronic Engineering, Muroran Institute of Technology,
27-1, Mizumoto-cho, Muroran 050-8585, Japan
e-mail: yokoyama@cbrl.csse.muroran-it.ac.jp

Yoshihumi Okada
College of Information and Systems, Muroran Institute of Technology,
27-1, Mizumoto-cho, Muroran 050-8585, Japan
e-mail: okada@cbrl.csee.muroran-it.ac.jp

V.-N. Huynh et al. (eds.), *Knowledge and Systems Engineering, Volume 2,*      155
Advances in Intelligent Systems and Computing 245,
DOI: 10.1007/978-3-319-02821-7_15, © Springer International Publishing Switzerland 2014

identifies biclusters in which the rating values for items are similar to those of a user requesting recommendation (hereinafter "an active user"). Finally, items unrated by the active user in those biclusters are recommended.

On the other hand, biclustering-based approach has a problem that enormous computation is required due to the combinatorial search for users, items and rating values. In addition, this approach needs to execute re-biclustering every time new users and items are added into database. The numbers of users/items are rapidly increasing day by day, thus it is crucially important to tackle those problems concerning such computational cost or scalability.

In this study, we propose a new recommendation method by query-based biclustering. This method is achieved by expanding the biclustering method based on closed itemsets which was developed by Kawahara et al. [3]. A major feature of this method lies in searching only for biclusters that include items specified in a query. This enables us to reduce the search space for biclusters drastically, and thus the computational time can also be considerably curtailed compared to the previous method. In addition, there is no need to perform biclutering for entire database in advance every time the database is updated. In this paper, we discuss the usefulness of the proposed method through comparing the performances with the previous method.

## 2   Recommendation of the Biclutering Based on a Closed Itemset

In this section, the biclustering-based method by Kawahara et al. [3] is outlined briefly. Suppose that a transaction database is given as shown in Fig.1. Each transaction corresponds to a user and includes one or more items. Namely, the transaction database is a database that consists of items rated by each user.

A closed itemset is a maximal set of items which appears concurrently in a subset of transactions, namely in a subset of users. A bicluster is a cluster that consists of a subset of items rated in common by multiple users. We can obtain a maximal bicluster by identifying a closed itemset and its supporting users. Maximal biclusters (hereinafter "bicluster" for short) are generated from the entire transaction database in advance. This method utilizes an efficient and fast closed itemset mining algorithm called LCM developed by Uno et al. [4] and can enumerate biclusters exhaustively under a pre-specified minimum support (i.e. "a minimum number of users") and a minimum number of items.

Recommendation to active user is executed according to the following steps. An input for this method is one or more items that are liked by the active user. First, we find biclusters that are similar to the query with regard to the ratings for items. Subsequently, items unrated by the active user out of items included in the bicluters are recommended (see [3] for details).

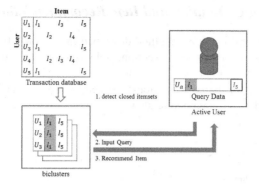

**Fig. 1** Biclutering-based recommendation by Kawahara et al.

## 3 Query-Based Biclutering Method

Figure 2 illustrates an item recommendation by query-based biclustering method.

### 3.1 Query Preparation

Here, suppose that an active user inputs a query with n items. First, these items are sorted in descending order of the number of users (hereinafter "user count") that rate those items in the transaction database. Second, n queries are created by eliminating item one by one in ascending order of the user counts. In the example of Fig.2, three items are inputted by the active user, and three queries are generated.

### 3.2 Transaction Database Reduction

For each query created by the above manner, the query-based biclustering method generates biclusters according to the two steps as described below. Note that a query is inputted in order of query with more user count. The aim here is to enumerate biclusters including all items of the inputted query. For this purpose, we extract only a set of users (rows) rating all items of the inputted query from the transaction database, in advance. The set of users is used as a new transaction database in the next bicluster generation step. This operation allows to downsizing drastically the transaction database because we can eliminate many unnecessary users that will never be included in the biclusters targeted here.

## 3.3 Biclustring Generation and Item Recommendation

Biclusters are extracted using the method described in section 2 from the reduced transaction database in the previous section. Subsequently, item recommendation is performed as follows. First, we find biclusters that are similar to the query with regard to the ratings for items. Next, items unrated by the active user out of items included in the bicluters are recommended.

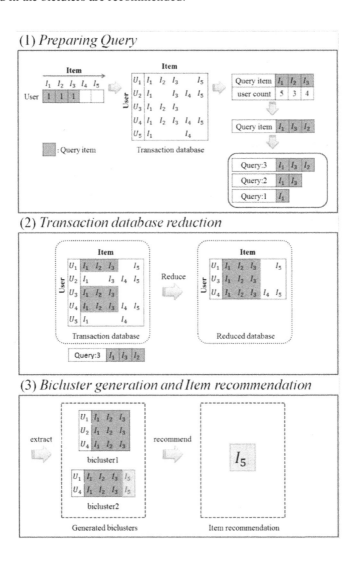

**Fig. 2** Query-based biclutering method

# 4 Evaluation Experiment

## 4.1 Dataset

This study uses a movie rating data set called MovieLens that was published by GroupLens [5]. MovieLens contains rating data from 943 users for 1,682 movies. Each movie is ranked from one to five. Each user rates at least 10 movies and there are a total of 100,000 ratings.

## 4.2 Transaction Data Creating

Transaction database for generating biclusters are constructed from MovieLens. A transaction and an item are regarded as a user and a movie, respectively. Ratings of four or five (highly-rated movies) are converted to one, and all other ratings (movies with lower ratings) are converted to zero. In this study, the transaction database is constructed using the top 900 users in the number of rating items.

## 4.3 Dataset for Evaluation

In this experiment, the computational times and the recommendation accuracies by increasing the size of the transaction database are evaluated on the basis of three-fold cross validation. First, the transaction database is divided into three datasets. The first two sets are training datasets, and the third is a test dataset.

The test dataset is a dataset for evaluating the recommendation accuracy and is composed of 300 users. Movies rated by each user are divided into movies for an evaluation and movies for a test query. For each user, 5 movies are used as a test query, and the remaining movies are used as movies for an evaluation. Note that each test query is divided into 5 queries using the step described in section 3.1. Training dataset is a dataset for generating biclusters and is composed of 600 users. In this experiment, we create six training datasets by increasing the size of the training dataset from 100 users to 600 users and evaluate the performances on those different sizes.

## 4.4 Computetional Time of Bicluter Generation

For each of the training datasets of the different sizes, we measure the computational time of bicluster generation by the following steps. First, we generate biclusters using 5 queries created from a user of the test dataset. This is repeated for the 300 users of the test dataset. Next, these computational times are averaged over the all queries. In this experiment, the minimum number of users is set to 10 and the minimum number of movies is set to 5.

## 4.5  Rating for Recommendation

Recommendation test is conducted for each user of test datasets by using the bicluters generated in section 4.4. Recommendation accuracy is estimated by precision, recall, and F-measure. Here precision and recall represent the correctness and completeness of the recommendation, respectively. F-measure is an index that takes the balance between the precision and recall into consideration. The precision, recall, and F-measure are calculated according to the following formulae:

$$Precision = \frac{|A \cap B|}{|B|}, \tag{1}$$

$$Recall = \frac{|A \cap B|}{|A|}, \tag{2}$$

$$F - measure = \frac{2 * Recall * Precision}{(Recall + Precision)}, \tag{3}$$

Where A is the set of movies rated highly by the user and B is the set of recommended movies.

## 5  Result

The results of the evaluation experiments are shown in Fig. 3. In these figures, "Entire" indicates the results of Kawahara's method that executes biclustering for an entire transaction database.

Figure 3 (a) is the computational time of biclusters generation calculated using the training datasets of the different sizes. Kawahara's method exhibits the exponential increase with an increase of the sizes of the training datasets. In contrast, our method shows the gradual changes against the sizes of the training datasets. This is because our method can drastically reduce the search space for biclusters by targeting only users rating movies included in the query.

Figure 3 (b) is the recommendation accuracies. In this figure, we show the averaged results on each of the 5 queries created in section 4.3. For precision, our method shows higher scores than Kawahara's method regardless of the number of items within queries. This is because our method focuses on only biclusters including items within queries. On other hands, recall shows substantially lower scores than Kawahara's method. This is because our method restricts the search space for items by conservative recommendation as described above. F-measure of our method is also inferior to Kawahara's method because it is affected by extremely low score of recall.

From these results, we can see that our method can recommend items accurately, and that the computational time is little-affected by increasing the number of users. On other hands, Completeness for items that should be recommended is lower than

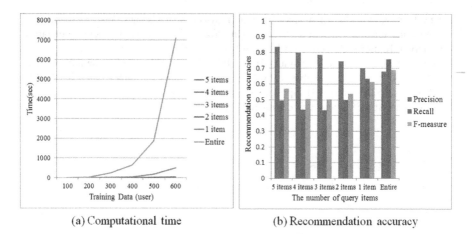

(a) Computational time                    (b) Recommendation accuracy

**Fig. 3** Result

Kawahara's method. In future, we need to develop a method to increase variations of recommended items for improving the recall.

## 6   Conclusion

In this study, we proposed a recommendation method based on a query-based bi-clustering and applied it to a movie rating dataset. As experiments, we evaluated computational time by changing the size of transaction database as well as recommendation accuracy by altering query size. A major feature of our method lies in searching only biclusters with items included in a query. By this approach, we can achieve fast biclustering regardless of the increase of the number of users because the size of transaction database can be drastically reduced. In addition, the precision (i.e. "correctness of recommendation") of our method substantially surpasses Kawahara's method that executes biclustering for entire transaction database.

In future, we will develop a method to increase variations of recommended items for improving Recall, namely completeness for items that should be recommended.

## References

1. Su, X., Khoshgoftaar, T.M.: A Survey of Collaborative Filtering Techniques. Advances in Artificial Intelligence (2009)
2. Symeonidis, P., Nanopoulos, A., Papadopoulos, A., Manolopoulos, Y.: Nearest Biclusters Collaborative Filtering. In: Nasraoui, O., Spiliopoulou, M., Srivastava, J., Mobasher, B., Masand, B. (eds.) WebKDD 2006. LNCS (LNAI), vol. 4811, pp. 36–55. Springer, Heidelberg (2007)

3. Kawahara, K., Okada, Y.: A recommendation system that narrows down recommendation results according to user's preference. In: Proc. of 14th Annual Conference of Japan Society of Kansei Engineering (in Japanese)
4. Uno, T., Asai, T., Uchida, Y., Arimura, H.: An Efficient Algorithm for Enumerating Closed Patterns in Transaction Databases. In: Suzuki, E., Arikawa, S. (eds.) DS 2004. LNCS (LNAI), vol. 3245, pp. 16–31. Springer, Heidelberg (2004)
5. GroupLens, http://www.grouplens.org

# A Cyber Swarm Algorithm for Constrained Program Module Allocation Problem

Peng-Yeng Yin and Pei-Pei Wang

**Abstract.** Program module allocation problem (PMAP) is an important application of the quadratic assignment problem (QAP), which has been shown to be NP-complete. The aim of the PMAP is to allocate a package of program modules to a number of distributed processors such that the incurred cost is minimal subject to specified resource constraints. We propose to employ a new metaheuristic, the cyber swarm algorithm (CSA), for finding the near optimal solution with reasonable time. The CSA has previously manifested excellent performance on solving continuous optimization problems. Our experimental results show that the CSA is more effective and efficient than modifications of genetic algorithm, particle swarm optimization, and harmony search in tackling the PMAP.

## 1 Introduction

The aim of the program module allocation problem (PMAP) is to allot a set of program modules to another set of computational processors such that the specified goal is optimized with respect to the resource constraints. PMAP has been formulated as a constrained combinatorial optimization problem and it can be viewed as a relaxed form of the well-known quadratic assignment problem (QAP) which stipulates a one-to-one assignment constraint. PMAP has been shown to be NP-complete, thus prohibiting the attempt for finding global optimal solutions [1].

There are, in literature, three main types of PMAP with different goals of optimization. The Type-I PMAP [2] searches for the module allocation with the least cost which may consist of dynamic cost (such as processor execution duration and communication cost) and fixed cost (processor purchase, overhead charges, etc). The Type-II PMAP [3] aims to maximize the overall system reliability for fulfilling the success of the module execution. The failure rate regarding to processors and

Peng-Yeng Yin · Pei-Pei Wang
Department of Information Management, National Chi Nan University, Nantou, Taiwan
e-mail: {pyyin,s2213522}@ncnu.edu.tw

V.-N. Huynh et al. (eds.), *Knowledge and Systems Engineering, Volume 2,*
Advances in Intelligent Systems and Computing 245,
DOI: 10.1007/978-3-319-02821-7_16, © Springer International Publishing Switzerland 2014

communication links should be taken into account. The Type-III PMAP [4] combines the previous two types of PMAP and creates a bi-objective formulation which intends to find the trade-offs between the least cost and the maximal reliability.

| | Mathematical programming | Heuristics | Genetic algorithms | Simulated annealing | Particle swarm optimization | Harmony search | Variable neighborhood search | Hybrid method |
|---|---|---|---|---|---|---|---|---|
| Chen and Yur, 1990 | ● | | | | | | | |
| Billionnet et al., 1992 | ● | | | | | | | |
| Ernst et al., 2001 | ● | | | | | | | |
| Pendharkar, 2012 | ● | | | | | | | |
| Lee and Shin, 1997 | | ● | | | | | | |
| Kafil and Ahmad, 1998 | | ● | | | | | | |
| Tripathi et al., 2000 | | | ● | | | | | |
| Pagea et al., 2010 | | | ● | | | | | |
| Hamam and Hindi, 2000 | | | | ● | | | | |
| Ho et al., 2008 | | | | | ● | | | |
| Zou, et al., 2010 | | | | | | ● | | |
| Lusa and Potts, 2008 | | | | | | | ● | |
| Yin et al., 2009 | | | | | | | | ● |
| Lin et al., 2009 | | | | | | | | ● |
| This study | | | | | | | | ● |

**Fig. 1** Categories of existing approaches for the PMAP problems

This paper focuses on the Type-I PMAP (referred to as PMAP hereafter), which is of the most interest to the research community. Many approaches have been proposed in the literature [5]. These methods can be divided into three categories as tabulated in Figure 1. (1) Mathematical programming. Ernst et al. [6] and Billionnet et al. [7] used column generation technique to solve the nonlinear integer programming formulation. Chen and Yur [8] applied the branch-and-bound-with-underestimates algorithm for solving the PMAP problem with module precedence constraint. Pendharkar [9] proposed several low bound estimates for facilitating the mathematical programming solutions. (2) Heuristics. Lee and Shin [10] and Kafil and Ahmad [11] have developed ad hoc heuristics for providing exact solutions in specific circumstances, such as the distributed systems with processor mesh and partial k-tree communication graph. (3) Metaheuristics. The metaheuristic algorithms, which can obtain near-optimal solutions within acceptable computational time, have become the prevailing solution methods for PMAP in the past decade. The most significant ones include genetic algorithms (GA) [12, 13], simulated annealing (SA) [14], particle swarm optimization (PSO) [15], harmony search (HS) [16], variable neighborhood search (VNS) [17], and hybrid methods [18, 19].

This paper presents a cyber swarm algorithm (CSA) [20] which marries PSO to the adaptive memory programming (AMP) in order to augment their synergy. The experimental results manifest that the proposed CSA algorithm is malleable against varying constraint requirements.

The remainder of this paper is organized as follows. Section 2 gives the formulation of the PMAP. Section 3 articulates the proposed CSA algorithm to the PMAP. Section 4 reports the performance evaluations. Finally, Section 5 concludes this work.

## 2  Problem Formulation

The PMAP problem addressed in this paper describes a distributed system which accommodates $n$ processors interconnected by communication links. A program with $m$ modules is to be executed in this distributed system and we seek to optimize the incurred execution cost and the communication cost for allotting the modules to processors. Moreover, the consumed memory capacity and communication load should be constrained by the characteristics of processors and communication links. Formally, the PMAP problem is formulated as follows.

$$\text{Min} \sum_{i=1}^{m} \sum_{k=1}^{n} e_{ik} x_{ik} + \sum_{i=1}^{m-1} \sum_{j=i+1}^{r} \sum_{k=1}^{n} \sum_{l \neq k} c_{ij} y_{ijkl}, \tag{1}$$

$$\text{subject to} \sum_{k=1}^{n} x_{ik} = 1, \forall i = 1, 2, \ldots, m \tag{2}$$

$$\sum_{i=1}^{m} r_i x_{ik} \leq R_k, \forall k = 1, 2, \ldots, n \tag{3}$$

$$\sum_{i=1}^{m} p_i x_{ik} \leq P_k, \forall k = 1, 2, \ldots, n \tag{4}$$

$$\sum_{l=1}^{n} y_{ijkl} = x_{ik} \forall i, j = 1, 2, \ldots, m; k = 1, 2, \ldots, n \tag{5}$$

$$\sum_{k=1}^{n} y_{ijkl} = x_{jl}, \forall i,j = 1,2,\ldots,m; l = 1,2,\ldots,n \qquad (6)$$

$$x_{ik}, y_{ijkl} \in \{0,1\}, \forall i, j, k, l \qquad (7)$$

The decision variables include $x_{ik}$ and $y_{ijkl}$. We define $x_{ik} = 1$ if module $i$ is assigned to processor $k$, and $x_{ik} = 0$ otherwise. We further let $y_{ijkl}$ be a binary variable and $y_{ijkl} = 1$ if and only if module $i$ is assigned to processor $k$ and module $j$ is assigned to processor $l$. Equation (1) describes the problem objective as to minimize the overall execution cost $e_{ik}$ and communication cost $c_{ij}$. Equation (2) states that each module should be assigned to exactly one processor. Equations (3) and (4) ensure that the demanded resources for memory ($r_i$) and processing ($p_i$) are less than capacity bounds $R_k$ and $P_k$ facilitated on each processor $k$. Equations (5) and (6) indicate the correspondence between $x_{ik}$ and $y_{ijkl}$. Equation (7) defines both $x_{ik}$ and $y_{ijkl}$ are binary decision variables.

## 3   Cyber Swarm Algorithm to the PMAP Problem

We previously proposed the Cyber Swarm Algorithm (CSA) [20] which enhances PSO by embedding search strategies from AMP. These strategies involve complex neighborhood concepts, memory structure, and adaptive search principles drawn from the Tabu Search and Scatter Search. The reader is referred to [21, 22]. We applied the CSA to the PMAP problem as follows.

### 3.1   Particle Representation and Fitness Evaluation

We employ a compact representation for the particle structure. Each particle contains $m$ variables and the value of the $i$th variable indicates the index of the processor to which the $i$th module is allotted. The initial swarm consists of $K$ particles generated at random. Each particle receives a fitness value estimating the quality of the solution the particle represents. As PMAP is a constrained optimization problem, the objective function (Equation (1)) is relaxed by including the constraints penalty, as follows.

$$\sum_{i=1}^{m}\sum_{k=1}^{n} e_{ik}x_{ik} + \sum_{i=1}^{m-1}\sum_{j=i+1}^{r}\sum_{k=1}^{n}\sum_{l\neq k} c_{ij}y_{ijkl}$$
$$+ \sum_{k=1}^{n}\left(\max\left(0, \sum_{i=1}^{m} r_i x_{ik} - R_k\right) + \max\left(0, \sum_{i=1}^{m} p_i x_{ik} - P_k\right)\right) \qquad (8)$$

Hence, the solution with a higher fitness is either having an expensive cost or violating the resource constraints.

## 3.2 Dynamic Social Networking

Along the CSA search history, we store quality solutions (i.e., good program module allocations to the PMAP problem) in the reference set, denoted $RefSol[i]$, $i = 1, \ldots,$ $R$, where $R$ is the size of the reference set. These solutions are sorted in decreasing quality order and they are used as a guiding solution in term to produce new solutions. In particular, the velocity in the $j$th dimension for the $i$th particle is updated as follows,

$$v_{ij}^m \leftarrow K \left( v_{ij} + (\varphi_1 + \varphi_2 + \varphi_3) \right.$$

$$\times \left. \left( \frac{\omega_1 \varphi_1 pbest_{ij} + \omega_2 \varphi_2 gbest_{ij} + \omega_3 \varphi_3 RefSol[m]_j}{\omega_1 \varphi_1 + \omega_2 \varphi_2 + \omega_3 \varphi_3} - p_{ij} \right) \right), m = 1, \ldots, R \qquad (9)$$

where $\varphi_i$ and $\omega_i$ are the acceleration and weighting parameters for the $i$th guiding solution.

And the position of the particle is updated by

$$P_i \leftarrow \text{best of} \{ P_i + v_i^m \mid m \in [1, R] \} \qquad (10)$$

## 3.3 Responsive Strategy

The tabu search exploiting longer term memory (LTM) is able to redirect the search when the current search trajectory loses its search efficacy. The LTM records the frequency of critical events and responsive strategies will be called upon when a specified condition on the frequency holds. We employ two responsive strategies for conducting convergence and diversity searches, respectively. The convergence strategy will regenerate the swarm of module allocation solutions if the whole swarm stagnates, while the diversity strategy will reposition an allocation solution if it fails to improve its best position within a time period. One of the effective longer term strategies is path relinking (PR). PR is a search process which constructs a link between two strategically selected solutions and the best solution observed on this link is considered as the local optimum. PR transforms the initiating solution into the guiding solution by generating moves that successively replace an attribute of the initiating solution that is contained in the guiding solution.

Our two responsive strategies proceed as follows. When the swarm overall best solution $gbest$ has not improved for $t_1$ successive search iterations, the convergence PR strategy, Convergence_PR($Rand_i$, $gbest$), $i = 1, \ldots, U$, reconstructs a new swarm of module allocation solutions within the neighborhood proximity of $gbest$ to improve the exploitation convergence, where $Rand_i$ is a random allocation solution drawn from the solution space. On the other hand, when a local region has been over-exploited (this phenomenon arises if when $pbest$ has not improved for $t_2$ successive search iterations), the corresponding particle should be repositioned in an

---

**Algorithm 5.** Pseudo codes of the CSA for the PMAP problem.

---

1. Initialization

    a. Randomly generate U module allocation solutions, $P_i = \{p_{ij}\}, 0 \le i < U, 0 \le j < m$

    b. Randomly generate U velocity vectors, $V_i = \{v_{ij}\}, 1 \le i \le U, 0 \le j < m$

    c. Evaluate the fitness value for each allocation solution. Update reference set.

2. Repeat until a stopping criterion is met

    a. For each allocation solution $P_i$, i = 1, …, U, Do

        i. **Guided moving with dynamic social networking:**

$$v_{ij}^m \leftarrow K \left( v_{ij} + (\varphi_1 + \varphi_2 + \varphi_3) \left( \frac{\omega_1 \varphi_1 pbest_{ij} + \omega_2 \varphi_2 gbest_{ij} + \omega_3 \varphi_3 RefSol[m]_j}{\omega_1 \varphi_1 + \omega_2 \varphi_2 + \omega_3 \varphi_3} - p_{ij} \right) \right)$$

           $P_i \leftarrow$ best of $\{ P_i + v_i^m \mid m \in [1,R] \}$

        ii. Update reference set

    b. **Convergence PR strategy:** If gbest has not been updated for $t_1$ iterations, restart each allocation solution by exploiting the region between a random solution $Rand_i$ and the gbest by

$$P_i \leftarrow Convergence\_PR(Rand_i, gbest), i = 1, …, U$$

    **Diversity PR strategy:** Else if a particular pbest has not been updated for $t_2$ iterations, restart this allocation solution by

$$P_i \leftarrow Diversity\_PR(Rand_1, Rand_2)$$

    Update reference set

3. Output the swarm's best allocation solution (gbest)

---

uncharted region to start a new search trajectory. We do this by performing the diversity PR strategy, Diversity_PR($Rand_1$, $Rand_2$), to explore under-exploited regions. The diversity PR strategy not only diversifies the search but also tunnels through different regions that have contrasting features. With the responsive strategies and PR search trajectories, the best allocation solution improves over time effectively and efficiently.

The details of the proposed CSA algorithm to the PMAP problem are presented in Algorithm 1.

## 4 Experimental Results

We denote the PMAP problem instance by three parameters, the number of modules ($r$), the number of processors ($n$), and the module interaction density ($d$). These are the key factors that affect the problem complexity degree. The testing dataset is generated as follows. We set the value of ($r$, $n$) to (5, 3), (10, 6), (15, 9) and (20, 12), respectively. For each pair of ($r$, $n$), we generate three different task interconnection graphs at random with density d equivalent to 0.3, 0.5, and 0.8. Ten problem instances with various resource demands and capacities are generated at random for each specification of $r$, $n$ and $d$. As such, we obtain a testing dataset of 120 problem instances for evaluating the comparative performances of the competing methods. All of the experiments are conducted on a 2.4 GHz PC with 256MB RAM.

A Lingo program based on the problem formulation introduced in Section 2 was codified to solve the problem instances to optimality. Three metaheuristic methods, a novel global harmony search (NGHS) [16], a hybrid particle swarm optimization with local search (HPSO) [23] and a Standard Genetic Algorithm (SGA) [24], were implemented for comparison. The competing algorithms use the same coding representation and fitness function as used by the CSA. In particular, the parameter setting used by HPSO is as follows: number of particles = 80, acceleration factors $c1 = c2 = 2$) and that by SGA is population size = 80, crossover rate = 0.7, and mutation rate = 0.1. The stopping criterion for all the compared methods are set to either executing 80,000 fitness evaluations or consuming 90 hours of CPU elapse time, depending on which is encountered first. Except the exact method (Lingo), the average performance over ten repetitive runs of each algorithm was gauged.

For our testing dataset with different problem size scales, the Lingo program is only able to solve optimally the small data instances with $(r, n)$ set to (5, 3) and (10, 6). For the remaining problem instances, no solution is obtained when Lingo has consuming 90 hours. Table 1 lists the numerical results for small problem instances obtained by various methods. It is observed that all competing methods can obtain exact optimal solutions as reported by Lingo for the smallest problem instances with $(r, n) = (5, 3)$. For problem instance with $(r, n) = (10, 6)$, only the proposed CSA method can solve optimally with the exact solutions as reported by Lingo. HPSO and NGHS obtain near-optimal results with the same number of fitness evaluations. However, SGA is the worst performer in terms of the obtained overall cost which is about 15% ˜ 20% worse than the global optimal cost.

**Table 1** The costs obtained by Lingo, GA, HPSO and CSA

| r | n | d | Lingo | SGA | HPSO | NGHS | CSA |
|---|---|---|-------|-----|------|------|-----|
| 5 | 3 | 0.3 | 180.97 | 180.97 | 180.97 | 180.97 | 180.97 |
|   |   | 0.5 | 202.40 | 202.40 | 202.40 | 202.40 | 202.40 |
|   |   | 0.8 | 215.63 | 215.63 | 215.63 | 215.63 | 215.63 |
| 10 | 6 | 0.3 | 374.68 | 436.32 | 376.11 | 375.77 | 374.68 |
|   |   | 0.5 | 403.15 | 501.05 | 403.15 | 403.15 | 403.15 |
|   |   | 0.8 | 715.61 | 806.21 | 719.55 | 715.61 | 715.61 |

For larger size problem instances, no solution is obtained when Lingo has consuming 90 hours, only metaheuristic methods (SGA, HPSO, NGHS, and CSA) can report feasible solutions within the specified CPU time limit. The performance difference between competing metaheuristic methods is more significant when the problem size increases. Figure 2 shows the average overall cost obtained by competing metaheuristic algorithms for problem instances with $(r, n) = (15, 9)$ and communication density $(d)$ set to 0.3, 0.5, and 0.8, respectively. It is observed that HPSO, NGHS, and CSA well surpass SGA in terms of the overall cost, independent of the

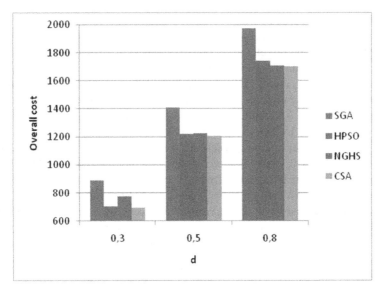

**Fig. 2** The costs obtained by SGA, HPSO, NGHS, and CSA for median-sized problem instances with (r, n) = (15, 9) and various values of d

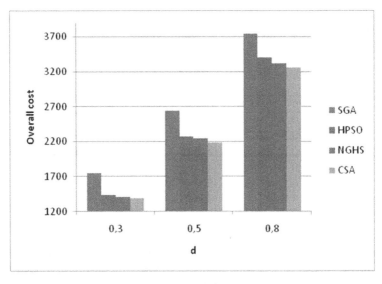

**Fig. 3** The costs obtained by SGA, HPSO, NGHS, and CSA for large-sized problem instances with (r, n) = (20, 12) and various values of d

communication density. The difference between HPSO, NGHS, and CSA is negligible, the reason could be due to that their results are already very close to the global optimal allocation solution. For problem instances with $(r, n) = (20, 12)$, HPSO, NGHS, and CSA still outperform SGA in terms of the overall cost as shown in Figure 3, and it seems that the performance gap is greater with larger value of $d$. Furthermore, CSA has a better performance against HPSO and NGHS with 3% ~ 5% lead in overall cost for various communication density, which means the proposed CSA algorithm is more scalable to problem complexity than the other compared methods.

## 5   Conclusions

In this paper, we have proposed a very effective solution method, namely the cyber swarm algorithm (CSA), for solving the program module allocation problem (PMAP), which has been known to be NP-complete. The CSA extends the notion of Scatter Search and Path Relinking, and marries them to particle swarm optimization (PSO) to create a more effective optimization method. The CSA has the following features. The dynamic social networking enables the particle swarm to learn from a historical set of elite solutions, in contrast to the constrained learning to best solutions as con-templated by traditional PSO. The responsive strategies enhance the CSA search capability in both convergence and diversification aspects. The experimental results on 120 PMAP problem instances show that the all compared metaheuristic methods can solve the small-sized problems to optimality verified by Lingo. For median- and large-sized problem instances, the CSA outperforms SGA, HPSO, NGHS and the difference is more significant as the problem size increases.

**Acknowledgments.** This research is partially supported by National Science Council of ROC, under Grant NSC 98-2410-H-260-018-MY3.

## References

1. Lo, V.M.: Task Assignment in Distributed Systems. Ph.D. dissertation, Dep. Comput. Sci., Univ. Illinois (1983)
2. Lee, C.H., Shin, K.G.: Optimal Task Assignment in Homogeneous Networks. IEEE Transactions on Parallel and Distributed Systems 8, 119–129 (1997)
3. Vidyarthi, D.P., Tripathi, A.K.: Maximizing Reliability of Distributed Computing System With Task Allocation Using Simple Genetic Algorithm. Journal of Systems Architecture 47, 549–554 (2001)
4. Yang, B., Hu, H., Guo, S.: Cost-oriented Task Allocation and Hardware Redundancy Policies in Heterogeneous Distributed Computing Systems Considering Software Reliability. Computers & Industrial Engineering 56, 1687–1696 (2009)
5. Semchedine, F., Bouallouche-Medjkoune, L., Assani, D.: Task Assignment Policies in Distributed Server Systems: A Survey. Journal of Network and Computer Applications 34, 1123–1130 (2011)

6. Ernst, A., Hiang, H., Krishnamoorthy, M.: Mathematical Programming Approaches for Solving Task Allocation Problems. In: Proc. of the 16th National Conf. of Australian Society of Operations Research (2001)
7. Billionnet, A., Costa, M.C., Sutter, A.: An Efficient Algorithm for a Task Allocation Problem. Journal of ACM 39, 502–518 (1992)
8. Chen, G.H., Yur, J.S.: A Branch-And-Bound-With-Underestimates Algorithm for the Task Assignment Problem With Precedence Constraint. In: Proc. of the 10th International Conf. on Distributed Computing Systems, pp. 494–501 (1990)
9. Pendharkar, P.C.: Lower Bounds For Constrained Task Allocation Problem in Distributed Computing Environment. In: IEEE Canadian Conference on Electrical & Computer Engineering (2012)
10. Lee, C.H., Shin, K.G.: Optimal Task Assignment in Homogeneous Networks. IEEE Transactions on Parallel and Distributed Systems 8, 119–129 (1997)
11. Kafil, M., Ahmad, I.: Optimal Task Assignment in Heterogeneous Distributed Computing Systems. IEEE Concurrency 6, 42–50 (1998)
12. Tripathi, A.K., Sarker, B.K., Kumar, N.: A GA Based Multiple Task Allocation Considering Load. International Journal of High Speed Computing, 203–214 (2000)
13. Pagea, A.J., Keanea, T.M., Naughton, T.J.: Multi-heuristic Dynamic Task Allocation Using Genetic Algorithms in a Heterogeneous Distributed System. Journal of Parallel and Distributed Computing 70, 758–766 (2010)
14. Hamam, Y., Hindi, K.S.: Assignment of Program Tasks To Processors: A Simulated Annealing Approach. European Journal of Operational Research 122, 509–513 (2000)
15. Ho, S.Y., Lin, H.S., Liauh, W.H., Ho, S.J.: OPSO: Orthogonal Particle Swarm Optimization and Its Application to Task Assignment Problems. IEEE Transactions on Systems, Man and Cybernetics, Part A: Systems and Humans 38, 288–298 (2008)
16. Zou, D., Gaoa, L., Li, S., Wua, J., Wang, X.: A Novel Global Harmony Search Algorithm for Task Assignment Problem. Journal of Systems and Software 83, 1678–1688 (2010)
17. Lusa, A., Potts, C.N.: A Variable Neighbourhood Search Algorithm for the Constrained Task Allocation Problem. Journal of the Operational Research Society 59, 812–822 (2008)
18. Yin, P.Y., Shao, B.M., Cheng, Y.P., Yeh, C.C.: Metaheuristic Algorithms for Task Assignment in Distributed Computing Systems: A Comparative and Integrative Approach. The Open Artificial Intelligence Journal 3, 16–26 (2009)
19. Lin, J., Cheng, A.M.K., Kumar, R.: Real-Time Task Assignment in Heterogeneous Distributed Systems with Rechargeable Batteries. In: International Conference on Advanced Information Networking and Applications, pp. 82–89 (2009)
20. Yin, P.Y., Glover, F., Laguna, M., Zhu, J.X.: Cyber Swarm Algorithms – Improving Particle Swarm Optimization Using Adaptive Memory Strategies. European Journal of Operational Research 201, 377–389 (2010)
21. Omran, M.: Special Issue on Scatter Search and Path Relinking Methods. International Journal of Swarm Intelligence Research 2 (2011)
22. Maquera, G., Laguna, M., Gandelman, D.A., Sant'Anna, A.P.: Scatter Search Applied to the Vehicle Routing Problem with Simultaneous Delivery and Pickup. International Journal of Applied Metaheuristic Computing 2, 1–20 (2011)
23. Yin, P.Y., Yu, S.S., Wang, P.P., Wang, Y.T.: A Hybrid Particle Swarm Optimization Algorithm for Optimal Task Assignment in Distributed Systems. Computer Standards & Interfaces 28, 441–450 (2006)
24. Goldberg, D.E.: Genetic Algorithms in Search, Optimization, and Machine Learning. Addison Wesley, Reading (1997)

# A Ray Based Interactive Method for Direction Based Multi-objective Evolutionary Algorithm

Long Nguyen and Lam Thu Bui

**Abstract.** Many real-world optimization problems have more than one objective (and these objectives are often conflicting). In most cases, there is no single solution being optimized with regards to all objectives. Deal with such problems, Multi-Objective Evolutionary Algorithms (MOEAs) have shown a great potential. There has been a popular trend in getting suitable solutions and increasing the convergence of MOEAs by considering by Decision Makers (DM) during the optimization process (interacting with DM) for checking, analyzing the results and giving the preference.

In this paper, we propose an interactive method for DMEA, a direction-based MOEA for demonstration of concept. In DMEA, the authors used an explicit niching operator with a system of rays which divide the space evenly for the selection of non-dominated solutions to fill the archive and the next generation. We found that, by using the system of rays with a niching operator, solutions will be convergence to the Pareto Front via the corresponding to the distribution of rays in objective space. By this reason, we proposed an interactive method using set of rays which are generated from given reference points by DM. These rays replace current original rays in objective space. Based on the new distribution of rays, a niching is applied to control external population (the archive) and next generation for priority convergence to DM's preferred region. We carried out a case study on several test problems and obtained quite good results.

## 1 Introduction

When solving multi-objective optimization problems (MOPs), we need to simultaneously optimize several objective functions [4]. As a result, we usually obtain *trade-offs*, which are called *Pareto optimal solutions* or *Pareto optimal Front* (POF).

Long Nguyen · Lam Thu Bui
Faculty of Information Technology, Le Quy Don Technical University
e-mail: {longit76,lam.bui07}@gmail.com

V.-N. Huynh et al. (eds.), *Knowledge and Systems Engineering, Volume 2*,                     173
Advances in Intelligent Systems and Computing 245,
DOI: 10.1007/978-3-319-02821-7_17, © Springer International Publishing Switzerland 2014

Methods for multi-objective optimization can be classified into several classes including *the Interactive method*. With the Interactive method, DM iteratively directs the searching process by indicating his/her preference information over the set of solutions until DM satisfies or prefers to stop the process[16]. An interesting feature of interactive methods is that during the optimal process DM is able to learn about the underlying problem as well as his/her own preference. To date, many interactive techniques have been proposed [6, 7, 8, 9, 10, 15, 17, 12, 13] for solving MOPs. It is worthwhile to note that the aim of the interactive method is to find most suitable solution in several conflicting objectives regarding the DM's preference. It requires a mechanism to support DM in formulating her/his preferences and identifying preferred solutions in the set of Pareto optimal solutions.

In this paper, we introduce an interactive method for DMEA[3], a direction-based multi-objective evolutionary algorithm. With this proposal, we allow DM to specify a set of reference points, with each point a ray is generated by the way to build the system of rays in DMEA. The rays are generated from control points and paralleled with the central line which starts from the ideal point to centre of the hyperquadrant containing POFs. In this way, DM has more flexibility to express his preference. Among several methods for taking set information, we propose to use some reference points to generate new rays which replace some ones in current system of rays. Then we use niching operator to control the population to be convergence to preferred region. We hypothesise that by the way altering the system of rays approach is a good way to express DM's preferences, since specifying a set of rays is considerably convenient for DM. After DM has specified a set of reference points, a set of new added rays are generated, they will be replace the farthest rays from the central of DM's points boundary. By using niching operator in DMEA the Pareto optimal solutions are found that best corresponds to preferred region in objective space. If DM is not satisfied, he/she can specify other reference points. In our research, we adjust the system of rays by replacing some rays in current system with new ones which are generated from specified reference points are given by DM.

In the remainder of the paper, section 2 briefly describes about multi-objective optimization interactive method using reference points. Thereafter, in Section 3 we show some related interactive MOEAs methods. In section 4 we have description for DMEA, section 5 we propose our methodology for an interactive with DMEA. Section 6 presents simulation results and discussion. Finally, the conclusion of this paper is outlined in section 7.

## 2  Reference Point Interactive Approaches

The reference point interactive method is suggested by Wierzbicki[1], this method is known as a classical reference point approach. The idea of the method to control the search by reference points using *achievement functions*. Here the achievement function is constructed in such a way that if the reference point is dominated, the optimization will advance past the reference point to a non-dominated solution. A reference point $z^*$ is given for an M-objective optimization problem of minimizing

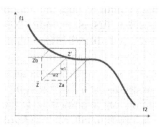

**Fig. 1** Altering the reference point, Here $Z_A$, $Z_B$ are reference points, $w$ is chosen weight vector used for scalarizing the objectives

$(f_1(x), \ldots, f_k(x))$ with $x \in S$. Then single-objective optimization one as following:
*minimize*

$$max_{i=1}^{M}[w_i(f_i(x) - z_i^*)] \tag{1}$$

subject to $x \in S$.

The algorithm for this method is described in five following steps:

**Step 1:** Present information to the DM. Set h=1

**Step 2:** Ask the DM to specify a reference point $z_*^h$

**Step 3:** Minimize achievement function. Present $z^h$ to the DM

**Step 4:** Calculate k other solutions with reference points
$\bar{z}(i) = \bar{z}^h + d^h e^i$ where $d^h = ||z_*^h - z^h||$ and $e_i$ is the $i^{th}$ unit vector

**Step 5:** If the DM can select the final solution, stop. Otherwise, ask the DM to specify $z_*^{h+1}$. Set $h = h + 1$ and go to **Step 3**.
Here $h$ is the number that DM specifies a reference point during process. By the way of using the series of reference points, DM actually tries to evaluate the region of Pareto Optimality, instead of one particular Pareto-optimal point. However DM usually deals with two situations:

1. The reference point is feasible and not a Pareto-optimal solution, the DM is interested in knowing solutions which are Pareto-optimal ones and near the reference point.
2. The DM finds Pareto-optimal solutions which is near supplied reference point.

## 3   Related Interactive MOEAs

In this section, we summarize several typical works on this area. In [8], authors proposed an Interactive MOEA using a concept of the reference point and finding a set of preferred Pareto optimal solutions near the regions of interest to a DM. The authors suggest two approaches: The first is to modify NSGA-II for effectively solving 10-objective. The other is to use hybrid-EMO methodology in allowing the DM to solve multi-objective optimization problems better and with more confidence.

The authors proposed in [15] a trade-off analysis tool that was used to offer the DM a way to analyze solution candidates. The ideas proposed here are directed to users of both classification and reference point based methods. The motivation here is that DM in certain cases miss additional local trade-off information so that they could get to know how values of objectives are changing, in other words, in which directions to direct the solution process so that they could avoid trial-and-error, that is, specify some preference information so that more preferred solutions will be generated.

In papers [12], [13] two reference point interactive methods are proposed to use single or multi reference points with multi-objective optimization based on decomposition-based MOEA (MOEA/D). In this method, a single point or a set of reference points are used in objective space to represent for DM's preferred region. The aggregated point from set of reference points (in case of multi-point) or the reference point is used in optimal process by two ways: replace or combine the current ideal point at the loop.

In paper [11] authors present a multiple reference point approach for multi-objective optimization problems of discrete and combinatorial nature. The reference points can be uniformly distributed within a region that covers the Pareto Optimal Front. An evolutionary algorithm is based on an achievement scalarizing function that does not impose any restrictions with respect to the location of the reference points in the objective space. Authors dealt with the design of a parallelization strategy to efficiently approximate the Pareto Optimal Front. Multiple reference points were used to uniformly divide the objective space into different areas. For each reference point, a set of approximate efficient solutions was found independently, so that the computation was performed in parallel.

# 4 DMEA

In DMEA two types of directional information are used to perturb the parental population prior to offspring production: convergence and spread (Fig. 2).

**Convergence Direction (CD).** In general defined as the direction from a solution to a better one, CD in MOP is a normalized vector that points from dominated to non-dominated solutions.

**Spread Direction (SD).** Generally defined as the direction between two equivalent solutions, SD in MOP is an unnormalized vector that points from one non-dominated solution to another.

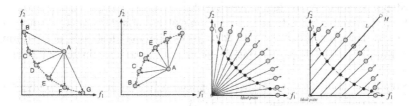

**Fig. 2** Illustration of convergence (black arrows in objective space- 1st figure) and spread (hollow arrows - 2nd figure in decision variable space). Two types of ray distribution (3rd, 4th figure).

## 4.1 Niching Information

A characteristic of solution quality in MOP is the even spread of non-dominated solutions across the POF [5]. In DMEA a bundle of rays are used to emit randomly from the estimated ideal point into the part of objective space that contains the POF estimate, (Fig. 2). The number of rays equals the number of non-dominated solutions wanted by the user. Rays emit into a "hyperquadrant" of objective space, i.e. the sub space that is bounded by the $k$ hyperplanes $f_i = f_{i,\min}$, $i \in \{1,2,\ldots,k\}$ and described by $f_i \geq f_{i,\min} \forall i \in \{1,2,\ldots,k\}$ where $f_{i,\min} \approx \min_{\text{all} A_1,A_2,\ldots} f_i$ with $A_1, A_2, \ldots$ being the solutions stored in the current archive. By their construction, the hyperquadrant contains the estimated POF.

A niching operator is used to the main population. From the second generation onward, the population is divided into two equal parts: one part for convergence, and one part for diversity. The first part is filled by non-dominated solutions up to a maximum of $n/2$ solutions from the combined population, where $n$ is the population size. This filling task is based on niching information in the decision space.

## 4.2 General Structure of Algorithm

The step-wise structure of the DMEA algorithm [3] with recent updates is as follows:

- **Step 1.** Initialize the main population $P$ with size $n$.
- **Step 2.** Evaluate the population $P$.
- **Step 3.** Copy non-dominated solutions to the archive $A$.
- **Step 4.** Generate an interim mixed population of the same size $n$ as $P$

    – Loop {
        · Select a random parent *Par*
        · Calculate $S_j$ [3].
        · Add $S_j$ to $M$.
    – } Until (the mixed population is full).

- **Step 5.** Perform the polynomial mutation operator [5] on the mixed population $M$ with a small rate.

- **Step 6.** Evaluate the mixed population $M$.
- **Step 7.** Identify the estimated ideal point of the non-dominated solutions in $M$ and determine a list of $n$ rays $R$ (starting from the ideal point and emitting uniformly into the hyperquadrant that contains the non-dominated solutions of $M$) [2]
- **Step 8.** Combine the interim mixed population $M$ with the current archive $A$ to form a combined population $C$ (i.e. $M + A \rightarrow C$).
- **Step 9:** Create new members of the archive $A$ by copying non-dominated solutions from the combined population $C$

  - Loop{
    - · Select a ray $R(i)$.
    - · In $C$, find the non-dominated solution whose distance to $R(i)$ is minimum.
    - · Select this solution and copy it to the archive.
  - } Until (all $n$ rays are scanned)

- **Step 10:** Determine the new population $P$ for the next generation.

  - Empty $P$.
  - Determine the number $m$ of non-dominated solutions in $C$. Using either niching in decision space ( function C1 in [3]) or in objective space (function C2 in [14]) to select $n/2$ solutions from C to P.
  - Apply a *weighted-sum scheme* to copy $\max\{n - m, n/2\}$ solutions to $P$.

- **Step 11:** Go to Step 4 if stopping criterion is not satisfied.

Function C1 and C2 are described as follow:

- Function C1 in Decision Space [3]:

  - Determine niching value (average Euclidean distance to other non-dominated solutions in decision space) for all non-dominated solutions in $C$.
  - Sort non-dominated solutions in $C$ according to niching values.
  - Copy the $n/2$ solutions with highest niching value to $P$.

- Function C2 in Objective Space [14]:

  - Determine density-based niching value for all non-dominated solutions in $C$ in objective space.
  - Sort non-dominated solutions in $C$ according to niching values.
  - Copy the $n/2$ solutions with highest niching value to $P$.

We use function C2 in [14] for experiments in this paper, for more detail please see [14].

In DMEA, the selection of non-dominated solutions to fill the archive and the next population is assisted by a ray based technique of explicit niching in the objective space by using a system of straight lines or rays starting from the current estimation of the ideal point and dividing the space evenly. Each ray is in charge of

locating a non-dominated solution, for that reason, a ray has an important role in optimal process. We suggest to adjust the distribution of ray according DM's preferred
region in objective space through interaction during process in the next section. The
proposed interactive EMO bases on system of ray so we called *The Ray based interactive method* for Direction based Multi-objective Evolutionary Algorithm.

In our experiments, the rays start from generated points and paralleled with the
central line of the top right hypequadrant.

## 5    Methodology

In DMEA, the rays are generated from control points and paralleled with the central
line which starts from the ideal point to centre of the hyperquadrant containing
POFs (Fig. 2). The number of rays equals the number of non-dominated solutions
wanted by the user. Base on this concept of DMEA, we proposed an reference point
interactive method on the second ray distribution as following steps:

- **Step 1:** Ask DM to input $n_p$ reference points which are their preferred regions
  in objective space.
- **Step 2:** Generate $n_p$ rays from reference points which paralleled with the central
  line ( In our experiment, we use the second distribution of rays (See Fig. 2)
- **Step 3:** Calculate the central point of DM's preferred region $P_c$.
- **Step 4:** Find $n_p$ rays which are farthest from $P_c$ by $n_p$ new ones are generated
  from Step 2.
- **Step 5:** Apply a niching to control external population (the archive) and next
  generation.

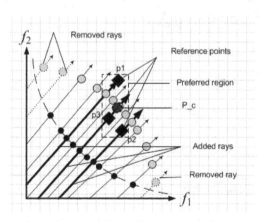

**Fig. 3** Illustration of proposed ray based interactive method for DMEA in a 2-dim MOP.
Three reference points are given by DM: p1, p2, p3. $p_c$ is the central point of DM's preferred
region, there three new rays (added rays) replace three ones (removed rays).

When DM interactive into the optimal process, we replace Step 7 in DMEA (See Section 4.2) with an interactive function is shown in Algorithm 1.

---

**Algorithm 6.** Ray-based interactive function.

---

**Input**: Number of reference points $n_p$
**Output**: New system of rays
1 For$i \leftarrow 0$ **to** $n_p$
- (1) Generate a ray $r_i$ from reference point $p_i$ ( $r_i$ through $p_i$ and paralleled with the central line (See Fig. 2).

- (2) Make a boundary of reference points (DM's preferred region) and find the central point $p_c$.

For$j \leftarrow 0$ **to** $n$ (The number of rays)

- (3) Calculate the Euclid distance from ray(j) to $p_c$.

- (4) Sort the index of rays in decrease of Euclid distance values in (3) (Using QuickSort).
- (5) Replace top $n_p$ rays in the Sorted ray indexes with $n_p$ ray from (1).

**return** $n$ *rays.*;

---

# 6  Case Studies and Discussion

## 6.1  *Test Functions*

In our experiments, we use 07 popular test problems designed by Zitzler, Deb and Thiele: [18]with two, three objectives:

- ZDT1: MOP with two objectives, convex.
$$\mathbf{f_1}(\mathbf{x}) = \mathbf{x_1}; \mathbf{f_2}(\mathbf{x}) = \mathbf{g}(\mathbf{x})\left[1 - \sqrt{\frac{x_1}{g(x)}}\right]$$
$$\mathbf{g}(\mathbf{x}) = 1 + 9\frac{\sum_{i=1}^{n}x_i}{n-1}$$

  $n = 30$, variables bound: $[0,1], x \in [0,1]$,
  $x_i = 0, i = 2,\dots,n$.
- ZDT2: MOP with two objectives, non-convex.
$$\mathbf{f_1}(\mathbf{x}) = \mathbf{x_1}; \mathbf{f_2}(\mathbf{x}) = \mathbf{g}(\mathbf{x})\left[1 - \left(\frac{x_1}{g(x)}\right)^2\right]$$
$$\mathbf{g}(\mathbf{x}) = 1 + 9\frac{\sum_{i=1}^{n}x_i}{n-1}$$

  $n = 30$, variables bound: $[0,1], x \in [0,1]$,
  $x_i = 0, i = 2,\dots,n$.
- ZDT3: MOP with two objectives, convex and disconnected.
$$\mathbf{f_1}(\mathbf{x}) = \mathbf{x_1}; \mathbf{f_2}(\mathbf{x}) = \mathbf{g}(\mathbf{x})\left[1 - \sqrt{\frac{x_1}{g(x)}} - \frac{x_1}{g(x)}sin?(10\pi\mathbf{x_1})\right]$$
$$\mathbf{g}(\mathbf{x}) = 1 + 9\frac{\sum_{i=1}^{n}x_i}{n-1}$$

$n = 30$, variables bound: $[0,1], x \in [0,1]$,
$x_i = 0, i = 2, \ldots, n$.

- ZDT4: MOP with two objectives, non-convex.

$$\mathbf{f_1}(\mathbf{x}) = \mathbf{x_1}; \mathbf{f_2}(\mathbf{x}) = \mathbf{g}(\mathbf{x}) \left[ 1 - \sqrt{\frac{\mathbf{x_1}}{\mathbf{g}(\mathbf{x})}} - \frac{\mathbf{x_1}}{\mathbf{g}(\mathbf{x})} sin(10\pi\mathbf{x_1}) \right]$$

$$\mathbf{g}(\mathbf{x}) = 1 + 10(\mathbf{n} - 1) + \Sigma_{\mathbf{i}=2}^{\mathbf{n}} \left[ \mathbf{x_1^2} - 10cos(4\pi\mathbf{x_i}) \right]$$

$n = 10$, variables bound: $x_1 \in [0,1], x_i \in [-5,5]$.

- ZDT6: MOP with two objectives, Non-convex,Non-uniformly spaced.

$$\mathbf{f_1}(\mathbf{x}) = 1 - exp(-4\mathbf{x_1}) \, sin^6(6\pi\mathbf{x_1}); \mathbf{f_2}(\mathbf{x}) = \mathbf{g}(\mathbf{x}) \left[ 1 - \left( \frac{\mathbf{f}(\mathbf{x_1})}{\mathbf{g}(\mathbf{x})} \right)^2 \right]$$

$$\mathbf{g}(\mathbf{x}) = 1 + 9 \left[ \Sigma_{\mathbf{i}=2}^{\mathbf{n}} \frac{\mathbf{x_i}}{\mathbf{n}-1} \right]^{0.25}$$

$n = 10$, variables bound: $[0,1], x \in [0,1]$,
$x_i = 0, i = 2, \ldots, n$.

- DTLZ3:High dimension objective space, many local Pareto of Fronts.

$$f_1(x) = (1 + g(X_m)) * cos(x_1 * \pi/2) \ldots cos(x_{m-1} * \pi/2)$$
$$f_2(x) = (1 + g(X_m)) * cos(x_1 * \pi/2) \ldots sin(x_{m-1} * \pi/2))$$
$$\ldots$$
$$f_m(x) = (1 + g(X_m)) * sin(x_1 * \pi/2)$$

*where*

$$g(X_m) = 100 * (|X_m| + \Sigma_{x_i \in X_m}(x_i - 0.5)^2 - cos(20\pi(x_i - 0.5)))$$
$$x = (x_1, \ldots, x_n) \in [0,1]^n$$

- DTLZ7: High dimension objective space.

$$f_1(x) = x_1, f_2(x) = x_2 \ldots f_{m-1}(x) = x_{m-1}$$
$$f_m(x) = (1 + g(X_m)) * h(f_1, f_2, \ldots, f_{m-1}, g)$$

*where*

$$g(X_m) = 1 + \frac{9}{|X_m|} \Sigma_{x_i \in X_m} x_i, h = m - \Sigma_{i=1}^{m-1} \frac{f_i}{1 + g(X_m)(1 + sin(3\pi f_i))}$$
$$x = (x_1, \ldots, x_n) \in [0,1]^n$$

## 6.2  *Results and Discussion*

At the step 7 of DMEA, the estimated ideal point of the non-dominated solutions are identified in $M$ and determine a system of $n$ rays $R$. We replace this step with a function in 1 to make new rays and replace corresponding number of current rays which are the farthest to the DM's preferred region. New rays are generated from given reference points are shown in Figures: 4, 5 and 6:

In figures ( 4 to  6): The big dots are reference points which is given by DM in objective space, the bolded lines are new rays which is generated from DM's reference points. The final obtained solutions are presented by small dots.

Through experiments with 07 test functions, we can see some features of the interactive method:

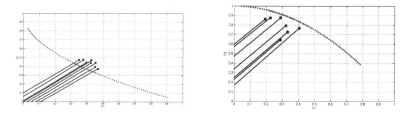

**Fig. 4** Results for the ray based interactive method on ZDT1(left) and ZDT2 (right)

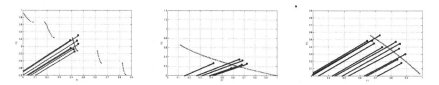

**Fig. 5** Results for the ray based interactive method on ZDT3(left), ZDT4 (middle) and ZDT6(right)

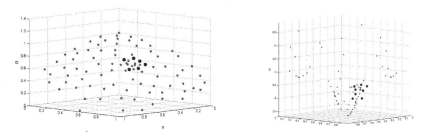

**Fig. 6** Results for the ray based interactive method on DTLZ3(left) and DTLZ7(right)

1. By Applying a niching to control external archive and next generation and replacing some rays in DM's preferred region, obtain solutions are strongly converged to DM's preferred region in objective space.
2. The final solutions are distributed uniformly outside DM's preferred region, except DM's unexpected region (region that contains farthest rays to DM's preferred region). It means DMEA with interactive still be balanced in maintaining two properties: convergence and spreading of population and indirectly balance between exploration and exploitation.

By using an interactive method with DMEA, where the system of rays are adjusted by replacing some rays which generated from reference points are given by DM. By applying a niching in step 9 and step 10, the final solutions will be strongly converged to the DM's preferred region. It ensures convergence and spreading of population and concept to use two kind of improvement directions in DMEA. With the interactive method help DM to get the most preferred solutions.

Our experiments were done at Software Technology Lab of Faculty of Information Technology, Le Quy Don Technical University.

## 7    Conclusion

In this paper, we proposed an interactive method using multi reference points with direction based multi-objective evolutionary optimization (DMEA). In our alternative method a set of new rays are generated from reference points given by DM in objective space. These rays will replace corresponding farthest rays to DM's preferred region. By applying a niching with new adjusted distribution of rays, the final solutions strongly converged to the DM's preferred region. It ensures convergence and spreading of population and concept to use two kind of improvement directions. With the interactive method help DM to get the most preferred solutions and concept of using two kind of improvement directions: Spread direction and Convergence direction.

**Acknowledgments.** We acknowledge the financial support from Vietnam's National Foundation for Science and Technology (Development Grant 107.04-2011.09).

## References

1. Wierzbicki, A.: The use of reference objectives in multi-objective optimisation. In: Proceedings of the MCDM Theory and Application. Lecture Notes in economics and mathematical systems, vol. 177, pp. 468–486 (1980)
2. Abbass, H.A.: An evolutionary artificial neural network approach for breast cancer diagnosis. Artificial Intelligence in Medicine 25(3), 265–281 (2002)
3. Bui, L.T., Liu, J., Bender, A., Barlow, M., Wesolkowski, S., Abbass, H.A.: Dmea: a direction-based multiobjective evolutionary algorithm. In: Memetic Computing, pp. 271–285 (2011)
4. Deb, K.: Multi-objective optimization using evolutionary algorithms. John Wiley & Sons, New York (2001)
5. Deb, K.: Multiobjective Optimization using Evolutionary Algorithms. John Wiley and Son Ltd., New York (2001)
6. Deb, K., Kumar, A.: Interactive evolutionary multi-objective optimization and decision-making using reference direction method. In: GECCO 2007, pp. 781–788 (2007)
7. Deb, K., Sinha, A., Korhonen, P.J., Wallenius, J.: An interactive evolutionary multi-objective optimization method based on progressively approximated value functions (2010)
8. Deb, K., Sundar, J.: Reference point based multi-objective optimization using evolutionary algorithms. In: GECCO 2006: Proceedings of the 8th Annual Conference on Genetic and Evolutionary Computation, pp. 635–642. ACM Press, New York (2006)
9. Gong, M., Liu, F., Zhang, W., Jiao, L., Zhang, Q.: Interactive moea/d for multi-objective decision making. In: GECCO 2011, pp. 721–728 (2011)
10. Branke, J.: Consideration of partial user preferences in evolutionary multi-objective optimization. In: Branke, J., Deb, K., Miettinen, K., Słowiński, R. (eds.) Multiobjective Optimization. LNCS, vol. 5252, pp. 157–178. Springer, Heidelberg (2008)

11. Talbi, E.-G., Wierzbicki, A.P., Figueira, J.R., Liefooghe, A.: A parallel multiple refer-
    ence point approach for multi-objective optimization. European Journal of Operational
    Research 205, 390–400 (2010)
12. Nguyen, L., Bui, L.T.: A decomposition-based interactive method formulti-objective
    evolutionary algorithms. The Journal on Information Technologies and Communications
    (JITC) 2(2) (2012)
13. Nguyen, L., Bui, L.T.: A multi-point interactive method for multi-objective evolutionary
    algorithms. In: The Fourth International Conference on Knowledge and Systems Engi-
    neering (KSE 2012), Da Nang, Vietnam (July 2012)
14. Nguyen, L., Bui, L.T., Abbass, H.: A new niching method for the direction-based multi-
    objective evolutionary algorithm. In: 2013 IEEE Symposium Series on Computational
    Intelligence, Singapore (April 2013)
15. Petri, E., Kaisa, M.: Trade-off analysis approach for interactive nonlinear multiobjective
    optimization. OR Spectrum, 1–14 (2011)
16. Thiele, L., Miettinen, K., Korhonen, P.J., Molina, J.: A preference based evolutionary
    algorithm for multi-objective optimization, pp. 411–436 (2009)
17. Belton, V., Branke, J., Eskelinen, P., Greco, S., Molina, J., Ruiz, F., Słowiński, R.: Inter-
    active multi-objective optimization from a learning perspective. In: Branke, J., Deb, K.,
    Miettinen, K., Słowiński, R. (eds.) Multiobjective Optimization. LNCS, vol. 5252, pp.
    405–433. Springer, Heidelberg (2008), OR Spectrum
18. Zitzler, E., Thiele, L., Deb, K.: Comparision of multiobjective evolutionary algorithms:
    Emprical results. Evolutionary Computation 8(1), 173–195 (2000)

# Phishing Attacks Detection Using Genetic Programming

Tuan Anh Pham, Quang Uy Nguyen, and Xuan Hoai Nguyen

**Abstract.** Phishing is a real threat on the Internet nowadays. According to a report released by an American security firm, RSA, there have been approximately 33,000 phishing attacks globally each month in 2012, leading to a loss of $687 million. Therefore, fighting against phishing attacks is of great importance. One popular and widely-deployed solution with browsers is to integrate a blacklist sites into them. However, this solution, which is unable to detect new attacks if the database is out of date, appears to be not effective when there are a lager number of phishing attacks created very day. In this paper, we propose a solution to this problem by applying Genetic Programming to phishing detection problem. We conducted the experiments on a data set including both phishing and legitimate sites collected from the Internet. We compared the performance of Genetic Programming with a number of other machine learning techniques and the results showed that Genetic Programming produced the best solutions to phishing detection problem.

## 1 Introduction

Genetic Programming (GP) [1, 2] is an evolutionary algorithm aimed to provide solutions to a user-defined task in the form of computer programs. Since its introduction, GP has been applied to many practical problems [1], producing a number of human competitive result [3]. GP has been used as a learning tool for solving

Tuan Anh Pham
Centre of IT, Military Academy of Logistics, Vietnam
e-mail: anh.pt204@gmail.com

Quang Uy Nguyen
Faculty of IT, Military Technical Academy, Vietnam
e-mail: quanguyhn@gmail.com

Xuan Hoai Nguyen
IT Research and Development Center, Hanoi University, Vietnam
e-mail: nxhoai@gmail.com

V.-N. Huynh et al. (eds.), *Knowledge and Systems Engineering, Volume 2,*                185
Advances in Intelligent Systems and Computing 245,
DOI: 10.1007/978-3-319-02821-7_18, © Springer International Publishing Switzerland 2014

some problems in network security [4, 5, 6]. However, to the best of our knowledge, there has not been any published attempt on the use of GP for learning to detect phishing web sites.

In the field of network security, phishing attack is one of the main threat on the Internet nowadays [7]. Phishing attackers attempt to acquire confidential information such as usernames, passwords, and credit card details by disguising as a trustworthy entity in an online communication. To achieve this, the attackers often use a set of web site spoofing techniques combining with social engineering to trick a user into revealing secret information with economic values. In a normal attack [7], a large number of spoofed (fake) emails are delivered to random Internet users. These emails are often masqueraded to appear coming from a legal business organization such as a bank. The receivers are provoked to update their personal information with the warning that the failure to comply with the request may result in the suspending of their online banking account. This kind of attack is very common and is an effective technique in convincing users.

Due to the simplicity, phishing attacks are very popular and become one of the major threat on the Cyber space today. According to a statistics reported by an American security firm, RSA, there have been nearly 33,000 phishing attacks globally each month in 2012, leading to a loss of $687 million [8]. Therefore, detecting and eliminating phishing attacks is very important for not only organizations but also individuals. One popular and widely-used solution with most web browsers is to integrate a blacklist sites into them. However, this solution, which is unable to detect a new attack if the database is out of date, appears to be not effective when there are a large number of phishing attacks carried out very day. In this paper, we propose a solution to this problem by applying Genetic Programming to phishing detection problem. We conduct the experiments on a data set including both phishing and legitimate sites collected from the Internet. We compare the performance of Genetic Programming with a number of other machine learning techniques.

The remainder of the paper is organised as follows. In the next section, we briefly review some previous research on detecting phishing attacks. In Section 3 we present our method using GP for solving the phishing detection problem. It is followed by a section detailing our experimental settings. The experimental results are shown and discussed in Section 5. The last section concludes the paper and highlights some potential future work.

## 2   Related Work

Since phishing attacks are very popular, there has been a number of anti-phishing solutions proposed to date. Some methods essay to solve the phishing problem at the email level by preventing users from visiting the phishing sites. That is, the emails containing phishing sites are filtered before being able to reach to the potential victims. Apparently, these techniques are closely related to anti-spam research and has been used by both Microsoft [9] and Yahoo [10]. Other solutions attempt to protect valuable information from being exposed to the phishers by replacing passwords with site-specific tokens, or by using novel authentication mechanisms. These methods have been used in some popular anti-phishing tools such as Pwd-Hash and AntiPhish. In PwdHash [11], a domain-specific password, that is rendered useless if it is submitted to another domain, is created (e.g., a password for www.gmail.com will be different if submitted to www.attacker.com). Conversely, AntiPhish [12] takes a different approach by keeping track of where confidential information such as a password is being submitted. That is, if it detects that a password is being entered into a form on an untrusted web site, a warning is generated and the current operation is destroyed.

In this paper, we will focus on the approaches that only use the information available from the URL and the pages source code. Currently, there are two main such approaches for identifying phising pages - based on URL blacklists; and based on the properties of the page and (sometimes) the URL.

*Blacklists*: Blacklists contain URLs that refer to malicious web sites. Whenever a browser loads a page, it sends a query to the blacklist to examine whether the current visiting URL is on the list. If it is, an appropriate countermeasure such as generating a warning can be taken. Otherwise, the page is considered benign. The blacklist can be stored at a central server or locally at the client.

Using blacklists as a mean to detect phishing is the most popular and widely-employed method. For example, Microsoft has used a blacklist-based anti-phishing solution in its Internet Explorer (IE) 7. The browser queries the lists of blacklisted sites from Microsoft servers and guarantee that the user is not accessing any phishing sites. Another browser that prevents phishing sites using blacklists is Google Safe Browsing [13]. When evaluating the efficiency of these methods, three important factors need to be considered [7]. The first one is the coverage, the phishing pages on the Internet are to be included in the list (recall). The second is the quality of the list. The quality indicates how many non-phishing sites (false positive) are incorrectly included into the list. Finally, the last factor that affects the effectiveness of a blacklist-based solution is the time for a new phishing site is included. In general, this approach is simple and widely used by many browsers, its disadvantage is that non blacklisted phishing sites are not recognized. Moreover, if there is new phising page on the internet (and the pace of creation of phising pages is high), it takes long time to include it in the updated version of the black listed sites.

*Page Analysis*: Page analysis techniques investigate properties of the web page and the URL to discriminate between phishing and legitimate sites. Page properties are typically extracted from the pages HTML source. For examples, some properties used in SpoofGuard [14] are the number of password fields, the number of links, or the number of unencrypted password fields. The effectiveness of page analysis approaches to determine phishing pages essentially depends on whether page properties exist that allow to differentiate between phishing and legitimate sites. Some previous research [14, 7, 15] evidenced that there are page properties that can support to distinguish the phishing from legitimate sites. In this paper, we follow the research by Ludl at el. [7] in identifying the list of pages properties (features) that are used by GP.

# 3 Methods

This section presents the methods used in this paper. The way to extract the features for each web site is presented first. After that, the GP system for phishing detection is described.

## 3.1 Features Extraction

The first step of using GP to tackle the phishing detection problem is features extraction/selection. This is a very important step since it may strongly affect the effectiveness of the learning algorithm (GP). The extracted features must contain information that helps to distinguish between phishing and legitimate sites. In this paper, we follow Ludl at el. [7] in selecting the following features from each site.

- **Form:** The number of forms (X1).
- **Input fields:** The number of input fields (X2).
- **Text fields:** The number of text fields (X3).
- **Password fields:** The number of password fields (X4).
- **Hidden fields:** The number of hidden fields (X5).
- **Internal images:** The number of internal image links (X6).
- **External images:** The number of external image links (X7).
- **Secure links:** The number of links (internal and external) over a secure connection (X8).
- **Secure images:** The number of image links (internal and external) over a secure connection (X9).
- **External references:** Sum of the external links and external images (X10).
- **Other links:** The number of links using $< link >$ tag (X11).
- **Script tags:** The number of JavaScript tags (X12).

## 3.2 System Description

The evolutionary learning process of GP for solving the problem of phishing detection is divided into two stages: training and testing. The objective of training stage is to evolve the model (the classifier) that can determine a site as either phishing or legitimate based on its feature values. In the testing stage, the learnt model is used to make predictions on the unseen data. The accuracy of these prediction is used as an indicator for the quality (effectiveness) of the model.

In the training stage, a set of training sites (both phishing and benign) with their labels (either as phishing or normal) are provided. The feature extraction process is called to convert every site to a feature vector. This vector is then served as the input for an individual in GP and the output of the individual is a real value. If this real value is greater than zero, this site is tagged as a phishing, otherwise it is considered as benign.

The next step in the training process is to measure the fitness of an individual in GP. In this paper, we use a simple way to measure the fitness of individual where the fitness is the percentage of sites in the training set that are correctly classified. This fitness, thought may not be a good indicator if the data is imbalance, is intuitive to identify the overall quality of a model.

## 4   Experimental Settings

This section outlines the settings used in our experiments. First, we present the way that data was collected for training and testing the systems. After that GP configurations for the experiments are described.

### 4.1   Data Collection

The data used for training and testing the system in this paper was collected from both phishing sites and legitimate sites on the Internet. To collect phishing data, we used web page *phishingtank.com*, which contains of an updating list of phishing URLs [7]. This site also provides a XML feed of verified and online phishing URLs. We extracted the first 2986 URLs from this XML file, and used a web crawling framework - Scrapy [16] to download HTML sources from these URLs. Because the phishing sites are often online for a short time, we need to remove all page sources that correspond to error pages. This process has finally resulted in 1450 page sources.

There are a large number of legitimate pages on the Internet. However, since phishers often focus on login pages [7], it is necessary that we collect these login pages to make the problem more difficult and realistic. These pages are achieved by manually making special queries search in Google. These queries were [7]: **inurl:login, inurl:logon, inurl:signin, inurl:signon,inurl:login.asp, inurl: login.php, inurl:login.htm**. We extracted 2800 URLs from the search results and downloaded page sources from them. After removing error pages, 2424 pages sources was obtained.

From the data set, twelve properties on each page were extracted to create the set feature vectors used for training and testing. We retained only one feature vector in case there is duplication in the data set. Moreover, if a feature vector presented in both phishing data and legitimate data, this vector was removed. As a result, 619 feature vectors for phishing and 244 feature vectors for legitimate data were retrieved.

Totally, we obtained 619+244=863 feature vectors of both phishing and legitimate sites. These vectors are mixed and divided into two sets: one for training (two third) and the other for testing (the rest). Finally, feature values were normalized to the range between $(0, 1)$, and the vectors extracted from phishing pages were labeled 1, otherwise labeled 0.

**Table 1** Run and Evolutionary Parameter Values

| Parameter | Value |
| --- | --- |
| Population size | 500 |
| Generations | 50 |
| Selection | Tournament |
| Tournament size | 7 |
| Crossover probability | 0.9 |
| Mutation probability | 0.1 |
| Initial Max depth | 6 |
| Max depth | 17 |
| Max depth of mutation tree | 5 |
| Non-terminals | +, -, *, / (protected version), |
|  | sin, cos, exp, iff, log (protected version) |
| Terminals | X1, X2, X3, X4, X5, X6, |
|  | X7, X8, X9, X10, X11, X12 |
| Raw fitness | percentage of correct classification |
| Trials per treatment | 100 independent runs for each value |

## 4.2  GP Parameters Settings

To tackle a problem with GP, several elements need to be clarified beforehand. These elements often depend on the problem and the experience of practitioners. The first and important element is the fitness function. As aforementioned, in this paper we use the percentage of correct classifications as the fitness measurement for each individual in the population.

Other factors that strongly affect the performance of GP are the set of non-terminals and terminals. The terminal sets include 12 variables ($X_1$, $X_2$,..., $X_{12}$) representing 12 features extracted from the sites. The non-terminal set include 9 functions (+, -, *, /, sin, cos, exp, iff, log). Here, we used the protected versions of division (/), meaning that if the denominator is zero, the returned value is 1. Similarly, the protected version of log was used: if the absolute of its parameter is zero, the returned value is 1. Other evolutionary parameters are presented in Table 1. These are typical values that are often used in the field of GP [2].

## 5  Results and Discussion

To determine quality of the models produced by GP, at the end of each run, we selected the best-of-the-run individual (the individual with the best fitness on the training set in the entire run). This model is then tested on the testing set and the output on the testing set is considered as the prediction error of the model.

We compare the results of GP with the results produced by a number of other machine learning techniques. These techniques include Decision Tree, Bayesian Networks and Artificial Neural Networks. They are detailed as follows.

*Decision Trees*: Decision trees are popular tools for classification and prediction [17]. The appealing property of decision trees is due to the fact that, in contrast to other methods, decision trees represent rules. Rules can easily be expressed so that humans can understand them or even directly use in a database access language like SQL. There have been a number of algorithms developed for decision trees induction. Among them, C4.5 is the most popular algorithm [18]. In this paper, we use an extension of C4.5 called J48 for constructing the decision tree for the problem. J48 has been implemented in Weka [19].

*Bayesian Networks*: Bayesian networks (BNs) are probabilistic graphical models for reasoning under uncertainty. In BNs the nodes represent random variables and arcs represent direct connections between them [20]. These direct connections are usually causal connections. In addition, BNs present the quantitative strength of the connections between variables. This allows the probabilistic beliefs about them to be updated automatically whenever new information is available. When constructing the network, two main components namely estimator and searching algorithm need to be considered. Estimator is a function used for evaluating a given network, and searching algorithm is once that searching through the space of possible networks. In our experiments, we used SimpleEstimator algorithm for estimator, and K2 algorithm for searching algorithm [19].

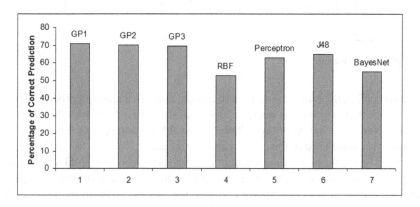

**Fig. 1** The percentage of correct prediction

*Artificial Neural Network*: An Artificial Neural Networks (ANN) is an information processing paradigm that is inspired by the way that the brain processes information [21]. The important component of this paradigm is the novel structure of the information processing system. It consists of a large number of highly interconnected processing elements (neurons) working together to solve problems. Similar to humans, ANNs learn by example. An ANN is constructed for a specific application,

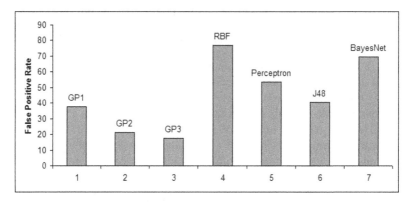

**Fig. 2** False positive rate

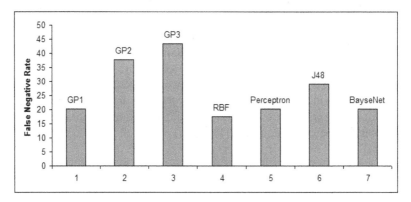

**Fig. 3** False negative rate

such as data classification or pattern recognition, through a learning process. There are several different kinds of neural network of these RBFNetwork (Radial Basis Function Network) and Multilayer Perceptron [19] are perhaps the most popular, therefore they were used in the experiments in this paper.

In order to use these machine learning techniques to solve the problem, we used their implementations in Weka. All parameters were selected as default settings in Weka and are described in [19]. We compare the results produced by these methods with three best results obtained by GP. The percentage of correct prediction, the false positive rate (the percentage of phishing sites that are misclassified) and the false negative rate (the percentage of legitimate sites misclassified) are presented in Figure 1, Figure 2, and Figure 3, respectively. In these figures, GP1, GP2 and GP3 are the three best models (classifiers) produced by GP. RBF is shorthanded for the Radial Basis Function Network while Perceptron stands for the Multilayer Perceptron network. Finally, J48 and BayseNet are the results of decision tree and Bayesian Network, respectively. It should be noted that in Figure 1, the greater values are better while in Figure 2 and Figure 3 the smaller values are better.

It can be seen from Figure 1 that three best models produced by GP are also the three best models among all models produced by all learning systems. Overall, the prediction accuracy of GP learnt models is about 70%. These values of other methods ranges from 55% to 65% with the lowest value is obtained by Radial Basis Function Network (RBF) while the highest value is obtained by decision tree (J48). This figure shows that GP can perform better than other machine learning techniques in solving this problem.

In terms of false positive rate, Figure 2 shows that the results of GP are often by far better than all other investigated machine learning techniques. The false positive rate of GP models are always smaller than 40% and even smaller than 20% with GP3. This result is quite significant in comparison with Firefox and Internet Explorer where it has been reported that based on blacklists data set [7], two these browsers have the false positive rate about 35% and 45%, respectively. In contrast to GP, all other machine methods produce worse results. The false positive rates of them are rather higher ranging from around 45% with J48 to nearly 80% with RBF.

Finally, the last figure shows the false negative rate of all methods. It can be observed from this figure that only GP1 is comparative with other methods while GP2 and GP3 are worse. In general, the results in this section show that GP can produce the better results in terms of the overall performance, the false positive rate while it is mostly equal to other techniques regarding to the false negative rate. Therefore, using the results of GP combined with blacklists-based browsers can potentially further enhance the ability of these browsers in protecting users from phishing attacks.

## 6 Conclusions and Future Work

In this paper, we conducted a first investigation on the use of Genetic Programming (GP) for solving the problem of detecting phishing attacks. The results were compared with three other machine learning techniques(Decision Trees, Baysian Networks and Artificial Neural Networks). The results show that GP is capable of producing the prediction models (classifiers) that are more accurate than other machine learning techniques. Moreover, GP also produces the models with rather low false positive rate compared to blacklists-based browsers and other machine learning methods. This result inspires us to get GP integrated with blacklists-based browsers to improve their ability in detecting phishing attacks.

In future, we are planning to extend the work in this paper in a number of ways. Firstly, we want to give GP more computational time (by increasing the population size) to see if it can help GP to find better models. Secondly, we want to use some recent advanced techniques, especially some technique for improving the generalization ability of GP [22] to see if they can further enhance the results. Last but not least, we want to make a more thorough analysis on the obtained models to get better understanding of the factors that affect the prediction accuracy.

**Acknowledgments.** The work in this paper was funded by The Vietnam National Foundation for Science and Technology Development(NAFOSTED), under grant number 102.01-2012.04.

# References

1. Poli, R., Langdonand, W., McPhee, N.: A Field Guide to Genetic Programming (2008), http://lulu.com
2. Koza, J.: Genetic Programming: on the Programming of Computers by Natural Selection. MIT Press, MA (1992)
3. Koza, J.: Human-competitive results produced by genetic programming. Genetic Programming and Evolvable Machines 11(3-4), 251–284 (2010)
4. Sen, S., Clark, J.A.: A grammatical evolution approach to intrusion detection on mobile ad hoc networks. In: WiSec 2009: Proceedings of the Second ACM Conference on Wireless Network Security, Zurich, Switzerland, March 16-19, pp. 95–102. ACM (2009)
5. Blasco, J., Orfila, A., Ribagorda, A.: Improving network intrusion detection by means of domain-aware genetic programming. In: International Conference on Availability, Reliability, and Security, ARES 2010, pp. 327–332 (February 2010)
6. Mabu, S., Chen, C., Lu, N., Shimada, K., Hirasawa, K.: An intrusion-detection model based on fuzzy class-association-rule mining using genetic network programming. IEEE Transactions on Systems, Man, and Cybernetics, Part C: Applications and Reviews 41(1), 130–139 (2011)
7. Ludl, C., McAllister, S., Kirda, E., Kruegel, C.: On the effectiveness of techniques to detect phishing sites. In: Hämmerli, B.M., Sommer, R. (eds.) DIMVA 2007. LNCS, vol. 4579, pp. 20–39. Springer, Heidelberg (2007)
8. RSA: Phishing in season: A look at online fraud in 2012 (2012), http://blogs.rsa.com/phishing-in-season-a-look-at-online-fraud-in-2012/
9. Microsoft: Sender id home page (2007), http://www.microsoft.com/mscorp/safety/technologies/senderid/default.mspx
10. Yahoo: Yahoo! antispam resource center (2007), http://antispam.yahoo.com/domainkeys
11. Ross, B., Jackson, C., Miyake, N., Boneh, D., Mitchell, J.C.: Stronger password authentication using browser extensions. In: Proceedings of the 14th USENIX Security Symposium, USENIX (August 2005)
12. Kirda, E., Krügel, C.: Protecting users against phishing attacks. Computer Journal 49(5), 554–561 (2006)
13. Schneider, F., Provos, N., Moll, R., Chew, M., Rakowski, B.: Phishing protection design documentation (2007), http://wiki.mozilla.org/PhishingProtection:DesignDocumentation
14. Chou, N., Ledesma, R., Teraguchi, Y., Mitchell, J.C.: Client-side defense against web-based identity theft. In: 11th Annual Network and Distributed System Security Symposium. The Internet Society (2004)
15. Blum, A., Wardman, B., Solorio, T., Warner, G.: Lexical feature based phishing URL detection using online learning. In: Greenstadt, R. (ed.) Proceedings of the 3rd ACM Workshop on Security and Artificial Intelligence, AISec 2010, pp. 54–60. ACM, Chicago (October 8, 2010)
16. Scrapy: Scrapy: web crawling framework, http://scrapy.org/
17. Quinlan: Learning decision tree classifiers. CSURV: Computing Surveys 28 (1996)

18. Quinlan, J.R.: C4.5: Programs for Machine Learning. Morgan Kaufmann, San Mateo (1993)
19. Witten, I.H., Frank, E.: Data Mining: Practical Machine Learning Tools and Techniques. Morgan Kaufmann (2005)
20. Heckerman, D.: Tutorial on learning in bayesian networks. Technical Report MSR-TR-95-06, Microsoft (1995)
21. Das, S.: Elements of artificial neural networks. IEEE Transactions on Neural Networks 9(1), 234–235 (1998)
22. Uy, N.Q., Hien, N.T., Hoai, N.X., O'Neill, M.: Improving the generalisation ability of genetic programming with semantic similarity based crossover. In: Esparcia-Alcázar, A.I., Ekárt, A., Silva, S., Dignum, S., Uyar, A.Ş. (eds.) EuroGP 2010. LNCS, vol. 6021, pp. 184–195. Springer, Heidelberg (2010)

# Solving Fuzzy Job-Shop Scheduling Problems with a Multiobjective Optimizer

Thanh-Do Tran*, Ramiro Varela, Inés González-Rodríguez, and El-Ghazali Talbi

**Abstract.** In real-world manufacturing environments, it is common to face a job-shop scheduling problem (JSP) with uncertainty. Among different sources of uncertainty, processing times uncertainty is the most common. In this paper, we investigate the use of a multiobjective genetic algorithm to address JSPs with uncertain durations. Uncertain durations in a JSP are expressed by means of triangular fuzzy numbers (TFNs). Instead of using expected values as in other work, we consider all vertices of the TFN representing the overall completion time. As a consequence, the proposed approach tries to obtain a schedule that optimizes the three component scheduling problems [corresponding to the lowest, most probable, and largest durations] all at the same time. In order to verify the quality of solutions found by the proposed approach, an experimental study was carried out across different benchmark instances. In all experiments, comparisons with previous approaches that are based on a single-objective genetic algorithm were also performed.

## 1 Introduction

Job-shop scheduling problems (JSPs) are known to be one of the hardest classes of combinatorial problems. They have formed an important body of research since the

Thanh-Do Tran · El-Ghazali Talbi
DOLPHIN Team, Inria Lille – Nord Europe and LIFL, Université Lille 1, France
e-mail: thanh-do.tran@inria.fr

Ramiro Varela
A.I. Centre and Department of Computer Science, University of Oviedo, Spain

Inés González-Rodríguez
Dept. of Mathematics, Statistics and Computing, University of Cantabria, Spain

\* TDT was supported by the ECSC Scholarship Program of the Master in Soft Computing and Intelligent Data Analysis at the European Centre for Soft Computing and, currently, is supported by the CORDI-S Doctoral Fellowship at Inria Lille.

late fifties, with multiple applications in industry, finance, and science [23]. In fact, JSPs are not only NP-complete but also among the worst NP-complete class members [21]. Therefore, they have usually been solved by using heuristic techniques, rather than exact methodologies.

During the past two decades, various proposals based on genetic algorithms have been introduced to solve large JSP instances: [4], [27], [22], and [15], to name but a few. Even though most proposals deal with crisp JSPs, i.e. all relevant information is assumed to be concrete, in many real-life situations, it is often the case that the exact duration of a task is not known in advance [10]. Instead, based on previous experience, an expert may have some knowledge about this duration and it is therefore possible to estimate this processing time. In such a situation, it is neither possible nor plausible to represent processing times with concrete numbers.

Depending on the available knowledge and the representation technique being used, the information about durations can be modeled by an interval for the possible processing times or its most typical values. When deeper knowledge of the problem is available, fuzzy intervals—which are considered as an alternative to probability distributions—can also be used; however, this technique usually requires complex computation [13]. When only little knowledge is available, we can use a confidence interval to represent the uncertain duration of a task. In this context, if some values in the interval appear to be more probable than others, it is natural to extend the representation to a fuzzy interval or fuzzy number [13].

In this work, following [10], triangular fuzzy numbers (TFNs) are employed to represent uncertain durations in a JSP. By this representation, only the *sum* and *maximum* operations are needed to calculate the *completion time of each task* in a job. Then, the *completion time of each job* is computed via a semi-active schedule builder [12]. This job completion time is also represented by a TFN. When completion times of all jobs have been determined, the *overall completion time of the JSP* (aka makespan) is taken as the maximum completion time over all the jobs. By taking the *expected value* of the overall completion time as an objective function, genetic and memetic algorithms have been successfully applied to search for a (near-) optimal schedule to the problem [11, 14, 24].

Being a single number, an expected value cannot fully represent the overall completion time that is expressed by a triangular fuzzy number. Consequently, using the expected value as an objective function implies an approximation of the problem to be solved. Such approximation might, to some extent, result in the loss of information; and therefore the obtained schedule might always be different from the true optimal one in some situations. To overcome this conspicuous drawback of the current techniques, we have investigated a new approach based on a multi-objective genetic algorithm.

This new approach enables us to take into account all three vertices of a TFN representing the overall completion time in the objective function. As a consequence, this approach considers at the same time three different scheduling problems corresponding to the lowest, most probable, and largest durations. With the use of a multi-objective genetic algorithm, the proposed approach tries to obtain a schedule that optimizes the three component scheduling problems all at the same time. In order to verify the quality of solutions found by the proposed approach, an experimental study has been carried out across different benchmark instances. Then a new proposal on analyzing the tolerance for the imprecision of knowledge representation is presented.

## 2   Job-Shop Scheduling with Uncertain Durations

### 2.1   Crisp Job-Shop Scheduling Problems

A general JSP [16] can be defined as scheduling a set of $n$ jobs $\{J_1, J_2, \cdots, J_n\}$ on a set of $m$ physical resources or machines $\{M_1, M_2, \cdots, M_m\}$, subject to a set of constraints. For a job $J_i$ to be completed, a series of tasks have to be done in a predefined order. Those tasks are enumerated by an index $j$, and, obviously, $j$ is at most equal to $m$. The job $J_i$ is then said to be composed of tasks $\theta_{ij}$'s, where $j = 1, 2, \cdots, m$. It is noteworthy that a task denoted by $\theta_{ij}$ does not imply that it will be processed on the $j$-th machine. The index $j$ is used only for task enumeration of the job $J_i$.

In such a general JSP, the predefined orders of tasks form *precedence constraints*; that is, $m$ tasks $\{\theta_{i1}, \theta_{i2}, \cdots, \theta_{im}\}$ of the job $J_i$, where $i = 1, 2, \cdots, n$, have to be sequentially scheduled. Also, there are *capacity constraints*; that is, each task $\theta_{ij}$ requires the uninterrupted and exclusive use of one of the machines for its whole processing time. A solution to this problem is a schedule $s$—which is an allocation of starting times for all the tasks. Such a solution, besides being feasible (i.e. all the constraints hold), has to be optimal according to some criteria, for instance, the makespan is minimal [10].

Without loss of generality, let us now consider a JSP instance of size $n \times m$ (i.e. $n$ jobs and $m$ machines), let $\mathbf{p}$ be a duration (aka processing time) matrix and $\mathbf{v}$ be a machine matrix such that $p_{ij}$ is the processing time of task $\theta_{ij}$ and $v_{ij}$ is the machine required by $\theta_{ij}$, where $i = 1, 2, \cdots, n$ and $j = 1, 2, \cdots, m$. Let $\sigma$ be a feasible *task processing order*, i.e. a lineal ordering of tasks which is compatible with a processing order of tasks that may be carried out such that all constraints hold. A feasible schedule $s$ may be derived from $\sigma$ using a semi-active schedule builder [17, 12]. Let $S_{ij}(\sigma, \mathbf{p}, \mathbf{v})$ and $C_{ij}(\sigma, \mathbf{p}, \mathbf{v})$ denote the starting and completion times of the task $\theta_{ij}$. According to the semi-active schedule builder, the starting and completion times can be computed as follows:

$$S_{ij}(\sigma, \mathbf{p}, \mathbf{v}) = C_{i(j-1)}(\sigma, \mathbf{p}, \mathbf{v}) \ \lor \ C_{rs}(\sigma, \mathbf{p}, \mathbf{v}), \tag{1}$$

$$C_{ij}(\sigma, \mathbf{p}, \mathbf{v}) = S_{ij}(\sigma, \mathbf{p}, \mathbf{v}) + p_{ij}, \tag{2}$$

where $\theta_{rs}$ is the task preceding $\theta_{ij}$ in the machine according to the processing order $\sigma$, $C_{i0}(\sigma, \mathbf{p}, \mathbf{v})$ is assumed to be zero and, analogously, $C_{rs}(\sigma, \mathbf{p}, \mathbf{v})$ is taken to be zero if $\theta_{ij}$ is the first task to be processed in the corresponding machine. The completion time of job $J_i$ will then be $C_i(\sigma, \mathbf{p}, \mathbf{v}) = C_{im}(\sigma, \mathbf{p}, \mathbf{v})$, and the makespan $C_{\max}(\sigma, \mathbf{p}, \mathbf{v})$ is the maximum completion time of any job under a given candidate schedule $\sigma$:

$$C_{\max}(\sigma, \mathbf{p}, \mathbf{v}) = \vee_{1 \leq i \leq n}[C_i(\sigma, \mathbf{p}, \mathbf{v})]. \tag{3}$$

For the sake of notation simplicity, we follow [10] to write $C_{\max}(\sigma)$ when the problem (hence $\mathbf{p}$ and $\mathbf{v}$) is fixed or even $C_{\max}$ when no confusion is possible.

## 2.2 Modeling Uncertain Durations with TFNs

In real-life applications, it is often the case that the exact duration of a task, i.e. the time it takes to be processed, is not known in advance. However, based on previous experience, an expert may have some knowledge, usually uncertain, about the duration. The most straightforward representation for uncertain durations would be a human-originated confidence interval. If some values appear to be more plausible than others, a natural extension is a fuzzy interval or fuzzy number [14]. The simplest model for this case is a triangular fuzzy number (TFN) [19], which use an interval $[p^1, p^3]$ of possible values and a modal value $p^2$ in between. For a TFN $A$, denoted by $A = (p^1, p^2, p^3)$, the membership function takes the following triangular shape:

$$\mu_A(x) = \begin{cases} 0, & x < p^1 \\ \frac{x-p^1}{p^2-p^1}, & p^1 \leq x \leq p^2 \\ \frac{x-p^3}{p^2-p^3}, & p^2 < x \leq p^3 \\ 0, & x > p^3. \end{cases} \tag{4}$$

In the JSP, we essentially need two operations on fuzzy quantities: the *sum* and the *maximum*. These operations are obtained by extending the corresponding operations on real numbers using the so-called *Extension Principle*. However, computing the resulting expression is cumbersome, if not intractable [10].

For the sake of simplicity and tractability of numerical calculations, we follow [7] to *approximate* the results of these operations by a TFN. In other words, we evaluate the sum and maximum operations on only three defining points of each TFN. The approximated sum coincides with the sum of TFNs as defined by the Extension Principle; thus, for any pair of TFNs $M$ and $N$, if $S = M + N$ denotes their sum, we have:

$$S = (m^1 + n^1, m^2 + n^2, m^3 + n^3). \tag{5}$$

Unfortunately, for the maximum of TFNs, there is no such simplified expression. For any two TFNs $M$ and $N$, let $F = N \vee M$ denote their [true] maximum, and $G = (m^1 \vee n^1, m^2 \vee n^2, m^3 \vee n^3)$ its *approximated value*, an illustration of the distinction between the maximum and its approximation is given in Fig. 1. It is interesting to note that such an approximated maximum can trivially be extended to the case of more than two TFNs [13].

Besides being simple, it is clear that this approximation possesses another nice property of preserving the support and modal value of the true maximum. It is however remarkable that, at all $\alpha$-cuts, the lower and upper bounds of the approximated maximum are either smaller than or equal to the lower and upper bounds of the true

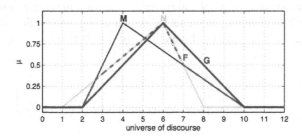

**Fig. 1** Exact (magenta) and approximated (red) maximum operations

maximum, respectively. More formally, let $[\underline{f}_\alpha, \overline{f}_\alpha]$ and $[\underline{g}_\alpha, \overline{g}_\alpha]$ be the $\alpha$-cuts of $F$ and $G$, respectively, it holds that:

$$\forall \alpha \in [0, 1], \quad \underline{f}_\alpha \leq \underline{g}_\alpha \text{ and } \overline{f}_\alpha \leq \overline{g}_\alpha. \tag{6}$$

In possibility theory, the membership function $\mu_Q$ of a fuzzy quantity $Q$ can be interpreted as a possibility distribution on real numbers; and this allows us to define the *expected value* of a fuzzy quantity. For a given TFN $A = (p^1, p^2, p^3)$, a typical model [20] for defining its expected value E[A] is given by:

$$E[A] = \frac{1}{4} \left( p^1 + 2p^2 + p^3 \right). \tag{7}$$

Importantly, the expected value coincides with the *neutral scalar substitute* of a fuzzy interval [28]. The neutral scalar substitute is among the most natural defuzzification procedures proposed in the literature [3]. The expected value can also be obtained as the center of gravity of its *mean value*, or using the *area compensation* method proposed by Fortemps and Roubens [8]. Most importantly, it induces a *total ordering* $\leq_E$ in the set of fuzzy intervals [3, 7]. For any two fuzzy intervals $M$ and $N$, $M \leq_E N$ if and only if $E[M] \leq E[N]$. Obviously, for any two TFNs $A = (a^1, a^2, a^3)$ and $B = (b^1, b^2, b^3)$, if $a^i \leq b^i \; \forall i$, then $A \leq_E B$; the reverse, however, does not hold.

## 2.3 Fuzzy Job-Shop Scheduling Problems

The fuzzy JSP considered in this study is the JSP with uncertain processing times (durations). Since processing times of operations are fuzzy intervals, the sum and maximum operations used to propagate constraints (in Eqs. 1 and 2) are taken to be the corresponding operations on fuzzy intervals, and approximated for the particular case of TFNs as explained in Sect. 2.2. The obtained schedule will be a fuzzy schedule in the sense that the starting and completion times of all tasks as well as the makespan are all fuzzy intervals. However, the task processing ordering $\sigma$ that determines the schedule $s$ is crisp—there is no uncertainty regarding the order in which the tasks have to be processed.

In order to demonstrate the graphical representation of a fuzzy JSP and a particular schedule for one of its instances, let's consider an instance with 3 jobs and 3
machines, having the following machine allocation and *fuzzy* processing time matrices:

$$\mathbf{p} = \begin{pmatrix} (3,4,6) & (2,3,4) & (1,2,5) \\ (1,2,4) & (2,3,4) & (1,2,3) \\ (2,3,5) & (2,3,4) & (1,2,4) \end{pmatrix}, \qquad \mathbf{v} = \begin{pmatrix} 1 & 2 & 3 \\ 1 & 3 & 2 \\ 2 & 1 & 3 \end{pmatrix}.$$

For a task processing order $\sigma = (\theta_{31}, \theta_{11}, \theta_{32}, \theta_{12}, \theta_{21}, \theta_{33}, \theta_{13}, \theta_{22}, \theta_{23})$, we
have the corresponding Gantt chart as shown in Fig. 2. This Gantt chart uses each
particular color for all tasks that belong to each particular job; tasks associated with
different jobs are accordingly colored differently.

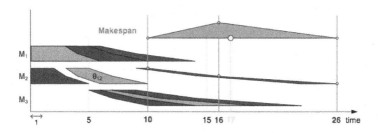

**Fig. 2** A sample Gantt chart for a fuzzy JSP instance

Since we could build a feasible schedule $s$ from a feasible task processing order
$\sigma$, we would therefore restate the goal of the fuzzy job-shop problem as finding an
*optimal* task processing order $\sigma$, in the sense that the makespan for the schedule
derived from that task processing order is *minimal*.

In [10], the authors employed a single-objective genetic algorithm that is enhanced by a local search to optimize the expected value of the makespan. However,
the expected value does not account for the width of a makespan. For instance,
the expected value cannot distinguish between two fuzzy makespans $A = (2,6,10)$
and $B = (4,6,8)$, since both have the same expected value 6, but one would think
$B = (4,6,8)$ is better because it is less uncertain and thus it provides more accurate
information on the possible values of the makespan.

On the other hand, the approximated maximum operator has identical support and
modal value to its exact version. As a consequence, the induced fuzzy makespan
also has identical support and modal value to the exact fuzzy makespan. Thanks
to this nice property, we could try to take into consideration the three *exact* vertices of a fuzzy makespan, instead of its *approximated* expected value, in searching
for an optimal schedule to the JSP. Such an idea naturally calls for the use of a
multi-objective optimization algorithm. As more information from the exact fuzzy
makespan is considered, the schedule obtained under this perspective is expected to
be more reliable and robust than those returned when considering only the modal
value or the expected value of the makespan.

## 3 An Evolutionary Approach to the Fuzzy JSPs

In order to apply a genetic algorithm to solving a combinatorial problem in general and a JSP in particular, we need to define a chromosome encoding strategy. Among different proposals available in the literature, the two most popular encoding schemes are the conventional permutations (CP) [2] and permutations with repetition (PR) [1]. In both cases, a chromosome expresses a total ordering of all operations of the problem [26]. For example, if we have a problem with $n = 3$ jobs and $m = 4$ machines, one possible ordering is given by the permutation $(\theta_{21}, \theta_{11}, \theta_{12}, \theta_{31}, \theta_{32}, \theta_{22}, \theta_{33}, \theta_{13}, \theta_{23}, \theta_{24}, \theta_{14}, \theta_{34})$, where $\theta_{ij}$ represents the task $j$ of the job $i$ (where $i = 1, 2, 3$; and $j = 1, 2, 3, 4$). In the CP scheme, operations are codified by the numbers $1, 2, \cdots, n \times m$, starting from the first job, so that the previous ordering would be codified by the chromosome $(5, 1, 2, 9, 10, 6, 11, 3, 7, 8, 4, 12)$. Whereas in the PR encoding scheme, an operation is codified just by its job number; hence the previous order would be given by $(2, 1, 1, 3, 3, 2, 3, 1, 2, 2, 1, 3)$ [26]. Also in [26], the authors have demonstrated that PR scheme is better than the CP. Taking that result, we will use the PR encoding scheme in this work.

After initialized randomly, each chromosome undergoes the crossover and mutation stages. New chromosome will then be created and selected to the next generation based on their fitness values—which are the expected makespans of the schedule they encode—by the well-known binary tournament selection strategy. To create new offspring, the job order crossover (JOX) [1, 12] and a simple swap mutation are employed. Specifically, given any two parents, the JOX selects a random subset of jobs from the first parent and copies the associated genes to the offspring at the same positions as they appear in that parent. Then, the remaining genes are taken from the second parent such that they maintain their relative ordering. The second offspring is produced in the same manner but considering the second parent first. Following the JOX, the simple swap mutation operator randomly selects and swaps two consecutive genes that encode two different jobs (i.e. having different values). For the single-objective GA, the generational-with-elitism survivor selection is employed; that is, the whole offspring population replaces the parent population except for the best chromosome in the parent population being copied directly to the offspring population.

The feasible search space for a JSP is usually very large; we therefore need to enrich GAs with some advanced algorithm that can somehow limit the feasible space to a narrower one but still guarantee the existence of optimal schedules. The algorithm proposed by Giffler and Thompson [9]—which is commonly referred to as the G&T algorithm—is the best known algorithm for that aim. In fact, this algorithm can be regarded as a transfer function that transforms a candidate schedule to a very similar yet better one in terms of makespan. In this sense, different candidate schedules might be transformed to a unique better schedule by this algorithm. In this work, for both single- and multi-objective GAs, we use the extended G&T algorithm for the fuzzy JSP as proposed in [13].

Our main aim in this work is to optimize simultaneously the three vertices of a makespan. Thus, we need to employ a technique in evolutionary multiobjective optimization (EMO) to handle these three objectives all at the same time, and to evolve a population of candidate solutions over generations in such a way that they get gradual improvements in all objectives. With the framework presented above for a single-objective GA, it is straightforward to extend the algorithm to a multiobjective version by the application of one of the existing EMO techniques. Among various EMO algorithms, the non-dominated sorting genetic algorithm (NSGA-II) [25, 5, 6] has gained a lot of popularity in the last few years, and becomes a landmark against which other EMO algorithms are often compared. In this work, we will therefore employed NSGA-II as a multiobjective optimizer to address the fuzzy JSP. The job order crossover and simple swap mutation described above will replace the simulated binary crossover and polynomial mutation in the original design of NSGA-II [6].

## 4  Analyzing the Tolerance for Knowledge Representation

A set of experiments has been conducted to examine the ability of the proposed approach to tolerate the imprecision inherent in representing the expert knowledge about task processing times by TFNs. This imprecision is due to the fact that the expert might not be completely sure about his specification of the fuzzy numbers. In other words, the specification of a fuzzy number is fuzzy itself. The location of the modal value as well as the support of a TFN is naturally imprecise. Accordingly, for a certain task, the processing time might be specified by various fuzzy numbers that are slightly different from each other. Another context can also be the case, that is, when more-than-one experts are jointly specifying the processing times of tasks. Obviously, they may have different knowledge about the tasks, and therefore the TFNs specified by each of them might be different from those specified by the others.

In the experiments, we have selected three typical benchmark instances from a famous library of 40 instances proposed by Lawrence [18]. On the other hand, three genetic algorithms that share common components have been implemented to enable a fairer comparative analysis. These algorithms are: (1) GAcrisp is the single-objective genetic algorithm considering only the most probable processing time of tasks, which is equivalent to a crisp JSP; (2) GAfuzzy is the single-objective genetic algorithm considering the expected value of the fuzzy makespan as the objective function; and (3) NSGAII is the multi-objective genetic algorithm (i.e. NSGA-II) considering at the same time the three vertices of a fuzzy makespan as its objectives. In all algorithms, we have used the binary tournament selection, the JOX with a crossover rate of 0.85, and the simple swap mutation with a mutation rate of 0.1; a randomly initialized population with 100 chromosomes will evolve across 100 generations.

The three benchmark instances that have been used are LA04 ($10 \times 5$), LA09 ($15 \times 5$), and LA18 ($10 \times 10$). For each benchmark, we randomly sample 10 fuzzy instances in the following manner. The modal value ($p^2$) for each fuzzy processing time ($p_{ij}$) is sampled uniformly at random in an interval having the crisp processing time as its center ($p$), with the lower and upper bounds being 95%$p$ and 105%$p$. Then, the left ($p^1$) and right ($p^3$) extremes of that fuzzy processing time is uniformly sampled at random in the intervals [50%$p$, 95%$p$] and [105%$p$, 150%$p$], respectively. An example to illustrate this procedure is given in Fig. 3. In this illustration, the red triangle is a randomly sampled fuzzy processing time, which has the three vertices being sampled in the blue, violet, and green intervals.

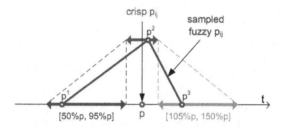

**Fig. 3** Sampling fuzzy processing time

For each benchmark, each of the ten sampled fuzzy instances is tested on the three algorithms, repeated 10 times with different initial populations in each repetition. With 3 algorithms tested on 3 benchmarks having 10 sampled fuzzy instances for each, and repeated 10 runs, we have done $3 \times 3 \times 10 \times 10 = 900$ runs.

For GAcrisp and GAfuzzy, we obtain a single best solution in each of the runs. This is however not the case for NSGAII; NSGAII returns a set of non-dominated solutions in the last generation. To facilitate the comparison with GAcrisp and GA-fuzzy, a single *best* solution from the obtained Pareto-optimal set must be nominated. For that purpose, we calculate the expected value of each solution in the Pareto-optimal set and select the one with the best (minimal) expected value of the makespan. In this way, we could reach a single best solution in each run of all the algorithms, and the results from the 10 runs can be summarized by box plots as presented in Fig. 4 for LA04. In this figure, each box contains 10 results from the 10 runs of a corresponding algorithm on a single sampled fuzzy instance. Also, the three algorithms referred to as Crisp, Fuzz, and NSGAII are the GAcrisp, GAfuzzy, and NSGAII, respectively. In addition, the ten sampled fuzzy instances are enumerated as Fuzz Samp 1 to Fuzz Samp 10.

Taking a look at Fig. 4, it is not difficult to realize that Crisp—which is the GA working with only the most probable task processing time in the sampled fuzzy instances—has makespans varying from instance to instance; whereas, the other two algorithms, i.e. Fuzz and NSGAII, are less sensitive to the random sampling. In addition, their expected makespans are much better (lower) than those of the Crisp on average. It should be noticed that, due to the space limitation, only the boxplot for LA04 is presented here; the similar plots for LA09 and LA18 also exhibit completely the same trend.

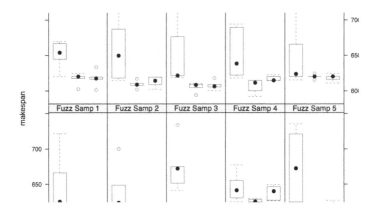

**Fig. 4** Results on asymmetrically sampled fuzzy versions of LA04 (10×5)

The comparison between Fuzz (i.e. the GA working with the expected fuzzy makespan) and NSGAII (i.e. the NSGA-II working with the three vertices of the fuzzy makespan) is however not intuitive. In fact, their results are close together for LA04 and LA18; and for LA09, their results look identical. (Notice again that the results for LA09 and LA18 are not shown here). Besides, the boxes for Crisp is larger than those for the other algorithms, which would suggest that under the current experimental setting of the algorithms (100 chromosomes evolved over 100 generations), Fuzz and NSGAII converge better than Crisp. Consequently, their results from different runs do not vary so much as those of the Crisp. Nonetheless, we have initialized the three algorithms by the same random seeds and they also share the same common structural components of the GA. In such a context, a more plausible explanation could be that, taking into consideration the triangular fuzzy processing time of tasks instead of only the most probable one, we could always gain benefits regardless of whether the expected value or all the three vertices of a makespan is utilized as the objective function(s).

What still remains interesting to know is how these algorithms perform on average in terms of the mean and variation of makespans over all the sampled fuzzy instances. To answer this question, we first take the mean (or median, alternatively) of each box, i.e. we are averaging the 10 runs. Then we calculate the mean and standard deviation of these ten means (or ten medians, alternatively). In other words, we are averaging the results of the 10 sampled fuzzy instances. As we have run each test only 10 times, the use of median could be a better estimate of the actual performance of the algorithms. Respective results from these calculations are fully presented in Table 1.

The results shown in Table 1 suggest that NSGA-II has a promising ability to tolerate the imprecision in representing the expert's knowledge about the fuzzy processing time. Under different randomly sampled fuzzy processing time, the final results of NSGA-II clearly exhibit less variation than those of GAcrisp on all the benchmarks. However, more extensive experiments are advocated in order to draw

**Table 1** Mean and standard deviation of makespans calculated on the mean/median of all runs over all sampled fuzzy instances

| Averaging over all sampled fuzzy instances | | **mean** over 10 runs | | **median** over 10 runs | |
|---|---|---|---|---|---|
| | | mean | std | mean | std |
| LA04 | GAcrisp | 655.9 | 11.14 | 642.3 | 19.49 |
| | GAfuzzy | 614.3 | 6.56 | 614.7 | 5.9 |
| | NSGAII | 617.4 | 8.34 | 616.8 | 9.34 |
| LA09 | GAcrisp | 980.9 | 11.04 | 975.1 | 14.23 |
| | GAfuzzy | 952.8 | 7.51 | 952.8 | 7.51 |
| | NSGAII | 952.8 | 7.51 | 952.8 | 7.51 |
| LA18 | GAcrisp | 958.8 | 13.3 | 956.3 | 21.28 |
| | GAfuzzy | 886.7 | 12.09 | 884.5 | 14.3 |
| | NSGAII | 894.2 | 8.16 | 891.6 | 9.74 |

a further conclusion about whether NSGA-II is better than the single-objective GA using the expected value of the makespan as its objective function. In fact, NSGAII has the same variation as GAfuzzy on the benchmark LA09, and a better (smaller) variation on LA18, but a worse (larger) variation on LA04, in comparison with GA-fuzzy.

## 5 Conclusions

In this work, we have investigated the application of a multi-objective genetic algorithm — the NSGA-II — to solving JSPs with uncertain durations, where uncertainty is modeled by triangular fuzzy numbers. The novelty of the investigation is that, we have considered the three vertices of a triangular fuzzy makespan all at the same time as three objectives of NSGA-II, rather than just using a representative which is the expected value as in previous work. Such a new proposal is often preferable to the existing approach in terms of offering the decision maker more options in selecting a scheduling strategy according to his preference to the earliest, most probable, or latest completion time. To validate the proposed multi-objective approach, a set of experiments has been performed. Even though the simulation results on a limited number of benchmarks do not strongly demonstrate the superiority of the proposal, they have provided some evidence for the imprecision tolerance ability of the obtained schedules with respect to the knowledge representation of the experts. The results have also suggested that a more comprehensive validation on a larger set of benchmark instances as well as a more extensive simulation would bring about a clear insight into the difference between the proposal and other available approaches.

# References

1. Bierwirth, C.: A generalized permutation approach to job shop scheduling with genetic algorithms. OR Spektrum 17(2-3), 87–92 (1995)
2. Bierwirth, C., Mattfeld, D.C.: Production scheduling and rescheduling with genetic algorithms. Evolutionary Computation 7(1), 1–17 (1999)
3. Bortolan, G., Degani, R.: A review of some methods for ranking fuzzy subsets. Fuzzy Sets and Systems 15(1), 1–19 (1985)
4. Cheng, R., Gen, M., Tsujimura, Y.: A tutorial survey of job-shop scheduling problems using genetic algorithms—I. representation. Computers & Industrial Engineering 30(4), 983–997 (1996)
5. Deb, K., Agarwal, S., Pratap, A., Meyarivan, T.: A fast elitist non-dominated sorting genetic algorithm for multi-objective optimization: NSGA-II. In: Schoenauer, M., Deb, K., Rudolph, G., Yao, X., Lutton, E., Merelo, J.J., Schwefel, H.-P. (eds.) PPSN 2000. LNCS, vol. 1917, pp. 849–858. Springer, Heidelberg (2000)
6. Deb, K., Pratap, A., Agarwal, S., Meyarivan, T.: A fast and elitist multiobjective genetic algorithm: NSGA-II. IEEE Transactions on Evolutionary Computation 6(2), 182–197 (2002)
7. Fortemps, P.: Jobshop scheduling with imprecise durations: a fuzzy approach. IEEE Transactions on Fuzzy Systems 5(4), 557–569 (1997)
8. Fortemps, P., Roubens, M.: Ranking and defuzzification methods based on area compensation. Fuzzy Sets and Systems 82(3), 319–330 (1996)
9. Giffler, B., Thompson, G.L.: Algorithms for solving production-scheduling problems. Operations Research 8(4), 487–503 (1960)
10. Gonzalez-Rodriguez, I., Puente, J., Vela, C.R., Varela, R.: Semantics of schedules for the fuzzy job-shop problem. IEEE Transactions on Systems, Man and Cybernetics, Part A: Systems and Humans 38(3), 655–666 (2008)
11. Gonzalez-Rodriguez, I., Vela, C.R., Puente, J.: A memetic approach to fuzzy job shop based on expectation model. In: IEEE Int. Conf. on Fuzzy Systems, pp. 1–6 (2007)
12. González, M.A., Vela, C.R., Varela, R.: Scheduling with memetic algorithms over the spaces of semi-active and active schedules. In: Rutkowski, L., Tadeusiewicz, R., Zadeh, L.A., Żurada, J.M. (eds.) ICAISC 2006. LNCS (LNAI), vol. 4029, pp. 370–379. Springer, Heidelberg (2006)
13. González, M.A., Vela, C.R., Puente, J.: A genetic solution based on lexicographical goal programming for a multiobjective job shop with uncertainty. Journal of Intelligent Manufacturing 21(1), 65–73 (2010)
14. González, M.A., Vela, C.R., Puente, J., Hernández-Arauzo, A.: Improved local search for job shop scheduling with uncertain durations. In: Nineteenth Int. Conf. on Automated Planning and Scheduling (ICAPS 2009), pp. 154–161 (2009)
15. Gonalves, J.F., de Magalhes Mendes, J.J., Resende, M.G.C.: A hybrid genetic algorithm for the job shop scheduling problem. European Journal of Operational Research 167(1), 77–95 (2005)
16. Jain, A.S., Meeran, S.: Deterministic job-shop scheduling: Past, present and future. European Journal of Operational Research 113(2), 390–434 (1999)
17. Jensen, M.T.: Improving robustness and flexibility of tardiness and total flow-time job shops using robustness measures. Applied Soft Computing 1(1), 35–52 (2001)
18. Lawrence, S.: Supplement to "Resource constrained project scheduling: An experimental investigation of heuristic scheduling techniques". Tech. rep., GSIA, Carnegie Mellon University, Pittsburgh PA (1984)

19. Lin, F.T., Yao, J.S.: Using fuzzy numbers in knapsack problems. European Journal of Operational Research 135(1), 158–176 (2001)
20. Liu, B., Liu, Y.K.: Expected value of fuzzy variable and fuzzy expected value models. IEEE Transactions on Fuzzy Systems 10(4), 445–450 (2002)
21. Nakano, R., Yamada, T.: Conventional genetic algorithm for job shop problems. In: Proceedings of ICGA, pp. 474–479 (1991)
22. Park, B.J., Choi, H.R., Kim, H.S.: A hybrid genetic algorithm for the job shop scheduling problems. Computers & Industrial Engineering 45(4), 597–613 (2003)
23. Pinedo, M.L.: Scheduling: Theory, Algorithms, and Systems, 3rd edn. Springer (2008)
24. Puente, J., Vela, C.R., González-Rodríguez, I.: Fast local search for fuzzy job shop scheduling. In: Proceedings of ECAI 2010, pp. 739–744. IOS Press (2010)
25. Srinivas, N., Deb, K.: Muiltiobjective optimization using nondominated sorting in genetic algorithms. Evolutionary Computation 2(3), 221–248 (1994)
26. Varela, R., Serrano, D., Sierra, M.R.: New codification schemas for scheduling with genetic algorithms. In: Mira, J., Álvarez, J.R. (eds.) IWINAC 2005, Part II. LNCS, vol. 3562, pp. 11–20. Springer, Heidelberg (2005)
27. Vázquez, M., Whitley, D.: A comparison of genetic algorithms for the static job shop scheduling problem. In: Schoenauer, M., Deb, K., Rudolph, G., Yao, X., Lutton, E., Merelo, J.J., Schwefel, H.-P. (eds.) PPSN 2000. LNCS, vol. 1917, pp. 303–312. Springer, Heidelberg (2000)
28. Yager, R.R.: A procedure for ordering fuzzy subsets of the unit interval. Information Sciences 24(2), 143–161 (1981)

# A Multi-objective Approach for Vietnamese Spam Detection

Minh Tuan Vu, Quang Anh Tran, Quang Minh Ha, and Lam Thu Bui

**Abstract.** In this paper, we propose a multi-objective approach for generating sets of feasible trade-off solutions for the Vietnamese anti-spam system (using SpamAssassin). The two objectives for considering are the Spam Detection Rate (SDR) and False Alarm Rate (FAR).The experiments were conducted based on Vietnamese spam data set through three scenarios with different numbers of SpamAssassin rules; and we used the non-dominated sorting genetic algorithm (version 2) – NSGA-II for finding the trade-off solutions. The result of each scenario was recorded to compare with the performance of the traditional approach (single objective optimization). According to the statistical results, the new approach not only achieved more efficient results but also created a set of ready-to-use rule scores which supports different levels of the trade-off between SDR and FAR.

## 1 Introduction

In recent years, when the spread of spams seems to be fierce and uncontrollable, researchers all around the world has managed to stop spammers from annoying email users by proposing a wide range of Anti-Spam solutions. For each solution with different approach, the pros and cons are various. There are also a number of factors to evaluate the efficiency of solutions. Among them, the Spam Detection Rate (SDR) and the False Alarm Rate (FAR) seems to be most obvious criteria to measure the effectiveness of a spam detection resolution.

The final purpose of any Anti-Spam approach is to maximize the SDR and to minimize the FAR as much as possible. The key point of problem is that the SDR

Minh Tuan Vu · Quang Anh Tran · Quang Minh Ha
Faculty of Information Technology, Hanoi University, Vietnam
e-mail: {minhtuan_fit,anhtq,minhhq_fit}@hanu.edu.vn

Lam Thu Bui
Le Quy Don Technical University, Vietnam
e-mail: lam.bui07@gmail.com

V.-N. Huynh et al. (eds.), *Knowledge and Systems Engineering, Volume 2,*
Advances in Intelligent Systems and Computing 245,
DOI: 10.1007/978-3-319-02821-7_20, © Springer International Publishing Switzerland 2014

is proportional to the FAR. Thus, the higher rate of detecting spam an approach brings the higher probability to alarm a ham (non-spam mail) as spam it gets and vice versa. An effective spam detection system is not expected to gain an absolute optimum which are 100% for SDR and 0% for FAR, but it is an acceptable trade-off between these criteria. Current approaches achieve the desired SDR (or FAR) by the following procedure:

1. A threshold at which an email is considered to be spam is predefined.
2. Model is built to train the system.
3. SDR (or FAR) is measured to evaluate the effectiveness of Anti-Spam solution at specific thresholds.

With this procedure, the only way to optimize the SDR and FAR without changing the model is to change the threshold. If email users' demand on the SDR and FAR are different, the threshold needs changing until matching their demands. For each time the threshold change, the whole training process is required to restart and consumes a lot of time.

In considering the concern of current Anti-Spam approach, the authors have applied the evolutionary multi-objective optimization algorithm –MOEA to solve the problem of SDR and FAR in Vietnamese spam detection. MOEAs have become popular as the solver for a number of multi-objective problems in different fields [1].By analyzing the nature of Anti-Spam problem and a wide range of MOEAs, authors figured out that NSGA-II [2] was suitable to build the framework and carry out the experiment. The performance of the algorithm was evaluated in [3] and said to outperform among other MOEAs.

The authors believe that the paper's contributions are two-fold. First of all, a set of Pareto is obtained. With this set of solutions, email users would have a list of SDR and FAR options for their different spam filtering demands. Each solution is available and ready-to-use without requiring retraining the dataset from the beginning. Secondly, Anti-Spam systems are provided a new approach to deal with the optimized tradeoff between SDR and FAR. The result of the paper illustrated that this approach was much more flexible and brought more satisfied results than single-objective optimization algorithms This paper is structured as follows: Section 2 introduced the background knowledge of the research. Section 3 explained the theoretical framework. Next, we presented the experiments and remarkable results in Section 5. Finally, the last section concluded the paper and talked about the future of our works.

## 2 Preliminaries

### 2.1 SpamAssassin Rules

SpamAssassin is one of the most popular mail filter developed by the Apache Software Foundation. It examines the message represented to it and assign a score to indicate the likelihood that the mail. SpamAssassin works basing on the predefined set of rules. A score is assigned to a rule. An email is marked as spam only when gaining enough the score which is greater than the threshold. Here is how a SpamAssassin looks like:

header FROM_STARTS_WITH_NUM From =~ /^\d\d/
describe FROM_STARTS_WITH_NUM From: starts with nums
score FROM_STARTS_WITH_NUM 0.390 1.574 1.044 0.579

The rule's name is FROM_START_WITH_NUMS. By applying the rule, SpamAssassin will examine whether the message's FROM header starts with at least two numbers against the regular expression. The score is added to the message's spam score if matching the rule. An anatomy of a rule was described in details by Schwartz (2004) [6].

## 2.2 NSGA-II

NSGA-II is an elitism algorithm introduced by Kaylyanmoy Deb in 2001[2]. The external set size (archive) equals to the initial population size. The current archive is determined based on the combination of the current population and the previous archive. The population is considered as a combination of several layers in such a way that the first layer is the best layer in the population by the dominance ranking.

The archive is formed against the order of ranking layers: Selecting the best ranking first. If the number of individuals in the archive is smaller than the size of population, the next layer will be taken into account and so on. A truncation operator is applied to that layer based on the crowding distance if adding a layer would increase the number of individuals in the archive to exceed the initial population size. Thus, the crowding distance of a solution x is the averaged total of objective-value differences between two adjacent solutions of the solution x, where the population is arranged according to each objective to find adjacent solutions and where also boundary solutions have infinite values. The truncation operator removes the individual with the smallest crowding distance.

An offspring population of the same size as the initial population is then created from the archive by using crowded tournament selection, crossover, and mutation operators. The crowded tournament selection rule is that the winner of two same-rank solutions is the one that has the greater crowding distance value [3].

## 3 Theoretical Framework

### 3.1 Problem

As mentioned in the introduction, the main concern of the traditional Anti-Spam approach is difficult and time-consuming to find out the optimized tradeoff between values of SDR and FAR if the threshold changes. If the set of spam detection rules remains unchanged, there is only one pair of values for SDR and FAR which are considered as the most wanted solution at a specific threshold. When the algorithm runs with different thresholds, the rule's scores (optimized for the pre-defined threshold) are no longer optimized for the current threshold which would cause the rate of spam detection and false alarm not optimized anymore. The training process must restart from the beginning to meet the email users' demand on various SDR and FAR.

## 3.2 Solution Design

This paper applied NSGA-II algorithm to solve the problem with two objectives: SDR and FAR. The first objectives SDR must be maximized while the second one FAR must be minimized.

**Step 1:** *Initialize the data input*
For the problem, the objective is also to find a set of ideal scores called x where

$$x = (x_1, .., x_m), \ m = (31, 51, 101), \ x_1 \in [2, 5], \ x_{2...m} \in [0, 2].$$

The set of x will be generated randomly with a random algorithm which is a part of NSGA-II. Each value inside the set is considered as a chromosome. The first value is set limitation from 2 to 5 because it is the threshold – the point at what an email is considered as spam. The other values are set from 0 to 2 which are the score of SpamAssassin rules. Experiments were carried out with three cases (three different numbers of x): 30 rules and 1 threshold (m = 31), 50 rules and 1 threshold (m = 51), 100 rules and 1 threshold (m = 101).

**Step 2:** *Create the objective function*
The objective function is designed to run on the spam dataset S (231 Vietnamese spam) and ham dataset H (251 Vietnamese ham).

$$S = \{s_1, s_2, .., s_K\}$$

$$H = \{h_1, h_2, .., h_L\}$$

The set of N rules is pre-designed based on the framework in [4].

$$R = \{r_1, r_2, .., r_N\}$$

Each rule might match with some spams or hams through the matching function.

$$m(r, e) = \begin{cases} 1 \ if\_r\_matches\_e \\ 0 \ otherwise \end{cases} \quad (1)$$

Where $r \in R, e \in \{S, H\}$
The effectiveness of the set of rules with randomly-generated scores (from step 1) is evaluated by SpamAssassin against the dataset S and H. Score sets bringing the best results would be selected as a solution for this multi-objective problem.

At threshold T, the function to detect spam is implemented as follows:

```
//Input is an email
//Out is 1 if e is spam else 0
is_spam(e){
    score = 0;
    for i= 0 to N
    score += m(r,e)*score_of_r
    if(score > T)
    then return 1
    else return 0
}
```

**Step 3:** *Compute two objectives*
The purpose of the objective function is to compute two objectives of the problem. Within the scope of this problem, two objectives SDR and FAR are compute against the formula:

$$SDR = \frac{\sum_{i=1}^{K} is\_spam(s_i)}{K} \qquad (2)$$

$$FAR = \frac{\sum_{i=1}^{L} is\_spam(h_i)}{L} \qquad (3)$$

However, all objectives of NSGA-II algorithms are minimized [2]. Therefore, the SDR objective of this specific problem should be reformulate as $(1 - SDR)$ to get the maximum.

**Step 4:** *Run NSGA-II algorithm*
After all data input and required parameters are ready, the NSGA-II program is called to run and figure out the best population. Based on that population, the final result would be evaluated and compared.

### 3.3 Algorithm Parameters

Due to the large number of parameters for the experiment, they were stored in a text file and passed into the program via the command line for each time the program called. The detailed descriptions of the parameters are shown in Table 1.

## 4 Experiments and Results

### 4.1 Experiment Settings

The experiments were carried out for three different numbers of rules' scores: 30, 50 and 100. Twenty simulation runs with twenty different random seeds are carried out for each set of rules. At the end of experiments for each set of rules, the results

**Table 1** Algorithm Parameters

| Algorthm Parameters | Values |
|---|---|
| Population size | 100 |
| Number of generations | 1000 |
| Number of objective functions | 2 |
| Number of constraints | 0 |
| Number of real variables | 31 or 51 or 101 |
| Lower limit of real variable 1 | 2 |
| Upper limit of real variable 1 | 5 |
| ... | |
| Lower limit of real variable n | 0 |
| Upper limit of real variable n | 2 |
| Probability of crossover of real variable | 0.9 |
| Probability of mutation of real variable | 1/number of real variables |
| Distribution index for crossover | 5 |
| Distribution index for mutation | 10 |

were recorded for analyzed and compared to that of the traditional approach with single objective optimization.

Results gained from the experiments of this paper were compared to that from the experiment using the single objective optimization carried out in [5].

## 4.2 Experiments with 30 Rules

According to statistical results (Figure 1) from the experiments with 30 rules, in term of minimizing the FAR (at 0%), the best solution recorded for SDR was 62.34% for SDR while that result with single objective optimization (Table 2) is only 40.3% for SDR. Among solutions which the FAR are around 10%, the SDR of new approach with multi-objective algorithm NSGA-II are also much better the single one. They are $\{(74.03\%, 7.79\%); (74.46\%, 8.66\%); (72.29\%, 6.93\%)\}$ in comparison to the best point $\{(67.1\%, 9.6\%)\}$. Further, the trade-off solutions found by NSGA-II were widely spread; this provides variety of good choices for the system.

## 4.3 Experiments with 50 Rules

According to statistical results (Figure 2) from the experiments with 50 rules, in term of minimizing the FAR (at 0%), the best solution recorded for SDR was 65.37% for SDR while that result with single objective optimization (Table 3) is only 43.7% for SDR. Among solutions which the FAR are around 10%, the SDR of new approach

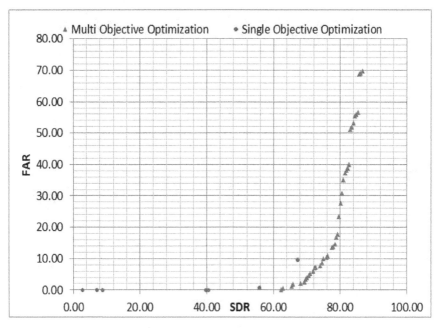

**Fig. 1** The result of experiments using NSGA-II with 30 rules

**Table 2** The result of experiments using single objective optimization with 30 rules

| Threshold | Spam Detection Rate | False Alarm |
|---|---|---|
| 0.5 | 67.1% | 9.6% |
| 1 | 67.1% | 9.6% |
| 1.5 | 55.8% | 0.8% |
| 2 | 55.8% | 0.8% |
| 2.5 | 40.3% | 0.0% |
| 3 | 39.8% | 0.0% |
| 3.5 | 8.7% | 0.0% |
| 4 | 6.9% | 0.0% |
| 4.5 | 2.6% | 0.0% |

with multi-objective algorithm NSGA-II are also much better the single one. They are {(83.98%, 9.96%); (83.55%, 8.66%); (82.68%, 7.36%)} in compare to {(68.8%, 9.6%)}.

Although the result of single objective optimization had improved, they were still far from feasible solutions obtained by NSGA-II.

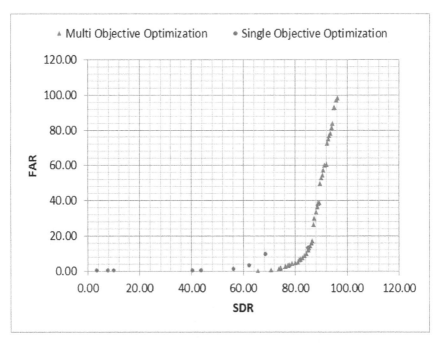

**Fig. 2** The result of experiments using NSGA-II with 50 rules

**Table 3** The result of experiments using single objective optimization with 50 rules

| Threshold | Spam    Detection Rate | False Alarm |
|-----------|------------------------|-------------|
| 0.5       | 84.8%                  | 13.1%       |
| 1         | 68.8%                  | 9.6%        |
| 1.5       | 62.3%                  | 3.2%        |
| 2         | 56.3%                  | 0.8%        |
| 2.5       | 43.7%                  | 0.0%        |
| 3         | 40.3%                  | 0.0%        |
| 3.5       | 10.0%                  | 0.0%        |
| 4         | 7.8%                   | 0.0%        |
| 4.5       | 3.5%                   | 0.0%        |

## 4.4   Experiments with 100 Rules

According to statistical results (Figure 3) from the experiments with 100 rules, the best solution recorded for FAR was 0.87% with SDR at 64.5% while that result with single objective optimization (Table 4) was50.6% and 0% for SDR and FAR namely. In this scenario, although the new approach could not eliminate the rate of false alarm, the result, in term of maximizing the SDR, were even better than the one with 50 rules. They are $\{(83.55\%, 8.23\%); (81.39\%, 6.06\%); (82.25\%, 6.93\%)\}$ in comparing to $\{(83.98\%, 9.96\%); (83.55\%, 8.66\%); (82.68\%, 7.36\%)\}$ of NSGA-II with 50 rules and $\{(78.4\%, 12\%)\}$.

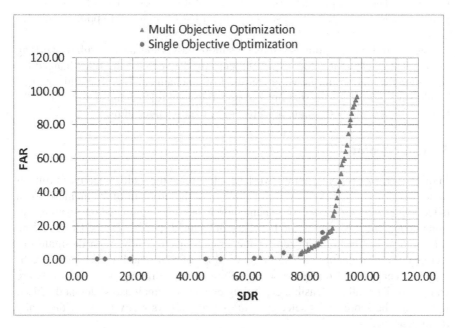

**Fig. 3** The result of experiments using NSGA-II with 100 rules

**Table 4** The result of experiments using single objective optimization with 100 rules

| Threshold | Spam Detection Rate | False Alarm |
|---|---|---|
| 0.5 | 86.1% | 15.9% |
| 1 | 78.4% | 12.0% |
| 1.5 | 72.7% | 4.0% |
| 2 | 62.3% | 0.8% |
| 2.5 | 50.6% | 0.0% |
| 3 | 45.5% | 0.0% |
| 3.5 | 19.0% | 0.0% |
| 4 | 10.0% | 0.0% |
| 4.5 | 7.4% | 0.0% |

## *Remarks*

Based on the statistical results of the experiments, it is undeniable that the application of multi-objective optimization algorithm to spam detection is reasonable and promising. The new approach not only figured out more effective solutions for the issue of SDR and FAR but it also suggested a list of optimized options ready for choosing.

The illustration also pointed out that the more set of rules the algorithms working on, the better results it achieved. However, only the score of the rule changed for each time the algorithm run while the rule kept unchanged. Therefore, this method would save more time for training and updating new rules than the way the traditional approach did with single objective optimization algorithms.

## 5 Conclusion

In this paper, we proposed a framework which applied the multi-objective optimization algorithms – NSGA-II by Deb [2] to solve the problem of Vietnamese spam detection. In fact, traditional anti-spam approaches have optimized the spam detection rate and the false alarm rate for years and gained specific results. However, the achievement has been optimized for the single objective only. With the-multi objective optimization approach, not only one pair of SDR and FAR for each threshold has been worked out but a set of solutions with different tradeoff levels are computed. They all are feasible depending on specific email users' demands. More important, the score set of selected solutions are always ready to use without any training needed.

Despite of being a promising approach, the proposed framework remains some issues which need more efforts to resolve in the future. Firstly, it is the problem of runtime. Currently, there is no measurement about the runtime of the system. Because conducted experiments were carried out against quite small dataset, it is not a big issue. However, when the dataset expands in the future, this concern should be analyzed seriously. Secondly, the result of the experiment strictly depends on the performance of NSGA-II algorithm. The framework should be tested on other evolutionary multi-objective optimization algorithms for more diverse results.

**Acknowledgement.** This research was supported by the Vietnam National Foundation for Science and Technology Development (NAFOSTED) under project number 102.01-2012.04

## References

1. Coello Coello, C.A., Veldhuizen, D.A.V., Lamont, G.B.: Evolutionary Algorithms for Solving Multi-Objective Problems. Kluwer Academic Publishers (2002)
2. Deb, K., Agrawal, S., Pratap, A., Meyarivan, T.: A Fast Elitist Non-dominated sorting genetic algorithm for multi-objective optimization: NSGA-II. In: Schoenauer, M., Deb, K., Rudolph, G., Yao, X., Lutton, E., Merelo, J.J., Schwefel, H.-P. (eds.) PPSN 2000. LNCS, vol. 1917, pp. 849–858. Springer, Heidelberg (2000)

3. Bui, L.T., Essam, D., Abbass, H.A., Green, D.G.: Performance analysis of evolution multi-objective optimisation algorithms in noisy environments. Complexity International 11, 29–39 (2005)
4. Tran, Q.A., Duan, H., Li, X.: Real-time statistical rules for spam detection. IJCSNS International Journal of Computer Science and Network Security 6(2B), 178–184 (2006)
5. Vu, M.T., Tran, Q.A., Jiang, F., Tran, V.Q.: Multilingual rules for spam detection. In: Proceedings of the 7th International Conference on Broadband and Biomedical Communications (IB2COM 2012), Sydney, Australia, pp. 106–110 (2012)
6. Schwartz: SpamAssassin. O'Reilly (2004)

# Risk Minimization of Disjunctive Temporal Problem with Uncertainty

Hoong Chuin Lau and Tuan Anh Hoang

**Abstract.** The Disjunctive Temporal Problem with Uncertainty (DTPU) is a fundamental problem that expresses temporal reasoning with both disjunctive constraints and contingency. A recent work by Peintner *et al* [6] develops a complete algorithm for determining Strong Controllability of a DTPU. Such a notion that guarantees 100% confidence of execution may be too conservative in practice. In this paper, following the idea of Tsamardinos [10], we are interested to find a schedule that minimizes the risk (i.e. probability of failure) of executing a DTPU. We present a problem decomposition scheme that enables us to compute the probability of failure efficiently, followed by a hill-climbing local search to search among feasible solutions. We show experimentally that our approach effectively produces solutions which are near-optimal.

## 1 Introduction

Expressive and efficient temporal reasoning is a significant task in planning and scheduling. A typical assumption is that all time points or temporal events such as starting and ending actions are under the complete control of the execution agent, and this can be modeled by the simplest and widely used Simple Temporal Network. The Simple Temporal Problem (STP) and the Disjunctive Temporal Problem (DTP) (that allows for disjunctive constraints) are concerned with checking temporal consistency of a given temporal network. While STP is polynomial-time solvable [1], solving DTP is known to be NP-hard [8] in general.

In real world practice, there are often needs to perform inference on events in the presence of exogenous factors (often referred as "Nature") which cannot be directly controlled by the agent, whose realization can only be observed. Due to these so-called "observable" time points, uncertainty is introduced into the above problems. Correspondingly, we have the Simple Temporal Problem with Uncertainty (STPU)

Hoong Chuin Lau · Tuan Anh Hoang
School of Information Systems, Singapore Management University

V.-N. Huynh et al. (eds.), *Knowledge and Systems Engineering, Volume 2*,
Advances in Intelligent Systems and Computing 245,
DOI: 10.1007/978-3-319-02821-7_21, © Springer International Publishing Switzerland 2014

and Disjunctive Temporal Problem with Uncertainty (DTPU), and the concept of temporal consistency is extended by varying notions of *controllability*, such as Strong, Weak and Dynamic Controllability, depending on how "observable" events affect controllable events.

Most works on temporal constraint problems with uncertainty focus on execution with 100% confidence. From the practical standpoint, this may lead to wasteful resource allocation or even infeasible networks. An interesting approach to deal with this is to take a probabilistic perspective. In [10], the Probabilistic Simple Temporal Problem (PSTP) was introduced by representing each uncontrollable event as a probability density function (PDF), where the goal was to find a schedule that maximizes the probability of execution under strong controllability (i.e. a one-size-fits-all schedule that maximizes the probability of successful execution against all possible realizations of uncertain events. The authors in [12] then presented heuristic techniques for determining the probability of executing a dynamically controllable strategy for PSTP. The concept of Robust Controllability was proposed in [3] to ensure the dynamic controllability of STPU within a specified degree of risk.

The Disjunctive Temporal Problem with Uncertainty (DTPU) was first defined in [13] for the purpose of modeling and solving planning and scheduling problems that feature both disjunctive constraints and contingency. In [6], the authors investigated the semantics of DTPU constraints and proposed a way to check if Strong Controllability holds. That work focused on checking if there exists a solution that ensures all constraints will be satisfied regardless of Nature's realization of uncontrollable events.

This paper is concerned with executing a DTPU from a risk minimization perspective as presented in [10]. Given a DTPU instance where the contingent constraints are modeled as random variables of known distributions, we are interested to find a solution that minimizes execution risk - or put in more positively, maximizes the probability of successful execution - that all temporal constraints will be satisfied. We propose a method for decomposing a DTPU into components so that the probability of success can be efficiently computed. Following that, we propose a computationally efficient hill-climbing local search that enables near-optimal solutions to be obtained.

## 2   Background and Literature Review

An STP [1] is defined as a pair $< V, S >$, where $V$ is a set of temporal variables representing temporal events or time points, and $S$ is a set of constraints between points, each taking the form $v_j - v_i \in [a_{ij}, b_{ij}]$, where $v_i, v_j \in V$ and $a_{ij}$ and $b_{ij}$ are some constants. Since STP contains only binary constraints, it can be represented by a weighted graph whose variables are represented as nodes and each directed edge $(v_i, v_j)$ is labeled by an interval $[a_{ij}, b_{ij}]$. An STP is consistent iff there exists at least one solution (an assignment to all temporal variables) such that all constraints in $S$ are satisfied, which can be determined in polynomial time [1].

A DTP [8] extends an STP by admitting disjunctive constraints. A DTP constraint consists of a disjunction of STP constraints of the form: $v_j - v_i \in [a_{ij1}, b_{ij1}] \bigvee v_j - v_i \in [a_{ij2}, b_{ij2}] ... \bigvee v_j - v_i \in [a_{ijk}, b_{ijk}]$. A DTP instance is consistent iff it contains at least a consistent component STP obtained by selecting one disjunct from each constraint. Efficient solvers for this NP-hard problem have been developed (e.g. Epilitis in [11]) to search through the meta space of component STPs. A number of pruning techniques are embedded in DTP solvers to reduce the search space, including conflict-directed backjumping, removal of subsumed variables, semantic branching, and no-good recording. In [4], the authors applied local search to DTP to generate solutions with minimal constraint violation and the operation is within the total assignment space of the underlying CSP rather than the partial assignment space of the related meta-CSP. In [9], the authors showed how their end-point ordering model can be used to express the qualitative interval algebra in a quantitative constraint solver with finite domains, and how to convert an interval algebra network into an equivalent non-binary CSP with finite integer domains. They observed that the relative positions of interval endpoints in an interval algebra network can be used to determine consistency.

To model uncertainty, two classes of temporal variables are defined: *executable variables* $V_e$ controlled by the execution agent, and *uncontrollable variables* $V_u$ controlled by Nature. In addition to the deterministic constraints $S$ defined above, $S_e$ and $S_c$ respectively denote the sets of *executable* and *contingent* constraints. $S_e$ model execution requirements in response to uncertain events (e.g.,"Activity can only be started (controllable) at least 30 minutes after the rain stops (uncontrollable)"). Hence, each executable constraint takes the form $y - x \in [a_{xy}, b_{xy}]$, where $y \in V_e$ and $x \in V_u$. $S_c$ model the temporal behavior of an uncontrollable event (e.g.,"The dinner will be ready (uncontrollable) between 20 and 30 minutes after cooking starts (controllable)"). They are used usually to model durational uncertainty whose values are controlled by Nature and can only be observed by the agent. In our paper, each contingent constraint takes the form $x - y = \tilde{d}$, where $y \in V_e, x \in V_u$ and $\tilde{d}$ is a random variable with a certain probability distribution.

Under different conditions of guaranteeing all constraints will be satisfied, three standard levels of controllability have been defined in the literature: Strong Controllability (i.e. existence of a universal solution), Weak Controllability (i.e. existence of a solution for each scenario), and Dynamic Controllability (i.e. existence of a solution that can always be built incrementally based on outcomes of contingent edges in the past). Tractable algorithms for checking them were provided in [14]. In [5], the authors proposed a pseudo-polynomial algorithm to handle dynamic controllability of STPUs based on constraint satisfaction. In [15], techniques were proposed to optimize the bounds on durations of contingent edges such that the resulting STPU is dynamic controllable.

Another line of work to deal with temporal uncertainty takes on a probabilistic context, i.e. allowing for probabilistic violation of constraints. In [3], the authors modeled contingent edges as random variables. Under some assumptions on the relationship between controllable and uncontrollable points, they provided an efficient polynomial-time approach based on second-order cone programming to check if an

STPU is Robust Controllable, i.e., can be executed dynamically within a given level of risk. In [10], the authors dealt with the problem of maximizing the probability of successful execution of PSTP (Probabilistic STPU) that models STPUs with continuous conditional PDF for each uncontrollable event. He formulated and solved the problem as a non-linear constrained optimization problem, and in general, the approach does not guarantee finding a globally optimal solution. In [12], the authors proposed heuristic techniques for approximating upper and lower bounds on the probability of executing a dynamically controllable strategy for PSTP.

## 3   Problem Definition

**Definition 1.** A Disjunctive Temporal Problem with Uncertainty is a tuple $<V_e, V_u, C, C_u, P>$ where

- $V_e = \{y_1, ..., y_n\}, V_u = \{x_1, ..., x_m\}$ respectively denote the sets of executable (or controllable) and observable (or uncontrollable) variables (or temporal events, time points) taking real values;
- $C$ is the set of controllable temporal constraints on $V_e$ and $V_u$. In the following, we describe five different types of constraints $\{S, S_e, D, D_e, D_{mix}\}^1$ where: $S$: set of standard STP constraints between controllable points; $S_e$: set of executable STPU constraints, each specifying the temporal requirement between an uncontrollable and a controllable point; $D$: set of DTP constraints, each being a disjunction of two or more STP constraints $S$; $D_e$ : set of disjunctions of two or more executable STPU constraints; and $D_{mix}$: set of disjunctions of a mix of STP and executable STPU constraints.
- $C_u$ is the set of contingent constraints, one for each uncontrollable time point in $V_u$, representing duration uncertainty determined by Nature. Each contingent constraint in a DTPU links an uncontrollable time point $x_j$ to a unique *parent* executable time point, denoted $pa(x_j)$, and is expressed as a disjunction of $K_j$ random variables taking nonnegative values. Hence, a contingent constraint can be expressed algebraically as $x_j - pa(x_j) = d_j^1 \vee ... \vee d_j^{K_j}$. In plain terms, this means that the point $x_j$ takes a value that Nature would determine, which is a realization of one of the random variables in $\{d_j^1, ..., d_j^{K_j}\}$; and this value gives the duration of the event whose start time is given by the executable time point $pa(x_j)$. For consistency of notations with controllable constraints, we write $C_u = \{S_c, D_c\}$, where $S_c$ and $D_c$ denote the set of contingent STPU and DTPU constraints respectively.
- $P$ is a set of continuous PDFs, one for each random variable $\tilde{d}$ providing the probability distribution over time duration of the uncontrollable event $x_j$ occurring after the executable point $pa(x_j)$ is started. The probability distributions are assumed to be independent of each other.

---

[1] For brevity, the reader may skip this and come back for more details later.

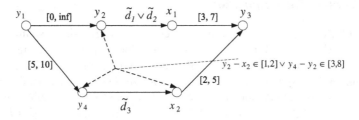

**Fig. 1** DTPU Example

Figure 1 gives an example of DTPU with four executable variables $y_1$ to $y_4$ (representing the start times of Tasks 1 to 4, where Task 1 is the dummy Time Reference point with no duration uncertainty), plus two observable variables $x_1$ and $x_2$ (representing the uncertain durations, i.e. uncontrollable events of Tasks 2 and 4's end times respectively). The constraint between executable events, such as that between $y_1$ and $y_4$, means that Task 4 is to start between 5 to 10 time units after $y_1$. The contingent constraint between $y_4$ and $x_2$ means the duration of Task 4 is controlled by Nature whose value follows a certain probability distribution. According to Definition 1, these constraints are classified as follows: $S = \{y_2 - y_1 \in [0,\infty], y_4 - y_1 \in [5,10]\}$; $S_e = \{y_3 - x_1 \in [3,7], y_3 - x_2 \in [2,5]]\}$; $D_{mix} = \{y_2 - x_2 \in [1,2] \vee y_4 - y_2 \in [3,8]\}$; $S_c = \{x_2 - y_4 = \tilde{d}_3\}$; and $D_c = \{x_1 - y_2 = \tilde{d}_1 \vee x_1 - y_2 = \tilde{d}_2\}$.

In this paper, we use probability distributions to model contingent durations. This frees us from the requirement for bounded intervals, which we believe is less realistic, since it is hard to provide bounds for the nature-controlled temporal events and there is always a non-zero probability that exogenous factors may cause an event to occur outside the bounds.

Given a DTPU, a *schedule* (or solution) is an assignment to all executable time points in $V_e$.

**Definition 2.** The Risk-Minimal DTP problem (RDTP) is defined as: given a DTPU instance, find a solution such that the probability that all constraints are satisfied is maximized.

This problem belongs to the class of "inequality constrained discrete MIN-MAX problem" [7] in optimization theory, which is known to be notoriously hard to solve computationally.

## 4 Solution Approach

First, we show how to decompose the original DTPU into different sub-problems. We then discuss how the success probability of each sub-problem can be computed. This leads us to the next section where we discuss our proposed local search scheme that makes use this computation.

## 4.1   Problem Decomposition

Note that for each disjunctive contingency constraint in $D_c$, since Nature will dictate which disjunct will ultimately be chosen and which value for that single disjunct will be realized during execution, the planner need to take into consideration ALL combinations of disjuncts that Nature may choose. On the other hand, since the planner has the right to decide on the values of the executable variables, an assignment that satisfies ANY of the disjunct of a disjunctive constraint in $\{D, D_e, D_{mix}\}$ is deemed to have satisfied that constraint. For this reason, one can imagine decomposing the DTPU into component DTPUs which can be solved independently and in parallel. Hence, we propose to split the constraint set into two categories $\{S, S_e, S_c, D, D_e, D_{mix}\}$ (termed the Canonical set) and $\{D_c\}$.

Intuitively, our idea is to decompose the problem as an MAX/MIN tree where we seek the maxima over the branches of the Canonical constraints, each of which is in turn computed as the minima over the branches of the disjunctive contingent constraint set $D_c$ (since Nature can pick any of these branches).

**Definition 3. Component DTPU (C-DTPU).** A C-DTPU is a DTPU with the set of temporal variables $V_e \cup V_u$, the set of disjunctive contingent constraints $D_c$, and a set of STPU constraints comprising the non-disjunctive constraints $\{S, S_e, S_c\}$ plus constraints obtained by selecting one disjunct from each constraint of the set $\{D, D_e, D_{mix}\}$. (In other words, a C-DTPU is a DTPU defined by $< V_e, V_u, S_{can}, D_c, P >$, where $S_{can}$ is an STPU comprising a unique disjunct combination derived from the *canonical* set of the original DTPU. To avoid ambiguity, we also term this STPU as a *canonical* STPU.)

Given a C-DTPU, in order to handle the disjunctive contingent constraints $D_c$, we further decompose into Component STPUs (as the MIN branches):

**Definition 4. Component STPU (C-STPU).** Given a C-DTPU $< V_e, V_u, S_{Can}, D_c, P >$, a C-STPU is an STPU with the set of temporal variables $V_e \cup V_u$, the entire set of $S_{Can}$ (defined above), a combination of contingent STPU constraints obtained by selecting one disjunct from each constraint of the set $D_c$, i.e. it is an STPU defined by $< V_e, V_u, S_{can}, S_{D_c}, P >$, where $S_{D_c}$ is an STPU comprising a unique contingent disjunct combination from $D_c$ of the original DTPU.

## 4.2   Computing Probability of Success

Consider a DTPU $T$ composed of $\varphi(T)$ number of C-DTPUs $T_l$ $(l = 1, ..., \varphi(T))$. For each $T_l$, we are interested to find an optimal schedule (denoted as $s_l^*$) that maximizes its success probability (i.e. the probability that all temporal constraints will be satisfied during execution if controllable points are executed following the schedule). Let $P(T_l|s_l)$ represent the success probability of $T_l$ when committing to schedule $s_l$ (i.e. an assignment to all executable points of $T_l$), an optimal schedule can then be defined as, $s^* = arg. \, max_s max_l(P(T_l|s))$.

Let $\theta(T_l)$ denote the total number of decomposed C-STPUs of $T_l$; $T_{lk}$ denote the $k^{th}$ C-STPU, and $P(T_{lk}|s_{lk})$ denote the success probability of $T_{lk}$ with schedule $s_{lk}$

$(l = 1, ..., \varphi(T)$ and $k = 1, ..., \theta(T_l))$. For simplicity, we will henceforth rewrite $s_l$ and $s_{lk}$ as $s$ where there is no ambiguity.

As noted above, since Nature will select the disjunct combination of constraints $D_c$ to realize, we adopt a worst-case approach in determining an optimal schedule $s^*$ for a given $T_l$. Given a schedule $s$, the worst-case probability for $T_l$ to be successfully executed when committing to $s$ is equivalently the minimum success probability of all its C-STPUs, i.e. $P(T_l|s) = min_k P(T_{lk}|s)$, where $k = 1, ..., \theta(T_l)$. Therefore, the optimization problem of determining an optimal schedule $s^*$ that maximizes the success probability of C-DTPU $T_l$ is:

$$P(T_l|s^*) = max_s P(T_l|s) = max_s min_k P(T_{lk}|s) \tag{1}$$

subject to the standard STP constraints.

We now show how we compute the success probability for a single C-DTPU. For simplicity, we will drop the subscript $l$ and write the C-DTPU as $T$ and the C-STPU $T_{lk}$ as $T_k$. The core computational part of our algorithm is to compute the success probability $P(T|s)$ of C-DTPU $T$ when committing to schedule $s$. We first consider the success probability of C-STPU $T_k$ and adopt the idea presented in (Tsamardinos, 2002) that deals with executing a PSTP so that the success probability is maximized. The difference is that in our work, the probability distribution is assumed with respect to a contingent constraint rather than an observable time point.

Recall that for a C-STPU $T_k$, three types of temporal constraints are involved: STP constraints, executable STPU constraints of the form $y_i - x_j \leq b_{x_j y_i}$ or $x_j - y_i \leq b_{y_i x_j}$[2], and contingent STPU constraints of the form $x_j - pa(x_j) = d_j$. STP constraints are standard constraints that must be satisfied as hard constraints. Contingent constraints are instantiations of random variables determined by Nature. Thus, given a schedule $s$, the success probability is equal to the probability that the executable STPU constraints are satisfied given the probability distributions associated with the random variables.

Consider first an arbitrary observable variable $x$ and its parent $pa(x)$. The success probability of C-STPU $T_k$ associated with this $x$ when committing to schedule $s$ can be computed as the probability that all the executable STPU constraints linked to $x$ are satisfied:

$$P(T_k|s) = P((\bigwedge_i y_i - x \leq b_{xy_i}) \wedge (\bigwedge_i x - y_i \leq b_{y_i x}) \tag{2}$$

Since $x - pa(x) = \tilde{d}$, we can rewrite the above as:

$$P(T_k|s) = P(L_s(d) \leq d \leq U_s(d)|s) \tag{3}$$

where $L_s(d) = Max_i(y_i - b_{xy_i}) - pa(x)$ is the lower bound for the contingent constraint to be satisfied for a given solution $s$, while $U_s(d) = Min_i(y_i + b_{y_i x}) - pa(x)$ is the upper bound.

---

[2] Note that the original form of the constraint $y_i - x_j \in [a_{x_j y_i}, b_{x_j y_i}]$ has been re-written by the two inequalities above for notational simplicity.

Extending to the general case when multiple observable variables are involved and letting $\tilde{d}_j$ denote the random variable associated with the $j^{th}$ contingent constraint, the above probability equation becomes:

$$P(T_k|s) = P(\bigwedge_j L_s(\tilde{d}_j) \leq \tilde{d}_j \leq U_s(\tilde{d}_j)|s) \tag{4}$$

Since all executable variables have been fixed with a given schedule $s$, the values of $L_s(\tilde{d}_j)$ and $U_s(\tilde{d}_j)|s$ are fixed for each contingent constraint $j$ and not dependent on any other random variables. Also, the probability distribution for different contingent durations represented with random variables $\tilde{d}_j$ are assumed to be independent of each other. Hence, we have,

$$P(T_k|s) = \prod_j P(L_s(\tilde{d}_j) \leq \tilde{d}_j \leq U_s(\tilde{d}_j)|s) \tag{5}$$

Thus, the success probability $P(T|s)$ of C-DTPU $T$ when committing to schedule $s$ can be represented as:

$$P(T|s) = min_k \prod_j P(L_s(\tilde{d}_{kj}) \leq \tilde{d}_{kj} \leq U_s(\tilde{d}_{kj})|s) \tag{6}$$

where $k = 1,...,\theta(T)$ and $\tilde{d}_{kj}$ is the $j^{th}$ contingent constraint of C-STPU $T_k$.

## 5  Local Search

In this section, we discuss our proposed local search algorithm to solve RDTP. Our search process is conducted on multiple C-DTPUs simultaneously and independently, and the one the gives the best solution will be returned. Note that such computations can occur in parallel without affecting the correctness of the solution. In the following, we therefore discuss our proposed local search algorithm to solve the problem associated with a single C-DTPU.

A C-DTPU is made up of an STPU plus the $D_c$ constraints. We perform search in the space of *feasible* solutions which comprise schedules that do not violate any of the STP constraints in $S$. Note that for a C-DTPU, every executable constraint is a simple (i.e. non-disjunctive) constraint. This means, for each executable variable, we can calculate its minimal domain using the algorithm in [1] in polynomial time, which is also can be used to find a feasible solution the minimal domain.

The initial solution denoted as $s^0 = \{s_1^0,...s_n^0\}$ is randomly selected from the feasible space, which is an assignment of all $n$ executable variables. From equation (6) given in the previous section, we see that the overall probability of success of a solution $s$ is a function of the probabilities that the individual random variables (representing the contingent constraints) will be realized within their respective interval bounds when committing to that schedule. Hence, the intuition for our local search neighborhood is to adjust the value of one executable variable of $s$ at a time in such a way that these probabilities might be increased while maintaining schedule feasibility.

Before discussing the neighborhood structure, we observe the following for any random variable $d$.

- For any interval $I \subseteq \mathscr{R}$, $P(d \in I)$ is not necessarily proportional to the length of $I$, and its value may vary, i.e., increase or decrease, when $I$ is slided to the left or right.
- For any two intervals $I_1$ and $I_2$, we have $P(d \in I_1) \geq P(d \in I_2)$ if $I_2 \subseteq I_1$.

Therefore, at each local search iteration, the intuition is to explore candidate neighbors that will either: (a) slide the interval $[L_s(\tilde{d}_j), U_s(\tilde{d}_j)]$ to one where the area under the probability density function of $\tilde{d}_j$ over that interval has an equal or higher value; or (b) expand the interval $[L_s(\tilde{d}_j), U_s(\tilde{d}_j)]$ as much as possible. In both cases, care must be exercised to ensure the resulting schedule is still feasible. This hill-climbing process will lead us to iteratively improve the overall value of the success probability.

Assume that $s = \{s_1, ... s_n\}$ is a feasible schedule a certain C-DTPU in question. For each $y_i \in V_e$, denote:

$$\delta^-(y_i) = max\{\delta > 0 : s' = (s_1, ..., y_i - \delta, ..., s_n) \text{ is a feasible solution}\}$$
$$\delta^+(y_i) = max\{\delta > 0 : s' = (s_1, ..., y_i + \delta, ..., s_n) \text{ is a feasible solution}\}$$

That is, $\delta^-(y_i)$ and $\delta^+(y_i)$ provide the limits, when all other executable variables are fixed, on how much the value of $y_i$ can be decreased or increased such that the resulting schedule is still feasible.

Since we must deal disjunctive contingent constraints (each of them comprising one or more random variables), rather than considering individual random variables, we instead focus on each uncontrollable variable. We say that two variables are said to be *related* if there is at least one constraint between them. For each uncontrollable variable $x_j$, the bound interval $[L_s(x_j), U_s(x_j)]$ of $x_j$ may be slided to the left or right by shifting the value of its parent variable $pa(x_j)$ along the interval $[pa(x_j) - \delta^-(pa(x_j)), pa(x_j) + \delta^+(pa(x_j))]$. Similarly, the bound interval may be expanded by shifting the values of the executable variables $y_i$ emanating from $x_j$. In order to understand which of these variables to shift, we first provide the following definitions.

Given a feasible solution $s$, the **Optimal Relatives for an uncontrollable variable** $x_j$ $(j = 1, ... m)$ are the controllable variables $y_j^l$ and $y_j^u$ *related* with $x_j$ that maximize the interval bounds of $x_j$ when committing to $s$, i.e.,

$$y_j^l = argmax_{y_i}\{(y_i - b_{x_j y_i}) | s\}, y_j^u = argmin_{y_i}\{(y_i + b_{y_i x_j}) | s\}$$

where $y_i$'s are the controllable variables *related* with $x_j$.

In other words, when $pa(x_j)$ is fixed, the expanded bounds of $x_j$ ($[L_s(x_j), U_s(x_j)]$) for the executable constraints emanating from $x_j$ to remain satisfied under the solution $s$ is given by $[y_j^l - b_{x_j y_j^l} - pa(x_j), y_j^u + b_{y_j^u x_j} - pa(x_j)]$.

Given a feasible solution $s$, the **Optimal Relatives for an executable (controllable) variable** $y_i$ $(i = 1,...n)$ are the controllable variables $Y_i^l$ and $Y_i^u$ *related* with $y_i$ that satisfy:

$$Y_i^l = argmax_{y_j}\{y_j - a_{y_i y_j}|s\}, Y_i^u = argmin_{y_j}\{y_j + a_{y_j y_i}|s\}$$

where $y_j$'s are the controllable points *related* with $y_i$.

Then, $\delta^-(y_i)$ and $\delta^+(y_i)$ for $y_i$ can be derived from its Optimal Relatives, i.e. $\delta^-(y_i) = y_i - Y_i^l + a_{y_i Y_i^l}$ and $\delta^+(y_i) = Y_i^u + a_{Y_i^u y_i} - y_i$.

For each $j = 1,...,m$ we obtain neighboring solutions by the two operators:

- Operator $\mathcal{O}_1$: Decrease or increase the value of the corresponding parent:
$$s_{\mathcal{O}_1}^l = (s_1, \cdots, pa(x_j) - \delta^-(pa(x_j)), \cdots, s_n)$$
$$s_{\mathcal{O}_1}^u = (s_1, \cdots, pa(x_j) + \delta^+(pa(x_j)), \cdots, s_n)$$
which are also feasible and may improve the probability of success as they slide the interval of $\tilde{d}_j$.

- Operator $\mathcal{O}_2$: Decrease or increase the value of the corresponding optimal relatives:
$$s_{\mathcal{O}_2}^l = (s_1, \cdots, y_j^l - \delta^-(y_j^l), \cdots, s_n)$$
$$s_{\mathcal{O}_2}^u = (s_1, \cdots, y_j^u + \delta^+(y_j^u), \cdots, s_n)$$
which are also feasible and may improve the probability of success since they expand the interval of $\tilde{d}_j$.

At each iteration, the neighbors of current solution $s$ is constructed by changing either the parent or an optimal relative of a uncontrollable variable to the value that guarantees the largest interval while keeping the other elements fixed, i.e. we apply either or both operators $\mathcal{O}_1$ and $\mathcal{O}_2$ to find the best neighbor from $s$.

The cost for computing the probability of success is $O(mK)$ where $K$ is the maximum number of disjunctions in a disjunctive contingent constraint (which is typically a small value such as 2). Since each solution keeps information about $Y_i^l$ and $Y_i^u$ for each $y_i$, and $y_j^l$ and $y_j^u$ for each $x_j$, the cost for updating one of them when we change the value of a $y_j$ is constant. Therefore, the cost for the optimal relative update step is $O(r)$ ($r$ is the maximum number of related variables). Consequently, the cost for a local search move is $O(mr)$. The complexity of our local search algorithm is $O((n+m)r) + O(Lm(r+K))$ where $L$ is the number of local search iterations.

## 6   Experimental Analysis

In this section, we present experimental results that verify the performance of our proposed local search algorithm. We implemented our algorithm in C# on a Core(TM) 2 Duo CPU 2.33GHz processor under Windows7 operating system with a main memory of 2GB.

Since there are no DTPU benchmark problems in the literature and it is impractical to optimally solve reasonably large instances, we resort to using reasonably tight lower bounds. More precisely, we generate problem instances which satisfy the condition that there is at least one feasible solution with the probability of success of at least $1 - \varepsilon$ for a given threshold value $\varepsilon$ between 0 and 1. On small values of $\varepsilon$

**Table 1** Parameters used in instance generation

|                          | Set 1 | Set 2 | Set 3 |
|--------------------------|-------|-------|-------|
| #Controllable variables  | 15    | 20    | 30    |
| #Uncontrollable variables| 5     | 10    | 10    |
| #DTP constraints         | 50    | 75    | 100   |
| #Executable constraints  | 10    | 20    | 20    |
| #Contingent constraints  | 4-6   | 6-12  | 6-12  |

(e.g., 0.2, 0.1, or 0.05), the value $1 - \varepsilon$ provides a tight lower bound for the optimal solution, and our aim is to show that our algorithm produces a solution very near to $1 - \varepsilon$. For our experiments, we restrict to two disjunctions per disjunctive constraint. This is a reasonable setting in modeling real-world uncertainty, where each disjunct is used to model either a rare event or a usual event.

Due to space constraints, we will not present how random DTPU instances are generated for our experiments. Interested readers may refer to the full length version of this paper [2] for details. We ran experiments on 3 sets of random DTPU instances generated with parameter settings given in Table 1. For each setting, we generate 100 instances whose random variables follow the normal distribution, and 100 instances follow the uniform distribution. For every instance, we generated random initial feasible solutions and applied three variants of our proposed local search: $\mathcal{O}_1$ denotes using only the operator $\mathcal{O}_1$ is applied to search for the best neighbor; $\mathcal{O}_2$ denotes using only operator $\mathcal{O}_2$; and $\mathcal{O}_1 + \mathcal{O}_1$ denotes using both operators. We then computed the mean and variance of the probabilities of success obtained by our three local search variants on each set of problem instances. And as a benchmark comparison, we also show the mean probability of success of the best random initial solutions.[3] In addition, to show the actual quality of solutions, we show the number of times the obtained probability exceeds the tight lower bound $1 - \varepsilon$ (for $\varepsilon = 0.2, 0.1$ and $0.05$). The results are summarized in Table 2. We observe that all local search variants performed well in general, since they generate solutions which are reasonably close to the value $1 - \varepsilon$ whereas the initial solutions are not so. This observation is consistent across all problem instance sets tested with different parameters and under both the normal and uniform distributions. Table 2 also clearly shows that, while the performance of $\mathcal{O}_1$ and $\mathcal{O}_2$ are not significantly different, the combined approach $\mathcal{O}_1 + \mathcal{O}_2$ always gives the best solutions. Moreover, the solutions obtained by $\mathcal{O}_1 + \mathcal{O}_2$ are more reliable, in the sense that they are closer to the optimal solutions (having the highest mean of probability of success among the three variants) and have the smallest variance.

The first three charts of Figure 2 show the scatter plot on the 100 solution values obtained from running $\mathcal{O}_1 + \mathcal{O}_2$ on Set 1 instances with contingent variables that

---

[3] We generated one random initial feasible solution for each C-DTPU, and these figures are based on the *best* solutions generated. It is noteworthy that many of the initial solutions have very low probability of success, almost near to zero.

**Table 2** Performance of Local Search

| | | Normal distribution | | | Uniform distribution | | |
|---|---|---|---|---|---|---|---|
| $\varepsilon$ | | **0.2** | **0.1** | **0.05** | **0.2** | **0.1** | **0.05** |
| **Set 1** | | | | | | | |
| Mean of best initial solutions | | 0.60 | 0.69 | 0.78 | 0.65 | 0.69 | 0.67 |
| $\mathcal{O}_1$ | Mean | 0.82 | 0.90 | 0.95 | 0.94 | 0.94 | 0.91 |
| | Variance | 0.07 | 0.11 | 0.06 | 0.08 | 0.10 | 0.13 |
| | #Instances exceeding $1-\varepsilon$ | 85 | 74 | 87 | 93 | 84 | 55 |
| $\mathcal{O}_2$ | Mean | 0.79 | 0.89 | 0.93 | 0.85 | 0.86 | 0.85 |
| | Variance | 0.12 | 0.10 | 0.12 | 0.16 | 0.15 | 0.18 |
| | #Instances exceeding $1-\varepsilon$ | 73 | 72 | 70 | 74 | 51 | 33 |
| $\mathcal{O}_1+\mathcal{O}_2$ | Mean | **0.86** | **0.94** | **0.97** | **0.97** | **0.97** | **0.97** |
| | Variance | **0.03** | **0.01** | **0.01** | **0.05** | **0.06** | **0.06** |
| | #Instances exceeding $1-\varepsilon$ | **98** | **100** | **99** | **95** | **89** | **80** |
| **Set 2** | | | | | | | |
| Mean of best initial solutions | | 0.57 | 0.69 | 0.80 | 0.50 | 0.49 | 0.50 |
| $\mathcal{O}_1$ | Mean | 0.80 | 0.88 | 0.93 | 0.81 | 0.80 | 0.81 |
| | Variance | 0.07 | 0.06 | 0.05 | 0.15 | 0.13 | 0.13 |
| | #Instances exceeding $1-\varepsilon$ | 64 | 55 | 40 | 61 | 24 | 8 |
| $\mathcal{O}_2$ | Mean | 0.77 | 0.88 | 0.93 | 0.72 | 0.73 | 0.72 |
| | Variance | 0.14 | 0.06 | 0.07 | 0.17 | 0.16 | 0.17 |
| | #Instances exceeding $1-\varepsilon$ | 63 | 58 | 54 | 35 | 15 | 5 |
| $\mathcal{O}_1+\mathcal{O}_2$ | Mean | **0.86** | **0.93** | **0.97** | **0.93** | **0.92** | **0.91** |
| | Variance | **0.02** | **0.01** | **0.01** | **0.10** | **0.09** | **0.08** |
| | #Instances exceeding $1-\varepsilon$ | **98** | **98** | **97** | **91** | **69** | **39** |
| **Set 3** | | | | | | | |
| Mean of best initial solutions | | 0.52 | 0.65 | 0.70 | 0.45 | 0.46 | 0.45 |
| $\mathcal{O}_1$ | Mean | 0.80 | 0.88 | 0.93 | 0.81 | 0.80 | 0.81 |
| | Variance | 0.07 | 0.06 | 0.05 | 0.15 | 0.13 | 0.13 |
| | #Instances exceeding $1-\varepsilon$ | 64 | 55 | 40 | 61 | 24 | 8 |
| $\mathcal{O}_2$ | Mean | 0.75 | 0.85 | 0.90 | 0.68 | 0.67 | 0.68 |
| | Variance | 0.14 | 0.11 | 0.16 | 0.19 | 0.17 | 0.16 |
| | #Instances exceeding $1-\varepsilon$ | 45 | 41 | 38 | 25 | 8 | 2 |
| $\mathcal{O}_1+\mathcal{O}_2$ | Mean | **0.86** | **0.93** | **0.96** | **0.91** | **0.90** | **0.89** |
| | Variance | **0.03** | **0.02** | **0.03** | **0.11** | **0.10** | **0.11** |
| | #Instances exceeding $1-\varepsilon$ | **98** | **96** | **93** | **90** | **59** | **39** |

follow the normal distribution. Similarly, the next three charts of the figure show the scatter plot on running $\mathcal{O}_1+\mathcal{O}_2$ on Set 3 instances whose contingent variables follow the uniform distribution. Both results illustrate that most solutions obtained by our algorithm $\mathcal{O}_1+\mathcal{O}_2$ are indeed very close to the optimal solutions (or more precisely, the respective tight lower bounds).

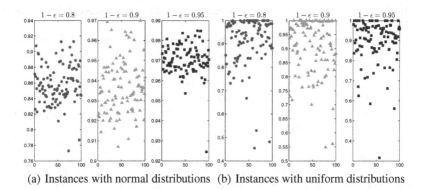

(a) Instances with normal distributions    (b) Instances with uniform distributions

**Fig. 2** Results for running $\mathcal{O}_1 + \mathcal{O}_2$ under normal and uniform distributions

## 7 Conclusion

In this paper, we presented what we believe to be the first efficient scheme to tackle the optimization problem associated with DTPUs, based on a decomposition mechanism and local search. Experimental results demonstrate the computational efficiency of our approach on a range of problem instances. We believe it is possible to make the search process more effectively with more powerful local search algorithms.

For future works, it is interesting to relax the assumption of independence among the random variables, which we inherited from [10], since such events are seldom independent. It is a challenging analytical and computational task to compute the probability of successful execution taking the correlation of random variables into consideration.

**Acknowledgments.** This research is supported by the Singapore National Research Foundation under its International Research Centre @ Singapore Funding Initiative and administered by the IDM Programme Office.

## References

1. Dechter, R., Meiri, I., Pearl, J.: Temporal constraint networks. Artif. Intell. 49 (May 1991)
2. Lau, H.C., Hoang, T.A.: Risk minimization of disjunctive temporal problems with unicertainty. Working paper, Singapore Management University (2013),
   http://www.mysmu.edu/faculty/hclau/dtpu-long.pdf
3. Lau, H.C., Li, J., Yap, R.H.: Robust controllability of temporal constraint networks under uncertainty. In: Proc. ICTAI (2006)
4. Moffitt, M.D., Pollack, M.E.: Applying local search to disjunctive temporal problems. In: Proc. IJCAI (2005)

 5. Morris, P., Muscettola, N.: Temporal dynamic controllability revisited. In: Proc. AAAI (2005)
 6. Peintner, B., Venable, K.B., Yorke-Smith, N.: Strong controllability of disjunctive temporal problems with uncertainty. In: Bessière, C. (ed.) CP 2007. LNCS, vol. 4741, pp. 856–863. Springer, Heidelberg (2007)
 7. Rustem, B., Nguyen, Q.: An algorithm for the inequality-constrained discrete Min–Max problem. SIAM J. Optimization 8(1) (1998)
 8. Stergiou, K., Koubarakis, M.: Backtracking algorithms for disjunctions of temporal constraints. Artif. Intell. 120 (June 2000)
 9. Thornton, J., Beaumont, M., Sattar, A., Maher, M.: A local search approach to modelling and solving interval algebra problems. J. Logic and Comput. 14 (February 2004)
10. Tsamardinos, I.: A probabilistic approach to robust execution of temporal plans with uncertainty. In: Vlahavas, I.P., Spyropoulos, C.D. (eds.) SETN 2002. LNCS (LNAI), vol. 2308, pp. 97–108. Springer, Heidelberg (2002)
11. Tsamardinos, I., Pollack, M.E.: Efficient solution techniques for disjunctive temporal reasoning problems. Artif. Intell. 151 (December 2003)
12. Tsamardinos, I., Pollack, M.E., Ramakrishnan, S.: Assessing the probability of legal execution of plans with temporal uncertainty. In: Proc. ICAPS 2004 Workshop (2004)
13. Venable, K.B., Yorke-Smith, N.: Disjunctive temporal planning with uncertainty. In: Proc. IJCAI (2005)
14. Vidal, T., Fargier, H.: Handling contingency in temporal constraint networks: from consistency to controllabilities. J. Experimental and Theoretical Artificial Intelligence 11 (1999)
15. Wah, B.W., Xin, D.: Optimization of bounds in temporal flexible planning with dynamic controllability. In: Proc. ICTAI (2004)

# Reference Resolution in Japanese Legal Texts at Passage Levels

Oanh Thi Tran, Bach Xuan Ngo, Minh Le Nguyen, and Akira Shimazu

**Abstract.** Sentences in the domain of legal texts are usually long and complicated. At the discourse level, they contains lots of reference phenomena which make the understanding of laws become more difficult. This paper investigates the task of reference resolution in the legal domain. The aim is to create a system which can automatically extracts referents for references in a real time. This is a new interesting task in the research of Legal Engineering. It does not only help readers in comprehending the law, support law makers in developing and amending laws, but also support in building an information system which works based on laws, etc. The main issues are to detect references and then resolve them to their referents. To detect references, we use a powerful machine learning technique rather than rule-based approaches as used in previous works. In resolving them, we design regular expressions to catch up the position of referents. We also build a corpus using Japanese National Pension Law to train and test our model. Our final system achieved 91.6% in the F1 score in detecting references, 96.18% accuracy in resolving them, and 88.5% in the F1 score in the end-to-end system.

## 1 Introduction

Legal documents contain lots of references which bring precious information. They are usually linguistic expressions that identify a specific act or a text partition referred to. References are used in many different ways in legal texts. One legal document may use a reference to import definitions from another one, or may limit the scope of applications of some laws. References may be used to change the status of the law that is referred to, i.e. enacting, altering, or amending laws. Due to high

Oanh Thi Tran · Bach Xuan Ngo · Minh Le Nguyen · Akira Shimazu
School of Information Science,
Japan Advanced Institute of Science and Technology,
1-1 Asahidai, Nomi, Ishikawa, 923-1292 Japan
e-mail: {oanhtt,bachnx,nguyenml,shimazu}@jaist.ac.jp

V.-N. Huynh et al. (eds.), *Knowledge and Systems Engineering, Volume 2,*                237
Advances in Intelligent Systems and Computing 245,
DOI: 10.1007/978-3-319-02821-7_22, © Springer International Publishing Switzerland 2014

frequency of references, a law will be difficult to comprehend if we do not read referenced texts within it. Detecting and resolving references, therefore, is an important task in the research on Legal Engineering [3, 4, 5], whose purposes are to help experts make complete and consistent laws; and to design an information system which works based on laws. For these purposes, it is vital to develop a system which can process legal texts automatically.

In this paper, we direct our attention to the task of reference detection and resolution in legal texts. This is a sequence of two steps: detection of the references inside documents, and resolution of those references to their referents. These referents usually correspond to articles, paragraphs of articles, items of paragraphs, or sub-items of items according to the naming rules used in the Japanese legal domain. They usually are a passage or a collection of passages. Once this task is solved, it will bring many benefits. For examples, it is used in legal information linking to link legal contents, in extending services in digital libraries such as querying, hypertext services (automatic hypertext creation). This also supports readers in comprehending laws or supports lawmakers in developing or even amending laws.

There exists some work related to this kind of researches, which was developed for many languages such as Italian [1, 14], Spanish [11], and Dutch [10]. In these studies, the authors propose rule-based approaches to recognize references. However, the disadvantage is that the created systems must be constantly extended in order to provide rules for yet unseen cases. In this paper, we consider this task for Japanese laws in which we propose to utilize a machine learning technique which results in the final system being automatically trainable from a corpus with a minimal amount of human intervention. Moreover, the machine learning technique can solve the ambiguous cases better than rule-based approaches. In legal texts, there are some linguistic expressions which conform to references' regular expressions, but it is not a real reference when using in some given contexts. For examples, in the case of *dai ni go hi hoken sha (the second insured person)*, the rule-based approach can wrongly detect out the reference *dai ni go* because it conforms to regular patterns. Using machine learning technique, however, we did not get this kind of errors because it is possible to integrate the context information in solving the problem. To train and test the system, we also build a Japanese National Pension Law corpus on reference resolution to conduct experiments.

Our main contributions can be summarized in the following points:

- Introducing the task of reference resolution in the domain of legal texts at the passage level.
- Analyzing the characteristics of references in the legal texts. Based on that, we propose a two-step framework in which we use machine learning techniques for the first step.
- Introducing an annotated corpus, the Japanese National Pension Law (JNPL) corpus on reference resolution.
- Performing experiments and evaluating our framework on the JNPL corpus.

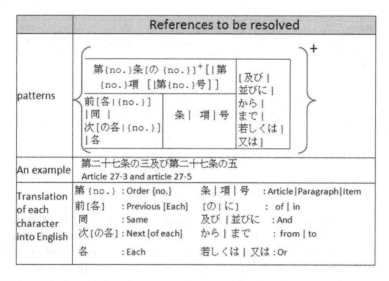

**Fig. 1** Some examples of different types of references

Experimental results on the annotated JNPL corpus showed promising results. We obtained 91.6% in the F1 score in detecting references, 96.18% accuracy in resolving them, and 88.5% in the F1 score in the end-to-end system.

The rest of this paper is organized as follows. Section 2 introduces some structures of references in legal texts. Section 3 presents our proposed method to solve this task. Next, we will show the experimental setup and results in Section 4. In this section, we also introduce our corpus of JNPL on reference resolution. In Section 5, we would like to analyze error cases caused by the final system. Finally, Section 6 presents our conclusion and future work.

## 2 Structures of References in Legal Texts

There are some structures of references in the legal domain that are worth consideration when dealing with automatic extraction and resolution of references. Most types of references in legal texts relate to terms, definitions or provision of articles. Naming references in legal texts have their own structures, which are different from references in the general domain. They mostly conform to several kinds of patterns. Figure 1 shows that a reference usually conforms to a regular expression.

In Figure 1, | means 'or', [ ] means 'optional', and + means 'repeat one or more times'. An example of a reference and its translation into English are also given in Figure 1.

By observing references of Japanese legal texts, we can see that references in legal texts mostly fall under one of the following types:

| No. | References | Types |
|-----|-----------|-------|
| 1 | 第七条第一項第二号　　--　　Article 7, paragraph 1, Item 1 | 1)　a) |
| 2 | 同項　　--　　The same paragraph<br>前項　　--　　The previous paragraph | 1)　b) |
| 3 | 次号及び第三号　　　　--　　The next and the third items | 2) |
| 4 | 第三条（第二項、第三項を除く。）<br>--　Article 3 (excluding Paragraph 2, and Paragraph 3) | 3) |
| 5 | 次の各号のいずれか　　--　　Each of the following items | 3) |

**Fig. 2** Some examples of different types of references

1. Single references: concerns a single referent.

   a. Well-formed references: These references comprise of a label, such as an
      article, a paragraph, and an item combined with a number. References usu-
      ally start with the broadest part and end with the narrowest part. These
      references contain all the information needed to identify the referred item.
   b. Anaphora and indirect references: They usually refer to an earlier reference.
      These references are always resolved to one of the former (label and num-
      ber). This requires some context information (documents referenced before
      and near the current reference, in a specific part of the same texts, ..) to
      solve the reference.

2. Co-ordinated references: Several referents are referred to in the same reference
   by using some linking expressions such as *and, or, from ... to*, etc.
3. Special cases: These are used when an element in the text contains a list that is
   preceded by a description of the list, without which the list does not make sense.
   Another special case is the use of the word *each time* that refers to sub-items of
   an article.

Figure 2 illustrates some examples of references in legal texts, which belong to
different types as distinguished above.

## 3　A Solution for Detecting and Resolving References in Legal Texts

To detect out references, most of previous work focus on rule-based approaches in
which they build regular expressions or context free grammars to recognize them
[1, 10, 11, 14]. However, the disadvantage is that the the created systems must be
constantly extended in order to provide rules for yet unseen cases. Moreover, using
machine learning techniques can solve ambiguous cases better as showed in the
introduction part.

In this work, we would like to make use of advanced machine learning techniques
to automatically detect out references. Figure 3 presents our framework. First, our
system learns a reference labeling model from an available corpus using a sequence

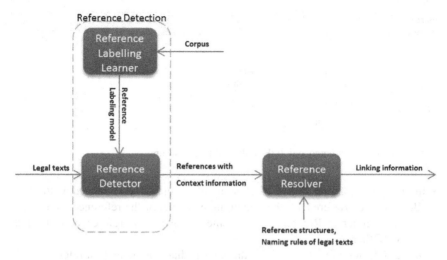

**Fig. 3** A framework for reference detecting and resolving in legal texts

labeling algorithm. After acquiring the model, it will be used to detect out references for each input legal text. This reference detector outputs all references found in the input. Each output reference also contains the context information if available to serve for later step. Context information can be the source document, fragment that contains the reference, and so on. The resolver step will use this information to match each reference to one or some legal passages. This step uses the knowledge of the reference structure as discussed in Section 2 as well as naming rules of legal texts to create linking information between the references and the referenced items.

Next, we will present the solution for each step of the framework in more detail.

## 3.1 Reference Detection

In the reference detection step, the goal is to extract out all references appearing in the input string. The input string is a sequence of words and the output is a collection of references that the input string contains.

Similarly to many classical natural language processing (NLP) tasks such as text chunking [15] and named entity recognition [2], we also formulate this task as a sequence labeling problem. In this step, each word is assigned a label indicating whether it starts a specific reference, is inside a specific reference, is the ending of a reference, or is outside any reference. Figure 4 illustrates an example of a Japanese law sentence in the IOB, IOE and FIL notations. In this example, the source sentence contains one reference that should be detected. The labels of this sentence using these three notations are described as follows:

| Source Sentence | 次に掲げる事項は、代議員会の議決を経なければならない。 The following matters will be decided through the decision of a representative board | | | | | | | | | | | | | |
|---|---|---|---|---|---|---|---|---|---|---|---|---|---|---|
| Word sequence | 次 | に | 掲げる | 事項 | は | 代議員 | 会 | の | 議決 | を | 経 | なければ | なら | ない。 |
| | The following matters | will be decided through the decision of a representative board. | | | | | | | | | | | | |
| IOB notation | B_R | I_R | O | O | O | O | O | O | O | O | O | O | O | O |
| IOE notation | I_R | E_R | O | O | O | O | O | O | O | O | O | O | O | O |
| FIL notation | F_R | L_R | O | O | O | O | O | O | O | O | O | O | O | O |

**Fig. 4** A Japanese law sentence in IOB, IOE, and FIL notations

- In the IOB notation, the first element of a reference is tagged with B_R (**B**eginning of **R**eference); the remaining elements of the reference are tagged with I_R (**I**nner of **R**eference); all elements outside the reference are tagged with O (**O**thers).
- In the IOE notation, the first and the intermediate elements of a reference are tagged with I_R (**I**nner of **R**eference); the last element of the reference is tagged with E_R (**E**nding of **R**eference); all elements outside the reference are tagged with O (**O**thers).
- In the FIL notation, the first element of a reference is tagged with F_R (**F**irst of **R**eference); the intermediate elements of the reference are tagged with I_R (**I**nner of **R**eference); the last element of the reference is tagged with L_R (**L**ast of **R**eference); all elements outside the reference are tagged with **O** (**O**thers).

In this task, we use Conditional Random Fields [8] as the learning method to learn the sequence labeling model. This method is an efficient and powerful framework for many sequence learning tasks [6, 18]. In extracting feature sets, we use a combination of n-gram ($n \leq 3$) of words, readings[1], part-of-speech tags, and the information of chunking. When doing experiments, we exploit three kinds of label settings which are the IOB notation, IOE notation and the FIL notation [9].

## 3.2 Reference Resolver

This step tries to locate the position of the referenced items of a given reference. The input is a reference, which is the output of the first step. The goal is to recognize which articles, which paragraphs, which items etc., a given reference refers to.

We use regular expressions which are carefully constructed to catch up and determine the scope of referenced items for a given reference. Some examples of position recognition results are given in Figure 5. The patterns of the references were described in Section 2.

---

[1] In Japanese, each character may have serveral ways of readings in different contexts. When it is used in a specific sentence, it has only one specific way of reading.

| No. | Position Parts of Mentions | Context Information | Position of Antecedents (Which article, which paragraph, which items) |
|---|---|---|---|
| 1 | 第九十条第一項第一号から第三号まで<br>From the 1st item to the 3rd item of paragraph 1 of article 90 | No need | --- *Referent 1* -----<br>( 90, 1, 1 )<br>--- *Referent 2* -----<br>( 90, 1, 2 )<br>--- *Referent 3* -----<br>( 90, 1, 3 ) |
| 2 | 同項<br>The same paragraph | -- *The current paragraph* --<br>第九十六条第四項<br>article 96, paragraph 4 | --- *Referent 1* -----<br>( 96, 4 ) |
| 3 | 前条第一項<br>The previous article, paragraph 1 | -- *The current article* --<br>第六条 ( article 6 ) | --- *Referent 1* -----<br>( 5, 1 ) |
| 4 | 同項<br>The same paragraph | -- *The previous paragraph* --<br>前条第一項<br>the previous article, paragraph 1 | --- *Referent 1* -----<br>( 5, 1 ) |

**Fig. 5** Some examples of the output of the position recognition step.

Resolving references is mostly simple when the reference belongs to the complete type (i.e. the first example in Figure 5). A reference is complete if it includes information of the complete document it refers to. We can find the article in the list. This list is built based on investigating the law and its structure. From this article, we will specify the precise location in it. Using regular expressions, we know that where it should be located.

We also use regular expressions and the common structure of references to catch up co-ordinated references and then resolve them.

Another group of references that is a little bit harder to resolve is anaphora. We will list out some typical cases as follows:

- References that refer to the current text: For example, this article, this paragraph, and so on. This type is feasible to resolve if we know the current location of the reference (i.e. the second example in Figure 5).
- References refer to an earlier point in the text, such as *the previous article*. These can be resolved using structure information of the law (i.e. the third example in Figure 5).
- References refer to an earlier reference, for example *dou kou (the same paragraph)*, it refers to not to the current paragraph, but to a paragraph that was previously mentioned in the text (i.e. the fouth example in Figure 5). It is usually the most recent documents referenced before and near the current reference in the text. To resolve this type, we need to keep a history of the references found so far.

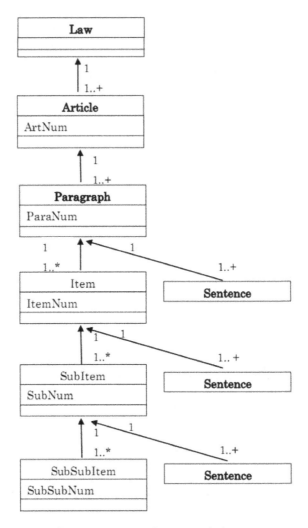

**Fig. 6** The architecture of JNPL corpus on reference resolution

## 4  Experiments

### 4.1  Building a Corpus on Reference Resolution

We used the Japanese National Pension Law (JNPL) corpus for reference resolution to conduct experiments[2]. In this corpus, all references that refer to the inside of JNPL are marked. It does not include the references that refer to other laws, or refer to ambiguous ranges.

---

[2] The corpus was manually built by a person who graduate a law school and a person who worked in the government and a student of a law school.

The architecture of JNPL is shown in Figure 6. The law consists of articles, an article consists of paragraphs, and a paragraph consists of sentences. A sentence may belong to an item, a sub-item, or a sub-sub-item of a paragraph. Bellows are some statistics about this corpus:

- # Articles: 99 articles
- # Paragraphs: 931 paragraphs
- # References: 748 references

## 4.2 Evaluation Method

We divided the JNPL corpus into 10 sets, and conducted 10-fold cross-validation tests for all experiments. For the reference detection task which we modeled as a sequence labeling problem, we evaluated the performance of our systems by precision, recall, and the F1 score as follows:

$$P = \frac{\#correctly\ detected\ references}{\#detected\ references},\ R = \frac{\#correctly\ detected\ references}{\#gold\ references},$$
$$F1 = \frac{2*Precision*Recall}{Precision+Recall}.$$

For the reference resolving step, because we conduct experiments using the gold references of the corpus, we use the accuracy score to evaluate the experimental results as follows:

$$Accuracy = \frac{\#correctly\ resolved\ references}{\#gold\ references}.$$

## 4.3 Results

### 4.3.1 Reference Detection

In this step, we chose a rule-based approach as a baseline model. This baseline is also the main approach of previous work. Based on the structure of references, we designed corresponding regular expressions to catch them up. When tested on the corpus of JNPL, we obtained 65.53% precision, 98.32% recall, and 78.4% in the F1 score. Using the rule-based approach, we obtained very high recall when adding more rules to the system. Along with this, however, the system also detected out much more referring expressions which seem to be references. Unfortunately, they are not real references when used in their specific contexts (i.e. the italic parts in: *Daiichigou* houtei, *dainigou* hihokensha, *korera* no, *sono*, etc.). This leads to low precision of the system.

To improve the performance of the reference detection system, we built another model using a machine learning approach. To learn the model, we used the CRF++[3] tool written by Kudo. In extracting feature sets, we used a combination of

---

[3] http://crfpp.googlecode.com/svn/trunk/doc/index.html

**Table 1** Experimental results for the reference detection task (%) (W - Word, R - Read, P - POS, C - Chunk)

| Label Settings | W | | | W + R | | | W + R + P | | | W + P + R + C | | |
|---|---|---|---|---|---|---|---|---|---|---|---|---|
| | P | R | F1 | P | R | F1 | P | R | F1 | P | R | F1 |
| IOB | 90 | 89.3 | **89.6** | 90.3 | 89.9 | **90** | 90.5 | 91 | **90.7** | 90.5 | 91.2 | **90.8** |
| IOE | 89.7 | 88.9 | **89.2** | 89.9 | 90 | **89.9** | 90.1 | 91 | **90.5** | 90 | 90.5 | **90.2** |
| FIL | 90.3 | 89 | **89.6** | 90.6 | 89.8 | **90.1** | 90.5 | 90.8 | **90.6** | 91.6 | 91.8 | **91.6** |

n-gram ($n \leq 3$) of words, part-of-speech tags, reading, and the information of chunking. These were taken from the output of the Cabocha[4]. Cabocha [7] is a Japanese dependency structure analyzer based on Support Vector Machine (SVM) [19]. When doing the experiments, we used three kinds of label settings which are the IOB, IOE, and FIL notations [9]. We also investigated the task using different combinations of feature sets to find out which feature sets yield better performances.

Experimental results of the reference detection step are given in Table 1. We realize that usually the more feature sets we use the better the results are. This is reasonable because these feature sets yield more benefits to the model. We obtained the highest performance of 91.6% in the F1 score using the FIL notation on feature sets which are based on words, readings, part-of-speech tags and chunking information. Among three notations, the FIL yields the best results, the IOE yields the lowest results, and the IOB yields the intermediate results.

These experimental results also showed that the machine learning approach yields the better performance than the rule-based approach in the context of Japanese national pension laws (from 78.4% to 91.6% in the F1 score).

### 4.3.2 Reference Resolving

By using regular expressions and context information as described in Section 2, we caught up to 96.18% of correct positions of all references (see results in Figure 2).

### 4.3.3 End-to-End Performance

The last experiment that was conducted is the end-to-end system in which the input of the current step will be the output of the previous step. In the whole task setting, we got the result of 88.83% recall, 88.27% precision, and 88.55% in the F1 score in reference resolving for references needed to be resolved in the corpus (see results in Figure 2). This performance is quite good to implement the system in real applications.

## 5 Error Analysis

This part analyzes cases that are not correctly caught by our framework. We show some examples of error cases in Figure 7.

---

[4] http://code.google.com/p/cabocha/

**Table 2** Experimental results of the reference resolver and the end-to-end system (%)

|  | Accuracy | | |
| --- | --- | --- | --- |
| Reference Resolver | **96.18** | | |
|  | P | R | **F1** |
| End-to-end system | 88.27 | 88.83 | **88.55** |

By observing the output of the final system, we realized that in the reference detection step, most of errors can be attributed to the following reasons:

- Detected out but they are not labeled as references in the corpus. The reason is that these references refer to the passages which are beyond the scope of our JNPL corpus (i.e. the first example).
- Only detected out a part of a very long reference (i.e. the second example), or a complex reference (i.e. the third example).

| No. | Examples |
| --- | --- |
| 1 | 第百三十九条<br>Article 139 |
| 2 | 第三十七条 、 第三十七条の二 、 第四十九条第一項 、 第五十二条の二第一項及び第五十二条の三第一項<br>Article 37, Article 37-2, Article 49 Para 1, Article 52-2Para 1 and Article 52-3 Para 1 |
| 3 | この項の本文若しくは次項又は他の法令<br>The texts of this para or next para or other laws and regulations |

**Fig. 7** Some examples of errors in detecting references

In resolving references, there are some cases where our system was unable to recognize the correct position (i.e. the third example). The reason is that they fall under some exceptions that do not conform to the regular expressions that were described.

## 6 Conclusion

In this paper, we investigated the task of reference resolution in legal texts. Based on the characteristics of reference phenomena in legal texts, we proposed a framework with two steps to solve the task. This framework can be easily derived to extend to other types of laws. Experiments on the Japanese National Pension Law corpus which was built manually showed good results. The results also showed that using a machine learning approach yielded better performance than a rule-based approach. Our work provides promising results for further studies on this interesting task.

**Acknowledgment.** This work was partly supported by Grant-in-Aid for Scientific Research, Education and Research Center for Trustworthy e-Society, and JAIST Research Grants.

We also would like to give special thanks to two people who made our corpus. They are a person who graduated from a law school and worked in the government and a student of a law school.

# References

1. Bolioli, A., Dini, L., Mercatali, P., Romano, F.: For the automated mark-up of italian legislative texts in xml. In: Proceedings of International on Legal Knowledge and Information Systems (Jurix), pp. 21–30 (2002)
2. Finkel, J.R., Manning, C.D.: Hierarchical Joint Learning: Improving Joint Parsing and Named Entity Recognition with Non-Jointly Labeled Data. In: Proceedings of the 48th Annual Meeting of the Association for Computational Linguistics (ACL), pp. 720–728 (2010)
3. Katayama, T.: The curent status of the art of the 21st coe programs in the information sciences field. verifiable and evolvable e-society - realization of trustworthy e-society by computer science. Information Processing Society of Japan 46(5), 515–521 (2005) (in Japanese)
4. Katayama, T.: Legal engineering - an engineering approach to laws in e-society age. In: Proceedings of International Workshop on Juris-informatics, JURISIN (2007)
5. Katayama, T., Shimazu, A., Tojo, S., Futatsugi, K., Ochimizu, K.: e-Society and legal engineering. Journal of the Japanese Society for Artificial Intelligence 23(4), 529–536 (2008) (in Japanese)
6. Kudo, T., Yamamoto, K., Matsumoto, Y.: Applying conditional random fields to japanese morphological analysis. In: Proceedings of Empirical Methods in Natural Language Processing (EMNLP), pp. 230–237 (2004)
7. Kudo, T., Matsumoto, Y.: Japanese Dependency Analysis using Cascaded Chunking. In: Proceedings of the 6th Conference on Natural Language Learning 2002 (COLING 2002 Post-Conference Workshops) (CoNLL 2002), pp. 63–69 (2002)
8. Lafferty, J., McCallum, A., Pereira, F.: Conditional random fields: Probabilistic models for segmenting and labeling sequence data. In: Proceedings of International Conference on Machine Learning, ICML, pp. 282–289 (2001)
9. Ludtke, D., Sato, S.: Fast base np chunking with decision trees experiments on different pos tag settings. In: Proceedings of Conferences on Computational Linguistics and Natural Language Processing (CICLing), pp. 139–150 (2003)
10. Maat, E., Winkels, R., Engers, T.: Automated detection of reference structures in law. In: Proceedings of International on Legal Knowledge and Information Systems (Jurix), pp. 41–50 (2006)
11. Martínez-González, M., de la Fuente, P., Vicente, D.-J.: Reference extraction and resolution for legal texts. In: Pal, S.K., Bandyopadhyay, S., Biswas, S. (eds.) PReMI 2005. LNCS, vol. 3776, pp. 218–221. Springer, Heidelberg (2005)
12. Ng, V.: Supervised noun phrase coreference research: The first fifteen years. In: Proceedings of Annual Meeting of the Association for Computational Linguistics (ACL), pp. 1396–1411 (2010)
13. Ng, V., Cardie, V.: Improving machine learning approaches to coreference resolution. In: Proceedings of Annual Meeting of the Association for Computational Linguistics (ACL), pp. 104–111 (2002)

14. Palmirani, M., Brighi, R., Massini, M.: Automated extraction of normative references in legal texts. In: Proceedings of International Conference on Artificial Intelligence and Law (ICAIL), pp. 105–106 (2003)
15. Pitler, E., Bergsma, S., Lin, D., Church, K.: Using Web-scale N-grams to Improve Base NP Parsing Performance. In: Proceedings of the 23rd International Conference on Computational Linguistics (Coling 2010), pp. 886–894 (2010)
16. Ponzetto, S., Strube, M.: Exploiting semantic role labeling, wordnet and wikipedia for coreference resolution. In: Proceedings of Human Language Technologies: Annual Conference of the North American Chapter of the Association for Computational Linguistics (HLT-NAACL), pp. 192–199 (2006)
17. Rahman, A., Ng, V.: Supervised models for coreference resolution. In: Proceedings of Empirical Methods in Natural Language Processing (EMNLP), pp. 968–977 (2009)
18. Sha, F., Pereira, F.: Shallow parsing with conditional random fields. In: Proceedings of Conference of the North American Chapter of the Association for Computational Linguistics (NAACL), pp. 213–220 (2003)
19. Vapnik, V.N.: Statistical Learning Theory. Wiley-Interscience (1998)
20. Yang, X., Su, J., Zhou, G., Tan, C.: An np-cluster based approach to coreference resolution. In: Proceedings of International Conference on Computational Linguistics (COLING), pp. 226–232 (2004)
21. Yang, X., Zhou, G., Su, J., Tan, C.: Coreference resolution using competitive learning approach. In: Proceedings of Annual Meeting of the Association for Computational Linguistics (ACL), pp. 176–183 (2003)

# Paragraph Alignment for English-Vietnamese Parallel E-Books

Quang-Hung Le, Duy-Cuong Nguyen, Duc-Hong Pham,
Anh-Cuong Le, and Van-Nam Huynh

**Abstract.** Parallel corpora are among the most important linguistic resources used in multilingual language processing such as statistical machine translation, cross-language information retrieval, and so on. Manually constructing such corpora takes a very high cost while there are many available parallel e-books containing a large number of parallel texts. This paper focuses on the task of aligning paragraphs of English-Vietnamese parallel e-books. A new method for this alignment is proposed. By doing an experiment we have collected an English-Vietnamese parallel corpus which contains nearly 40,000 sentence pairs aligned at the paragraph level.

## 1 Introduction

There are different levels of parallel items in a parallel corpus, these items include document, paragraph, phrase, and word. Statistical machine translation [1] uses parallel sentences as the input for the alignment module to produce word translation probabilities. Cross language information retrieval [2] uses parallel texts for determining corresponding information in both questioning and answering. Parallel texts are also used for acquisition of lexical translation. However, the available parallel corpora are not only in relatively small size, but also unbalanced even in the major languages [3]. Many studies focused on extracting bilingual texts from the Web, such as [4, 5]. However, obtaining a corpus with high accuracy is still a challenge because the texts presented in the Internet are very "noisy". Meanwhile, there are

Quang-Hung Le
Faculty of Information Technology, Quynhon University, Vietnam

Duy-Cuong Nguyen · Duc-Hong Pham · Anh-Cuong Le
University of Engineering and Technology, Vietnam National University, Hanoi, Vietnam

Van-Nam Huynh
School of Knowledge Science, Japan Advanced Institute of Science and Technology,
Ishikawa, Japan

V.-N. Huynh et al. (eds.), *Knowledge and Systems Engineering, Volume 2,*            251
Advances in Intelligent Systems and Computing 245,
DOI: 10.1007/978-3-319-02821-7_23, © Springer International Publishing Switzerland 2014

many available parallel e-books containing a large number of parallel texts which are carefully translated.

The task of paragraph alignment is to match corresponding paragraphs in a text from one language to paragraphs in a translation of that text in another language [6]. Different methods have been proposed for finding paragraph alignments between parallel texts [7, 8, 9]. In our opinion, these methods can be classified into two major approaches: statistical based [10, 11], and linguistic knowledge based (also can call content based) [12, 13]. The first approach exploits the correlation of length of text units (paragraphs or sentences) in different languages and tries to establish the correspondence between the units of the expected size [14]. The size can be measured by using a number of words or characters. However, texts (in the parallel e-books) are often heavily reformatted in translation, so they don't only contain one-to-one paragraph mappings, i.e. one English paragraph can match to two or more Vietnamese paragraphs, and vice versa. In this case, statistical methods which rely on structure correspondence such as words or characters may not perform well. The second approach uses linguistic data (usually, dictionaries) for establishing the correspondence between structural units. However, dictionary-based methods coverage limitations [6] mean that we will often encounter problems establishing a correspondence in non-general domains.

In this paper, we follow the second approach by extending Gelbukh's idea [14] and combining a Statistical Machine Translation System (SMTS) to paragraph alignment. The contribution of our work includes two main points: (1) we propose a method to paragraph alignment for the English-Vietnamese parallel e-books; (2) we will extract a large number of parallel paragraphs with a high accuracy.

The idea of using a machine translation system to the alignment problem is also mentioned in [15, 16] by Sennrich et al. They have used BLEU-based heuristics or the length-based algorithm by Gale and Church [17] to sentence alignment for parallel texts. Note that we differ from the Sennrich's works. Here, we focus on aligning paragraphs for the parallel e-books. Our method has advantages that it can detect some kinds of mapping between parallel paragraphs and reduce the search space[1] in comparison with those existing methods.

For the remaining of the paper, the section 2 will present our method for the task of paragraph alignment. Section 3 shows our experimental results. Finally, the conclusion is presented in Section 4.

## 2   The Proposed Method

### 2.1   The Problem

Given an English book $E$ consisting of $I$ paragraphs $pe_1, ..., pe_I$ and a Vietnamese book $V$ consisting of $J$ paragraphs $pv_1, ..., pv_J$, we define a link $l = (i, j)$ to be existed if $pe_i$ and $pv_j$ are the translations (or partly translated) of each other. Then,

---

[1] By using some predefined patterns which will be described in the next sections.

an alignment $A$ (between $E$ and $V$) is defined as a subset of the Cartesian product of the paragraph positions:

$$A \subseteq \{(i,j) : i = 1,...,I; j = 1,...,J\} \tag{1}$$

The problem of this study is finding the alignment $A$, which connects paragraphs in the English book to their translation paragraphs in the corresponding Vietnamese book.

## 2.2  Paragraph Alignment

In this section, we first describe the work called pre-processing step which has to be processed before paragraph alignment. Second, we present how to measure the similarity between two paragraphs. Finally, we show the algorithm to paragraph alignment for parallel e-books.

### 2.2.1  Pre-processing

The e-books are originally in PDF format then they must be converted into plain texts. However, the information about paragraph boundaries is lost during this conversion. Therefore we must recover the paragraph boundaries. To do this work we first use an available toolkit to convert e-books from PDF format to text, and then restoring the boundaries of paragraphs by our own codes. Table 1 shows an example about recovering the structures of original paragraphs.

Next, we use a statistical machine translation system to translate English books into Vietnamese ones, which will be used for measuring the similarity between paragraphs in an English book and its corresponding Vietnamese book. By using statistical machine translation we can reduce ambiguity of lexical translation which usually occurs in previous works using bilingual dictionaries.

In our method, anchor points are used to limit the effect of cascaded errors during paragraph alignment. With this, if an error occurs, it will not affect alignment of the whole text. The algorithm to paragraph alignment will process the units between the anchor points. Some previous works [14, 15, 16] have used a similarity score between the translated source text and the target text to identify the anchor points (e.g., using the BLEU score). These anchor points are short units with very high similarity. However, by this way, we cannot achieve the desired results. For example, the positions of two adjacent anchor points too far or too near. In some cases, the anchor points are determined incorrectly. All these problems may cause a low accuracy of the alignment.

From our observation, in almost English-Vietnamese parallel e-books, some units can be used as anchor points: "Part" $\leftrightarrow$ "Phn", "Chapter" $\leftrightarrow$ "Chng", etc. As we can see in the Figure 1, the anchor points $(ae_1, av_1)$, $(ae_2, av_2)$, ..., $(ae_n, av_n)$ in an English-Vietnamese parallel e-book pair is determined by the units: "Chapter 1" $\leftrightarrow$ "Chng 1", "Chapter 2" $\leftrightarrow$ "Chng 2", ..., "Chapter n" $\leftrightarrow$ "Chng n". Here, an unit between two anchor points can be the whole text of a chapter or a section in an e-book. Note that the number of anchor points in the e-book are not much. So that, we can be easy to detect them by manually. In our method, the units as described above will be used to automatically detect anchor points. To do this work, we implement two steps:

1. We define the list $L$, which includes units (predefined) such as patterns.
2. For each unit/pattern $l \in L$, if $l$ matches with an unit in the e-book then an anchor point is defined.

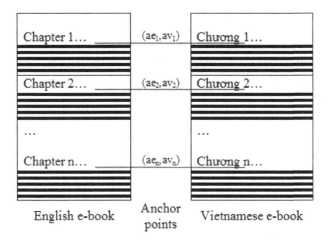

**Fig. 1** An example of the anchor points $(ae_1, av_1)$, $(ae_2, av_2)$, $\ldots$, $(ae_n, av_n)$ in an English-Vietnamese parallel e-book pair

### 2.2.2 Similarity Measure

Suppose that we are working with a parallel e-book pair of English and Vietnamese. The English book $E$ consists of $I$ paragraphs $pe_1, \ldots, pe_I$ and the Vietnamese book $V$ consists of $J$ paragraphs $pv_1, \ldots, pv_J$. Let $T$ is a Vietnamese translation of the English book $E$ and $pt_i$ is a Vietnamese translation of the paragraph $pe_i$ (in the English book). Let $S_n(pt_i)$ and $D_n(pv_j)$ are sets of n-gram of paragraph $pt_i$ and $pv_j$, respectively. Now, similarity between $pt_i$ and $pv_j$ are defined as follows.

$$
\begin{aligned}
Score_n(pt_i, pv_j) &= Similarity(pt_i, pv_j) \\
&= \frac{|S_n(pt_i) \cap D_n(pv_j)|}{|S_n(pt_i) \cup D_n(pv_j)|}
\end{aligned} \tag{2}
$$

In this equation, $score_n$ is the resemblance metric of two paragraphs when segmented by $n$, $0 \le score_n \le 1$.

**Table 1** An example about recovering the structure of paragraphs. In the first paragraph, the information about paragraph boundaries is lost. The paragraph structure is recovered in the second paragraph.

| **The information about paragraph boundaries is lost (when converting from PDF format to plain text).** |
|---|
| Your Highnesses, as Catholic Christians, and princes who love and promote the holy Christian faith, and are enemies of the doctrine of Mahomet, and of all idolatry and heresy, determined to send me, Christopher Columbus, to the above-mentioned countries of India, to see the said princes, people, and territories, and to learn their disposition and the proper method of converting them to our holy faith; and furthermore directed that I should not proceed by land to the East, as is customary, but by a Westerly route, in which direction we have hitherto no certain evidence that anyone has gone. |
| **The paragraph structure was restored.** |
| Your Highnesses, as Catholic Christians, and princes who love and promote the holy Christian faith, and are enemies of the doctrine of Mahomet, and of all idolatry and heresy, determined to send me, Christopher Columbus, to the above-mentioned countries of India, to see the said princes, people, and territories, and to learn their disposition and the proper method of converting them to our holy faith; and furthermore directed that I should not proceed by land to the East, as is customary, but by a Westerly route, in which direction we have hitherto no certain evidence that anyone has gone. |

### 2.2.3 The Algorithm to Paragraph Alignment

From our observation, the alignment of paragraphs in the source and target texts is not only 1-to-1. That mean, there are other kinds of mapping between parallel paragraphs. However, it commonly belongs to five kinds: 1-to-1, 1-to-2, 1-to-3, 3-to-1, or 2-to-1. The following is an example of a mapping between one paragraph in the English e-book and two paragraphs in the Vietnamese e-book:

- English text (page 65, Steve Jobs, English book).

  – *There was another reason that Joanne was balky about signing the adoption papers. Her father was about to die, and she planned to marry Jandali soon after. She held out hope, she would later tell family members, sometimes tearing up at the memory, that once they were married, she could get their baby boy back.*

- Vietnamese text (page 21, Steve Jobs, Vietnamese book).

  – *Cng cn mt l do khc khin Joanne lc u khng khng khng k giy chuyn nhn con nui l v cha b sp cht v b d nh s kt hn vi Jandali ngay sau .*
  – *B hi vng rng sau khi ci nhau, h s thuyt phc dn c gia nh v nhn li con.*

The objective of this study is finding the alignment $A$, which connects paragraphs in the English book to paragraphs in the corresponding Vietnamese book. For each connection, we now must detect parallel units (includes one or many paragraphs). To do this work, we will calculate the similarities for all the matching pattern (1-to-1, 1-to-2, 1-to-3, 2-to-1, and 3-to-1) by using the function $Similarity(pt_i, pv_j)$ in the Equation (2). Then the pattern $(s,t)$ with the highest similarity score will be selected as in the Equation (3). And consequently we will obtain parallel unit texts $(u_s, u_t)$ from the best pattern $(s,t)$.

$$(s,t) = \arg\max \begin{cases} Similarity(pt_i, pv_j) \\ Similarity(pt_i, pv_j pv_{j+1}) \\ Similarity(pt_i, pv_j pv_{j+1} pv_{j+2}) \\ Similarity(pt_i pt_{i+1}, pv_j) \\ Similarity(pt_i pt_{i+1} pt_{i+2}, pv_j) \end{cases} \qquad (3)$$

Where, $i = 1, \ldots, I - 2, j = 1, \ldots, J - 2$. The algorithm to paragraph alignment is implemented in two stages. First, it finds for anchor points, and then it determines aligned units between the anchor points. The inputs to the algorithm are a parallel e-book pair of English-Vietnamese $(E,V)$ and the Vietnamese translation $T$ of the English book $E$. The output identifies the alignment $A$ between paragraphs. The algorithm to paragraph alignment is described as follows:

- Input: $E,V,T$
- Output: $A$

  1. $A = \emptyset$
  2. Finding the positions of anchor point pairs $(ae_1, av_1)$, $(ae_2, av_2)$, ..., $(ae_n, av_n)$.
  3. For each text unit between two anchor point pairs $(ae_i, av_i)$ and $(ae_{i+1}, av_{i+1})$:
     a. Using the Equation (2) to calculate the similarities for all the patterns 1-to-1, 1-to-2, 1- to-3, 2-to-1, and 3-to-1.
     b. Selecting the pattern $(s,t)$ with the highest similarity score defined by the Equation (3).
     c. Adding the pattern/link $l = (s,t)$ to $A$.

## 3 Experiments

### 3.1 Data Preparation

The data, which is used in our experiments, is extracted from four parallel e-books of English-Vietnamese. The English titles of them are: *Steve Jobs* (by Walter Isaacson), *The Open Society And Its Enemies* (by Karl R. Popper), *The World is Flat* and *The Lexus and the Olive Tree* by Thomas L. Friedman). These e-books were collected from the Internet, which are carefully translated by famous translators

**Table 2** Some parallel e-books of English-Vietnamese

| # | Author | English title | Paragraphs | Vietnamese title | Paragraphs |
|---|--------|---------------|------------|------------------|------------|
| 1. | Walter Isaacson | Steve Jobs | 1.968 | Steve Jobs | 1.948 |
| 2. | Karl R. Popper | The Open Society And Its Enemies | 950 | X Hi M V Nhng K th ca N | 904 |
| 3. | Thomas L. Friedman | The World is Flat | 1.114 | Th gii Phng | 1.348 |
| 4. | Thomas L. Friedman | The Lexus and the Olive Tree | 1.349 | Chic Lexus v Cy Liu | 1.391 |

in Vietnam[2]. The table 2 presents their detail information. The data size is about 10.1 MB (only text data). They consist of 5,381 English paragraphs and 5,591 Vietnamese paragraphs. In the pre-processing step, we first convert e-books from PDF format to plain texts. To do this work, we first used the freely available toolkit *PDF to Text*[3], and then recover paragraph boundaries (using our own codes). In the next step, we use Google translator as a statistical machine translation system to translate English books into Vietnamese ones. Finally, anchor point pairs are manually detected manually for both English and Vietnamese books.

## 3.2 Experimental Results

To measure the similarity between *pe* and *pv*, we use Equation (2) with $n = 1$. We randomly select 200 patterns of paragraphs from the set of the experimental data to evaluate the performance of our method. The experimental results are presented in Table 3. The precision of the method in this experiment was 97%. These results have shown that the our proposed method is so effective for the task of extracting

**Table 3** Alignment results for 200 patterns of paragraphs

| Pattern | Correct | Incorrect |
|---------|---------|-----------|
| 1-1 | 158 | 2 |
| 1-2 | 16 | 2 |
| 1-3 | 0 | 0 |
| 2-1 | 16 | 2 |
| 3-1 | 4 | 0 |
| Total | 194 | 6 |

---

[2] *Steve Jobs* was translated by Bookstore Alezaa.com, *The Open Society And Its Enemies* was translated by Nguyen Quang A, *The World is Flat* and *The Lexus and the Olive Tree* were translated by Nguyen Quang A, Cao Viet Dung, Nguyen Tien Phong.

[3] http://www.pdf-technologies.com/pdf-library-pdf-to-text.aspx

**Table 4** Some parameters of corpus

| Parameter | English | Vietnamese |
|-----------|---------|------------|
| Words | 771,565 | 1,035,358 |
| Sentences | 39,066 | 36,104 |
| Paragraphs | 5,042 | 5,042 |

parallel paragraphs. From these obtained parallel paragraphs we have generated nearly 40,000 sentence pairs. And we can obtain a bigger parallel corpus if we have more parallel e-books.

## 4   Conclusion

This paper presents our work on aligning paragraphs for parallel English and Vietnamese e-books. We have proposed a method for this task, which is based on using predefine anchor points and using a new way of measuring similarity between two paragraphs in source and target languages (using a statistical machine translation system, and using a set of matching patterns between paragraphs).

The obtained results have shown the effectiveness of the proposal. We believe that parallel e-books are absolutely the valuable resources of knowledge containing bilingual texts with high quality (they are carefully translated and are not "noise" as other resources, e.g., the Web). And by using our method we can obtain enough parallel sentences for many natural language processing tasks such as statistical machine translation.

**Acknowledgment.** This work is partly supported by the project QGTD.12.21 "Studying Methods for Analyzing and Summarizing Opinions from Internet and Building an Application" funded by VNU-Vietnam National University of Hanoi.

## References

1. Brown, P.F., Cocke, J., Pietra, S.A.D., Pietra, V.J.D., Jelinek, F., Lafferty, J.D., Mercer, R.L., Roossin, P.S.: A statistical approach to machine translation. Comput. Linguist. 16(2), 79–85 (1990)
2. Davis, M.W., Dunning, T.: A trec evaluation of query translation methods for multilingual text retrieval. In: TREC (1995)
3. Hung, L.Q., Cuong, L.A.: Extracting parallel texts from the web. In: Proceedings of the 2010 Second International Conference on Knowledge and Systems Engineering, KSE 2010, pp. 147–151. IEEE Computer Society, Washington, DC (2010)
4. Zhao, B., Vogel, S.: Adaptive parallel sentences mining from web bilingual news collection. In: Proceedings of the 2002 IEEE International Conference on Data Mining, ICDM 2002, pp. 745–748. IEEE Computer Society, Washington, DC (2002)

5. Resnik, P., Smith, N.A.: The web as a parallel corpus. Comput. Linguist. 29(3), 349–380 (2003)
6. Collier, N., Ono, K., Hirakawa, H.: An experiment in hybrid dictionary and statistical sentence alignment. In: Proceedings of the 17th International Conference on Computational Linguistics, vol. 1, pp. 268–274. Association for Computational Linguistics (1998)
7. Tay, R., Ibrahim, T.: Research on paragraph alignment technology in chinese-uighur bilingual corpus. Journal of Xinjiang University (Natural Science Edition) 1, 021 (2010)
8. Rasooli, M.S., Kashefi, O., Minaei-Bidgoli, B.: Extracting parallel paragraphs and sentences from english-persian translated documents. In: Salem, M.V.M., Shaalan, K., Oroumchian, F., Shakery, A., Khelalfa, H. (eds.) AIRS 2011. LNCS, vol. 7097, pp. 574–583. Springer, Heidelberg (2011)
9. Gupta, A., Pala, K.: A generic and robust algorithm for paragraph alignment and its impact on sentence alignment in parallel corpora (2012)
10. Brown, P.F., Lai, J.C., Mercer, R.L.: Aligning sentences in parallel corpora. In: Proceedings of the 29th Annual Meeting on Association for Computational Linguistics, ACL 1991, pp. 169–176. Association for Computational Linguistics, Stroudsburg (1991)
11. Gale, W.A., Church, K.W.: A program for aligning sentences in bilingual corpora. Comput. Linguist. 19(1), 75–102 (1993)
12. Chen, S.F.: Aligning sentences in bilingual corpora using lexical information. In: Proceedings of the 31st Annual Meeting on Association for Computational Linguistics, ACL 1993, pp. 9–16. Association for Computational Linguistics, Stroudsburg (1993)
13. Meyers, A., Kosaka, M., Grishman, R.: A multilingual procedure for dictionary-based sentence alignment. In: Farwell, D., Gerber, L., Hovy, E. (eds.) AMTA 1998. LNCS (LNAI), vol. 1529, pp. 187–198. Springer, Heidelberg (1998)
14. Gelbukh, A., Sidorov, G., Vera-Félix, J.Á.: Paragraph-level alignment of an english-spanish parallel corpus of fiction texts using bilingual dictionaries. In: Sojka, P., Kopeček, I., Pala, K. (eds.) TSD 2006. LNCS (LNAI), vol. 4188, pp. 61–67. Springer, Heidelberg (2006)
15. Sennrich, R., Volk, M.: Mt-based sentence alignment for ocr-generated parallel texts. In: The Ninth Conference of the Association for Machine Translation in the Americas (AMTA 2010), Denver, Colorado (2010)
16. Sennrich, R., Volk, M.: Iterative, mt-based sentence alignment of parallel texts (2011)
17. Gale, W.A., Church, K.W.: A program for aligning sentences in bilingual corpora. Computational Linguistics 19(1), 75–102 (1993)

# Part-of-Speech Induction for Vietnamese

Phuong Le-Hong and Thi Minh Huyen Nguyen

**Abstract.** This paper presents a method for automatically inducing the parts-of-speech of the Vietnamese language from a large text corpus. We first build a class-based bigram language model using several statistical algorithms assigning words to classes based on their ability to combine with neighbouring words. We then show that this model is able to extract word classes that have the flavor of either syntactically based or semantically based groupings of Vietnamese words, which are the long disputed approaches among the Vietnamese linguistic community. Finally, the quality of word clusters is quantitatively evaluated when word cluster features are used to improve the accuracy of a statistical part-of-speech tagger for Vietnamese.

## 1 Introduction

In linguistics, a part-of-speech (POS) or a lexical category is a linguistic category of words (e.g. noun, verb) which is generally defined by its morpho-syntactic behaviour. POS tagging is the process of labeling each word in a text with a particular part-of-speech tag based on both its definition and its context. This task is important as it is very useful for a wide variety of text processing areas such as syntactic and semantic parsing, information retrieval, text to speech, word sense disambiguation, *etc.*

In the traditional POS tagging problem, we have a predefined tagset, *i.e.* the list of all possible categories for each word. The POS induction problem, discussed here, concerns the determination of a tagset which is suitable for a given language. In recent years, POS induction has been one of the most popular tasks in research on unsupervised natural language processing (NLP), and is also of theoretical interest in language acquisition and learnability. It is said that syntactic category information is part of the basic knowledge about language that children must learn before they can acquire more complicated structures [1].

Phuong Le-Hong · Thi Minh Huyen Nguyen
University of Science, Vietnam National University, Hanoi
e-mail: {phuonglh,huyenntm}@vnu.edu.vn

V.-N. Huynh et al. (eds.), *Knowledge and Systems Engineering, Volume 2,*          261
Advances in Intelligent Systems and Computing 245,
DOI: 10.1007/978-3-319-02821-7_24, © Springer International Publishing Switzerland 2014

The categorization of Vietnamese words has been greatly discussed and debated among Vietnamese linguists. Although most linguists agree to use both semantic and syntactic criteria to define POS categories in Vietnamese, they do not agree upon the actual classification criteria and the resulting word categories [2]. The common drawback is that they are still subjective in the determination of word classes based on their personal perspective or experience, for example the tagsets defined in [3], [4] and [5]. As a compromise, the computer scientists and linguists of the Vietnamese Language and Speech Processing (VLSP) project suggested a practical tagset composed of only 17 different parts-of-speech [6]. This tagset includes rather coarse categories; for example, there is no sub-categories of verbs or adjectives.

This lack of consensus among linguists motivates us to investigate a principled method for inducing Vietnamese word classes so that the different views of linguists can be objectively evaluated. Our motivation also stems from the fact that we try to improve the accuracy of a number of Vietnamese processing tools that we have developed over the years.

In this paper, we present a statistical model for automatically inducing Vietnamese word classes from a large text corpus. We first build a class-based bigram language model using statistical algorithms for assigning words to classes based on their ability of combination with neighbouring words. We then show the efficiency of this model in extracting word classes that have the "flavor" of either syntactically based or semantically based groupings. Our result quantitatively justifies the dominant approach in the linguistic community and suggests a way to objectively classify Vietnamese words.

The remainder of this paper is structured as follows: Section 2 describes the class-based bigram language model and three algorithms to induce word classes from a corpus; Section 3 presents empirical results and discussions; finally, Section 4 concludes the paper and gives some directions for future work.

## 2 POS Induction Algorithms

Many POS induction systems have been developed in recent years as unsupervised learning has become an active area in NLP. Most systems use sophisticated machine learning approaches. In a recent survey, Christodoulopoulos et al. [7] did a comprehensive comparison and evaluation of seven different POS induction systems spanning nearly 20 years of research. They found the surprising result that, when using today's hardware and large corpora, some of the oldest POS induction systems using relatively simple models and algorithms work as well as or better than systems using newer and far more sophisticated and time-consuming machine learning methods.

This interesting result encourages us to develop a system using simpler algorithms to induce word classes for Vietnamese. We choose the class-based $n$-gram language model originally proposed by Brown et al. [8]. In comparison with Brown's system, we use a slightly different procedure to create initial word classes. In this section, we describe in detail the model and the algorithms used in our system.

## 2.1 Class-Based Bigram Language Model

Suppose that we partition a vocabulary of $V$ words into $C$ clusters using a function $\pi$ that maps a word $w_i$ to its class $c_i$. We say that a language model is an *n-gram class model* if it is an $n$-gram language model and for $1 \leq k \leq n$:

$$p(w_k|w_1, w_2, \ldots, w_{k-1}) = p(w_k|c_k)p(c_k|c_1, c_2, \ldots, c_{k-1}),$$

That is, in this model, the probability of a word given the word-history is defined by its probability given its class and the probability of its class given the class-history.

Brown *et al.* [8] propose a bottom-up algorithm to derive a hierarchical clustering of words[1]. Each word appearing in the input sequence of $T$ word occurrences $t_1, t_2, \ldots, t_T$ is considered as a cluster. At each stage, the algorithm merges two clusters so as to maximize the quality of the resulting cluster, defined as the log-probability of the input text normalized by the text length:

$$Q(\pi) = \frac{1}{T} \log p(t_1 t_2 \ldots t_T).$$

This quantity in a bigram class model can be approximated as

$$Q(\pi) = \frac{1}{T} \log \prod_{i=1}^{T} p(\pi(w_i)|\pi(w_{i-1}))p(w_i|\pi(w_i))$$

$$= \frac{1}{T} \sum_{i=1}^{T} \log[p(\pi(w_i)|\pi(w_{i-1}))p(w_i|\pi(w_i))],$$

where $t_0$ is the dummy word added to the start of the text. Let $C(w)$ be the number of occurrences of a word $w$ in the text and $C(w, w')$ the number of occurrences of a bigram $(w, w')$. We define the number of occurrences of a cluster $c$ to be $C(c) = \sum_{w \in c} C(w)$ and that of a cluster pair $(c, c')$ to be $C(c, c') = \sum_{w \in c, w' \in c'} C(w, w')$. We empirically have $p(w) = \frac{C(w)}{T}$, $p(c) = \frac{C(c)}{T}$ and $p(c'|c) = \frac{C(c, c')}{\sum_{c'} C(c, c')}$. Then

$$p(c, c') = p(c)p(c'|c) = \frac{C(c)}{T} \frac{C(c, c')}{\sum_{c'} C(c, c')}$$

$$= \frac{C(c, c')}{T} \frac{C(c)}{\sum_{c'} C(c, c')}.$$

Since $C(c)$ and $\sum_{c'} C(c, c')$ are the numbers of words for which the class is $c$ in the strings $t_1 t_2 \ldots t_T$ and $t_1 t_2 \ldots t_{T-1}$ respectively, the quantity $\frac{C(c)}{\sum_{c'} C(c, c')}$ tends to 1 as $T$ tends to infinity. Thus $p(c, c')$ tends to the relative frequency of $c, c'$ as consecutive classes in the training text. The quality of the model can be rewritten as

---

[1] In this algorithm, each word belongs to exactly one cluster. In other words, the algorithm generates a hard clustering.

$$Q(\pi) = \frac{1}{T} \sum_{i=1}^{T} \log[p(c_i|c_{i-1})p(w_i|c_i)]$$

$$= \frac{1}{T} \sum_{i=1}^{T} \log\left[\frac{p(c_i|c_{i-1})}{p(c_i)} p(w_i|c_i)p(c_i)\right]$$

$$= \sum_{w_0,w_1} \frac{C(w_0,w_1)}{T} \log\left[\frac{p(c_1|c_0)}{p(c_1)} p(w_1|c_1)p(c_1)\right]$$

$$= \sum_{w_0,w_1} \left[\frac{C(w_0,w_1)}{T} \log\frac{p(c_1|c_0)}{p(c_1)} + \frac{C(w_0,w_1)}{T} \log[p(w_1)]\right].$$

Note that

$$\sum_{w_0,w_1} \frac{C(w_0,w_1)}{T} \log\frac{p(c_1|c_0)}{p(c_1)} = \sum_{c_0,c_1} \frac{C(c_0,c_1)}{T} \log\frac{p(c_1|c_0)}{p(c_1)}$$

and

$$\sum_{w_0,w_1} \frac{C(w_0,w_1)}{T} \log[p(w_1)] = \sum_{w_1} \frac{\sum_{w_0} C(w_0,w_1)}{T} \log[p(w_1)]$$

$$= \sum_{w_1} \frac{C(w_1)}{T} \log[p(w_1)]$$

$$= p(w_1) \log[p(w_1)],$$

we obtain

$$Q(\pi) = \sum_{c_0,c_1} \frac{C(c_0,c_1)}{T} \log\frac{p(c_1|c_0)}{p(c_1)} + \sum_{w_1} p(w_1) \log[p(w_1)]$$

$$= \sum_{c_0,c_1} p(c_0,c_1) \log\frac{p(c_0,c_1)}{p(c_0)p(c_1)} + \sum_{w} p(w) \log[p(w)]$$

$$= I(c_0,c_1) - H(w),$$

where $H(w)$ is the entropy of the unigram word distribution and $I(c_0,c_1)$ is the average mutual information of adjacent classes. As $H(w)$ is not affected by $\pi$, the partition that maximizes $I(c_0,c_1)$ is the one that maximizes $Q(\pi)$:

$$Q(\pi) \rightarrow \max \Leftrightarrow I(c_0,c_1) \rightarrow \max.$$

There is no exact method for finding a clustering that maximizes the average mutual information. However, we can use an approximate search procedure to make optimizations.

## 2.2 Naive Algorithm

We can use a naive approach to find the approximately optimal clustering as follows:

1. Initially, each word is assigned to a distinct class;
2. Compute the average mutual information between adjacent classes;
3. Merge the pair of classes for which the loss in average mutual information is least;
4. Repeat two steps above. After $V - C$ merges, $C$ classes remain.

At the $i$-th merge, there are $V - i$ classes. So, there are approximately $(V - i)^2/2$ merges that we must investigate. To compute $I(c_0, c_1)$ we need sum over $(V - i)^2$ terms, each of which involves a logarithm. In consequence, picking the best merge takes $O(V^4)$ time. Altogether, we need to make $V - C$ merges. Therefore, the time complexity of this algorithm is of order $O(V^5)$, making its computation impossible except for small values of $V$.

## 2.3 Dynamic Programming Algorithm

Using the dynamic programming technique, Brown *et al.* [8] present an optimization that reduces the time from $O(V^5)$ to $O(V^3)$. It can compute the average mutual information remaining after a merge in constant time, independent of $V$. The optimized algorithm maintains a table containing the change in clustering quality due to each of the $O(V^2)$ merges. With the table, picking the best merge takes $O(V^2)$ time instead of $O(V^4)$ time. Since we use a variant of this algorithm, we present it here in detail to allow better understanding of the following section.

Suppose that we have already made $V - k$ merges resulting in $k$ classes $c_k(1), c_k(2), \ldots, c_k(k)$ and now want to investigate the merge of $c_k(i)$ with $c_k(j)$ for $1 \le i < j \le k$. Let $p_k(u, v) = p(c_k(u), c_k(v))$ be the probability that a word in class $v$ follows a word in class $u$. Let $l_k(u)$ and $r_k(v)$ be marginal probabilities of classes: $l_k(u) = \sum_v p_k(u, v)$, $r_k(v) = \sum_u p_k(u, v)$ and let $q_k(u, v)$ be the mutual information between two classes $u$ and $v$:

$$q_k(u, v) = p_k(u, v) \log \frac{p_k(u, v)}{l_k(u) r_k(v)}.$$

The average mutual information of the clustering remaining after $V - k$ merges is $I_k = \sum_{u,v} q_k(u, v)$.

Using the notation $i + j$ to represent the cluster obtained by merging $c_k(i)$ and $c_k(j)$, we have

$$p_k(i + j, v) = p_k(i, v) + p_k(j, v)$$

$$q_k(i + j, v) = p_k(i + j, v) \log \frac{p_k(i + j, v)}{l_k(i + j) r_k(v)}.$$

The merging of two clusters $c_k(i)$ and $c_k(j)$ results in two changes in clustering: 1) the deletion of clusters $c_k(i)$ and $c_k(j)$ and 2) the addition of cluster $i + j$. The first change decreases the average mutual information by $s_k(i) + s_k(j)$ where

$$s_k(i) = \sum_u q_k(u,i) + \sum_v q_k(i,v) - q_k(i,i).$$

The second change increases the average information by

$$q_k(i,j) + q_k(j,i) + q_k(i+j,i+j) + \sum_{u \neq i,j} q_k(u,i+j) + \sum_{v \neq i,j} q_k(i+j,v).$$

Therefore, after merging two clusters $c_k(i)$ and $c_k(j)$, the average mutual information is

$$\begin{aligned} I_k(i,j) = I_k - [s_k(i) + s_k(j)] + q_k(i,j) + q_k(j,i) \\ + q_k(i+j,i+j) + \sum_{u \neq i,j} q_k(u,i+j) + \sum_{v \neq i,j} q_k(i+j,v). \end{aligned}$$

By investigating all pairs, we can find a pair $(i^*, j^*), i^* < j^*$ for which the loss in average mutual information $L_k(i,j) = I_k - I_k(i,j)$ is minimal. We complete the step by merging $c_k(i^*)$ and $c_k(j^*)$ to form a new cluster $c_{k-1}(i)$. If $j^* \neq k$, we rename $c_k(k)$ as $c_{k-1}(j^*)$ and for all $u \neq i^*$ and $u \neq j^*$ we set $c_{k-1}(u)$ to $c_k(u)$. The values of $p_{k-1}(u,v)$ can be easily obtained from $p_k(\cdot,\cdot)$ as follows:

$$p_{k-1}(u,v) = \begin{cases} p_k(u,v), & \text{if } \{u,v\} \cap \{i,j\} = \emptyset \\ p_k(u,i) + p_k(u,j), & \text{if } v = i+j \\ p_k(i,v) + p_k(j,v), & \text{if } u = i+j. \end{cases}$$

New values of $l_{k-1}, r_{k-1}$ and $q_{k-1}$ can be easily computed from $p_{k-1}(u,v)$ as their evaluation in the previous iteration. If $u$ and $v$ both denote indices neither of which is equal to either $i$ or $j$, we have:

$$\begin{aligned} L_{k-1}(u,v) = L_k(u,v) - [q_k(u+v,i) + q_k(i,u+v)] \\ - [q_k(u+v,j) + q_k(j,u+v)] \\ + [q_{k-1}(u+v,i) + q_{k-1}(i,u+v)]. \end{aligned}$$

Finally, the losses $L_{k-1}(u,i), \forall u \neq i,j$ are evaluated as follows: [2]

$$\begin{aligned} L_{k-1}(u,i) &= I_{k-1} - I_{k-1}(u,i) \\ &= I_k(i,j) - I_{k-1}(u,i). \end{aligned}$$

---

[2] Note that for each $k$, we need to compute only $L_k(u,v)$ for all $u < v$ since obviously $L_k(u,v) = L_k(v,u), \forall u,v$. Therefore, to speed up the computation, we compute $L_{k-1}(u,i), \forall u < i$ and $L_{k-1}(i,u), \forall u > i$.

The equation computing $L_k(i,j)$ shows that if we know $I_k, s_k(i)$ and $s_k(j)$ then most the time involved in computing $I_k(i,j)$ is devoted to computing the sums over all $u \neq i,j$ and $v \neq i,j$. Each of these sums has approximately $V - k$ terms and so we have reduced the evaluation of $I_k(i,j)$ from one of order $V^2$ in the naive algorithm to one of order $V$. We can improve this further by keeping track of those pairs $u,v$ for which $p_k(u,v) \neq 0$ (by convention, $q_k(u,v) = 0$ when $p_k(u,v) = 0$). The evaluation of $s_k(i)$ for all possible $i$ has complexity $O(V^2)$. Thus, to determine the best merge, we need $O(V^2)$ computations. After a merge, the computation of $s_{k-1}(u)$ and $L_{k-1}(u,v)$ can be done in constant time, independent of $V$. We thus reach a complexity of $O(V^3)$.

## 2.4 Incremental Algorithm

Although the dynamic programming algorithm significantly reduces the complexity of the clustering, it is still too slow for realistic values of $V$ (above about 8000 in our experiments). In an effort to obtain clusters for large corpora, Brown et al. use an improved version of the algorithm above which proceeds in an incremental fashion: first, the algorithm arranges the words in order of decreasing frequency and each of the first $C$ words is assigned to its own cluster $c_1, c_2, \ldots, c_C$. Then considering one by one each of the remaining words, a new cluster $c_{C+1}$ for the word is created and a pair of clusters from $c_1, c_2, \ldots, c_{C+1}$ is chosen to be merged: pick the merge that gives the maximum value for $Q(\pi)$ and the clustering of $C$ clusters is restored.

For efficiency reason, we propose the following incremental algorithm to obtain clusters for large vocabularies of size $V$.

1. Sort the vocabulary by decreasing frequency;
2. Run the Brown algorithm on the $N$ most frequent words ($N \ll V$) to obtain $C$ clusters $c_1, c_2, \ldots, c_C$.
3. For each $i$ from $N + 1$ to $V$:

   a. assign the $(i)$–th most frequent word to a new cluster $c_{C+1}$;
   b. find the cluster $c_j, j = 1, 2, \ldots, C$ such that the merge of $c_j$ and $c_{C+1}$ gives the maximum value for $Q(\pi)$;
   c. merge clusters $c_j$ and $c_{C+1}$ — back to $C$ clusters.

It is easy to verify that at the end of the algorithm, each of the words in the vocabulary will have been assigned to one of $C$ clusters. The running time of the algorithm is $O(VN^2)$.

The main difference between this algorithm and Brown's improved algorithm lies in the way subsequent words are distributed into clusters. In Brown's algorithm, each new word assignment needs to consider $O(C^2)$ pairs of existing clusters to find a merge, while in our algorithm, the number of pairs to be considered is only $C$ since the cluster $c_{C+1}$ is fixed. This significantly improves the efficiency of the clustering process, especially when it involves many expensive operations like the logarithm and the merge of large structures. In addition, this approach does not hurt the clustering quality if the number of clusters is reasonable, as we see in the experimental results.[3]

## 3 Experimental Results

### 3.1 Corpus

We have used the word clustering algorithms presented in the previous section on a large corpus developed by the Vietnamese national project VLSP (Vietnamese Language and Speech Processing). This corpus is collected from the social and political sections of the *Tui Tr* on-line daily newspaper. The corpus contains 69,482 sentences (two million syllables) segmented into about 1.5 million occurrences of 36,441 distinct words [10]. Word segmentation is first done using an automatic tokenizer [11] and then manually checked.

### 3.2 Results

We have use the incremental algorithm to find word clusters in the corpus. In the first stage, we use $C = 100$ and $N = 500$ to run step 2 of the algorithm—using Brown's algorithm to find 100 clusters of the 500 most frequent words. Table 1 shows twenty of those clusters.

For ease of interpretation, these clusters are listed into groups. We see that the clustering result is quite interesting when the words in a cluster share both syntactical and semantic properties. In particular, the first cluster includes common pronouns found in the corpus. The second group (clusters 2–5) contains nouns. Each cluster includes nouns that are semantically related, for example cluster 3 indicates a common noun which usually collects a community of people (hospital, mountain village, center, school, market-place, village); cluster 4 includes nouns which indicate administrative organization units (district, commune, province, town). The third and fourth groups contain adjectives and adverbs respectively. There are two clusters of adverbs whose words are different in use, both syntactically and semantically. The next group includes clusters (9–10) of determiners. Clusters 11-14 contain different verbal types. It is interesting to note that clusters 13 and 14 include only two-syllable verbs. Clusters 15 and 16 also include (intransitive) verbs but they are all related to actions or movements of a human being (to run, to go, to stand, to sit, to think, to tell, to speak...). The next cluster includes words which signify directions (end, begin, south, north, east). Finally, the last group are clusters containing numerals and scientific measurement units.

---

[3] We thank an anonymous reviewer for pointing out that the algorithm reported in [9] has the same complexity as ours.

**Table 1** Some clusters of the most frequent words

| | |
|---|---|
| 1. | ch, chng ta, ti, bc, chng, anh, h, mnh, chng ti, b, n, ng |
| 2. | g, my, c, tu, xe, gi |
| 3. | BV, bnh vin, bn, trung tm, trng, ch, lng |
| 4. | ngi dn, mi trng, nhn dn, b con, x hi, dn, kinh t |
| 5. | Tui Tr, huyn, x, Si Gn, a phng, tnh, thnh ph, TP |
| 6. | r, kh, nng, kh khn, cao, ngho, tt, vi phm, nhanh, nh, ln, di |
| 7. | c th, ch, chng, khng, cha |
| 8. | ang, va, u, s, |
| 9. | mi, nhiu, nhng |
| 10. | vi, ton b, tt c |
| 11. | , nhm |
| 12. | s dng, dng, thu, , k, thuł, mua, tr, bn, ng, chu |
| 13. | ngh, bo co, ch o, ułu cu |
| 14. | nghiłn cu, qun l, phc v, tham gia, thi cng, kim tra, pht trin, bo v, iu tra, thanh tra, h tr |
| 15. | chy, ng, ngi, nh, sng, ch, nm, i |
| 16. | ngh, bo, k, ni, nh, tin |
| 17. | nam, cui, bc, u, ng |
| 18. | ba, hai, my, bn, vi, su, mi, mt |
| 19. | 15, 11, 12, 3, 2, 20, 1, 10, 7, 30, 6, 5, 4, 9, 8, 100, 50 |
| 20. | g, tui, m2, m, km, ha, gi |

This clustering is used in the next stage to incrementally classify remaining words of the vocabulary as in step 3 of the algorithm. If the previous stage is a time-consuming task (it took 8.5 hours on a modern personal computer), this stage is much faster. We have been able to find clusters of about 12,000 words of the vocabulary (whose count is greater than a cut-off threshold fixed at 5) in a reasonable time. Due to space restriction, we do not present example clusters given by this stage.

## 3.3 Evaluation

We have given results of our model by showing syntactically and semantically plausible clusterings of Vietnamese words on a large text corpus. However, these results are only qualitative. Quantitative evaluation of POS induction result is a difficult task: there are many different measures which have been proposed but there is still no consensus on which is best. For this reason, we propose to evaluate the clustering results using an application-oriented approach rather than using an intrinsic measure. We have integrated word clustering features into a statistical POS tagger of Vietnamese. The word clustering information helps improve the accuracy of the tagger further even when simple clustering features are used.

The tagger that we experiment employs a maximum entropy Markov model (MEMM) to predict part-of-speech sequence of a sentence. MEMM is a discriminative counterpart of hidden Markov model (HMM) which offers many advantages and usually outperforms HMM in the sequence prediction problems [12]. In MEMM for POS tagging, we model the conditional probability of a tag sequence given a word sequence (a sentence) as follows:

$$P(\mathbf{y}\,|\,\mathbf{o}) = P(y_1, y_2, \ldots, y_T\,|\,o_1, o_2, \ldots, o_T)$$
$$= \prod_{t=1}^{T} P(y_t\,|\,y_1, \ldots, y_{t-1}, \mathbf{o})$$
$$\approx \prod_{t=1}^{T} P(y_t\,|\,y_{t-2}, y_{t-1}, \mathbf{o}) \approx \prod_{t=1}^{T} P(y_t\,|\,y_{t-1}, \mathbf{o}),$$

where the observations $o_t, t = 1, 2, \ldots, T$ are words and $y_t$ are their corresponding tags. The first approximation of the probability is used in a second order MEMM while the second approximation is used in a first order MEMM. Let $h_t = \langle \mathbf{o}, t, y_{t-2}, y_{t-1} \rangle$ denote the tagging context at position $t$, each local model of MEMM is defined by a maximum entropy model (a.k.a multinomial logistic regression model):

$$P(y_t\,|\,h_t) = \frac{\exp(\theta \cdot f(h_t, y_t))}{\sum_{s \in S} \exp(\theta \cdot f(h_t, s))},$$

where $f(h_t, s) \in \mathbb{R}^D$ is a feature vector and $\theta \in \mathbb{R}^D$ is the parameter to be estimated from training data. A detail presentation of this model for POS tagging can be found in [13].

The accuracy of a first order MEMM using standard feature templates in POS tagging, trained and tested on the Vietnamese treebank [10] is shown in the first half of Table 2. By including a simple cluster feature for each current word, the accuracy of the model is significantly improved from 90.08% to 91.85%. These results on the one hand show a net benefit of using word cluster features in predicting Vietnamese POS tagging and on the other hand, in some aspect, quantitatively justify the quality of the induced word clusters.

**Table 2** The accuracy of a first order MEMM

| Feature templates | Accuracy |
|---|---|
| $o_t, o_{t-1}, y_{t-1}$ | 88.75% |
| $o_t, o_{t-1}, y_{t-1}$, word form of $o_t$ | 90.08% |
| $o_t, o_{t-1}, y_{t-1}$, word form of $o_t$, cluster of $o_t$ | 91.85% |

# 4 Conclusion and Future Work

We have presented a statistical model and some algorithms for inducing words classes from a large text corpus. The words are grouped together using the statistical similarity of their surroundings. Although only distributional information is used in the clustering, the obtained classes include words that are both syntactically and semantically similar. This result quantitatively justifies both lines of linguistic research related to POS classification for Vietnamese. Based upon this result, we confirm that a good classification of Vietnamese words should be based not only on syntactic but also on semantic features of each word. The resulting clusters can contribute to the determination of word categories in linguistic studies. In particular, we have shown that by using simple clustering features, we can significantly improve the accuracy of a Vietnamese statistical POS tagger.[4]

Although there is no consensus on intrinsic measures used to evaluate a clustering result, the entropy-based V-measure [15] has recently been shown to be more stable and useful than other measures. An appealing property of this metric is that it is less sensitive both to the number of induced clusters and to the size of the data set. We plan to perform a quantitative evaluation of the model using $V$-measure and its comparison to other clustering metrics in a future work. In addition, a comparison of the derived clusters of words with other pre-defined POSs using V-Measure should be useful for tagset design.

For many occidental languages, the use of morphological information can substantially improve POS induction for rare words [16]. Vietnamese being an isolating language which does not use morphology, the question whether adding word form information may improve the POS induction for Vietnamese is an interesting one that we intend to investigate.

Another important direction resides in the study of clustering methods on very large corpora to induce more stable word classes. The major drawback of the Brown clustering technique is that it is a hard-clustering algorithm and could not reflect the traditional POS-based cluster which is basically a soft-clustering, in which a word may belong to many POS clusters. Techniques such as co-clustering of words and contexts [17], topic-based clustering [18] and distributed word representation approaches [19, 20] seem promising. These methods will be explored in our future work.

**Acknowledgment.** The authors would like to thank the University of Science, Vietnam National University for grants number TN-12-02. The first author is also partly supported by the FPT Technology Research Institute. We are grateful to anonymous reviewers for helpful comments on the draft.

---

[4] An anonymous reviewer brings our attention to the work [14] where the authors also used word cluster features to improve the accuracy of a statistical tagger for Vietnamese.

# References

1. Schütze, H.: Part-of-speech induction from scratch. In: Proceedings of ACL, pp. 251–258 (1993)
2. Con, N.H.: On the determination of Vietnamese word classes. Journal of Language, Vietnamese Institute of Linguistics, 36–46 (2003) (in Vietnamese)
3. Vietnam Social Science Committee (ed.): Vietnamese Grammar. Social Sciences Publisher, Hanoi (1983) (in Vietnamese)
4. Diep, Q.B., Hoang, V.T.: Vietnamese Grammar. Vietnam Education Publisher, Hanoi (1999) (in Vietnamese)
5. Doan, T.T., Nguyen, K.H., Pham, N.Q.: A Concise Vietnamese Grammar (For Non-native Speakers). World Publishers, Ha Noi (2003) (in Vietnamese)
6. Bao, H.T.: Building basic resources and tools for Vietnamese language and speech processing (VLSP). Technical report, The KC/01/06-10 project (2010)
7. Christodoulopoulos, C., Goldwater, S., Steedman, M.: Two decades of unsupervised POS induction: How far have we come? In: Proceedings of ACL (2010)
8. Brown, P.F., deSouza, P.V., Mercer, R.L., Pietra, V.J.D., Lai, J.C.: Class-based $n$-gram models of natural language. Computational Linguistics 18, 467–479 (1992)
9. Liang, P.: Semi-supervised learning for natural language. Master's thesis. MIT (2005)
10. Nguyen, P.T., Xuan, L.V., Nguyen, T.M.H., Nguyen, V.H., Le-Hong, P.: Building a large syntactically-annotated corpus of Vietnamese. In: Proceedings of the 3rd Linguistic Annotation Workshop, ACL-IJCNLP, Singapore (2009)
11. Le-Hong, P., Nguyen, T.M.H., Roussanaly, A., Ho, T.V.: A hybrid approach to word segmentation of Vietnamese texts. In: Martín-Vide, C., Otto, F., Fernau, H. (eds.) LATA 2008. LNCS, vol. 5196, pp. 240–249. Springer, Heidelberg (2008)
12. McCallum, A., Freitag, D., Pereira, F.: Maximum entropy Markov models for information and segmentation. In: Proceedings of ICML (2000)
13. Le-Hong, P., Roussanaly, A., Nguyen, T.M.H., Rossignol, M.: An empirical study of maximum entropy approach for part-of-speech tagging of Vietnamese texts. In: Proceedings of Traitement Automatique des Langues Naturelles (TALN 2010), Montreal, Canada (2010)
14. Minh, N.L., Bach, N.X., Cuong, N.V., Minh, P.Q.N., Shimazu, A.: A semi-supervised learning method for Vietnamese part-of-speech tagging. In: KSE, pp. 141–146 (2010)
15. Rosenberg, A., Hirschberg, J.: V-measure: a conditional entropy-based external cluster evaluation measure. In: Proceedings of EMNLP-CoNLL, pp. 410–420 (2007)
16. Clark, A.: Combining distributional and morphological information for part-of-speech induction. In: Proceedings of EACL (2003)
17. Leibbrandt, R.E., Powers, D.M.W.: Robust induction of parts-of-speech in child-directed language by co-clustering of words and contexts. In: Proceedings of the Joint Workshop on Unsupervised and Semi-Supervised Learning in NLP, Avignon, France, pp. 44–54 (2012)
18. Chrupała, G.: Hierarchical clustering of word class distributions. In: Proceedings of the NAACL-HLT Workshop on the Induction of Linguistic Structure, Montréal, Canada, pp. 100–104 (2012)
19. Turian, J., Ratinov, L., Bengio, Y.: Word representations: A simple and general method for semi-supervised learning. In: Proceedings of ACL, Uppsala, Sweden, pp. 384–394 (2010)
20. Huang, E.H., Socher, R., Manning, C.D., Ng, A.Y.: Improving word representations via global context and multiple word prototypes. In: Proceedings of the ACL, pp. 873–882 (2012)

# Resolving Named Entity Unknown Word
# in Chinese-Vietnamese Machine Translation

Phuoc Tran, Dien Dinh, and Linh Tran

**Abstract.** Vocabulary of natural language is an open set. So we cannot collect all words of a language. Therefore, arising unknown word (UKW) in statistical machine translation (SMT) is unavoidable. Named entity is the most common UKW. In this paper, we will present a new approach based on the meaning relationship in Chinese and Vietnamese to re-translate named entity UKW. Applying this approach to Chinese-Vietnamese SMT, experimental results show that our approach has significantly improved machines performance.

## 1  Introduction

New words are frequently arised from activities such as explaining a new concept, a new invention, named newborn children, a new established organization, etc. In SMT problem, we found that even if the training corpus of machine translation system are large, there will not be able to cover all the words of a language. Therefore, instead of finding the way to translate all the words of a language in order to not arising UKWs, here, we see UKWs as obvious part of machine translation (MT) and try to re-translate these UKWs to improve the overall quality of MT.

Phuoc Tran
University of Food Industry, Faculty of Information Technology,
140 Le Trong Tan, Tan Phu dist., HCMC, Vietnam
e-mail: phuoctt@cntp.edu.vn

Dien Dinh
University of Science, Faculty of Information Technology,
270 Nguyen Van Cu, dist. 5, HCMC, Vietnam
e-mail: ddien@fit.hcmus.edu.vn

Linh Tran
University of Science, Foreign Language Center,
270 Nguyen Van Cu, dist. 5, HCMC, Vietnam
e-mail: tranletamlinh@yahoo.com.vn

V.-N. Huynh et al. (eds.), *Knowledge and Systems Engineering, Volume 2*,          273
Advances in Intelligent Systems and Computing 245,
DOI: 10.1007/978-3-319-02821-7_25, © Springer International Publishing Switzerland 2014

A Chinese word usually includes many meaningful characters. When translating it into Vietnamese, its meaning is usually divided into three cases. Firstly, the meanings of Chinese characters are their Sino-Vietnamese meanings, usually 1-1 respectively. Secondly, the meanings of the Chinese characters are similar or related to the meanings of the Chinese word containing those characters. Lastly, the meanings of Chinese characters are not relevant to the meaning of the Chinese word containing them.

In the first case, Vietnamese words are largely borrowed from Chinese words, often called Sino-Vietnamese and about 65% of the total number of Vietnamese words. The Chinese itself is pronounced differently, even in China, depending on the area where there are many different voices or pronunciations such as Cantonese, Hokkien, Beijing and so on. Neighboring countries such as Korea has its own reading of Korea, known as the Sino-Korean (); Japanese have their own Chinese reading of the Japanese people, called Sino-Japanese (); and the Vietnamese have their private reading, called Sino-Vietnamese (Ł). Thus, Sino-Vietnamese is a reading way of Vietnamese people. For example, Chinese word (bank), Chinese (Pinyin) will pronounce yin hang and Vietnameses pronunciation is ngn hng. A Chinese character may be pronounced by many Sino-Vietnamese words, but in a specific context, one Chinese character is only corresponding to one Sino-Vietnamese. As the above example , corresponding Sino-Vietnamese pronunciation of character  is ngn, and pronunciation of  is hnh or hnh or hng or hng. However, when  and  combined into a unique word , we only pronounce ngn hng.

In the second case, the meaning of the Chinese word is a combination of Pure-Vietnamese meanings of Chinese characters forming that word. For example, word (trong nc: inland), the corresponding meaning of each character is / nc, /trong.

In the other cases, the words which their meanings are not related to the characters forming them.  (ng: right) is a typical instance. The corresponding Sino-Vietnamese meanings of characters are ho, ch and their Pure-Vietnam meanings are tt (good), ca (of). Clearly, ho ch and tt ca (good of) are not relevant to the proper meaning of (right).

Most UKWs in Chinese-Vietnamese SMT are Named Entities (NEs). These NEs are divided into the following categories: person name (PN) (1), organizations name (OG) (2), location name (LC) (3) and Factoid (date, time, percentage, number, phone number) [4] [5]. Chinese words in (1), (2) and (3) are often translated into Vietnamese to be their Sino-Vietnamese. The Factoids are translated into Vietnamese based on grammatical transformation, the translation method is completely based on rule. In this paper, we are only interested in identifying and re-translating UKWs in (1), (2) and (3).

This paper is presented as follows: in section 2, we will present related works about handling UKWs problem in machine translation as well as some approaches for Chinese NER. Section 3 will present knowledge background of Chinese NE. The recognizing as well as re-translating NE-UKWs will be presented in section 4. Meanwhile, in section 5, we will present experiments and some discussion will be presented in section 6. The conclusion will be presented in section 7.

## 2 Related Work

In this section, we will present some works related to the named entity recognition and re-translating UKW.

Chinese NER approaches in recent times are mainly the hybrid approaches, combining of statistic and rule. The key point of recognizing Chinese NEs is based on the keywords of each individual NE (rule based approach). For example, to recognize a PN, Jianfeng *et al* [8] analyzed the Chinese PN in the form of ¡Family name¿ (F) + ¡Given name¿ (G) (rule based approach). In particular, F and G have a length of one or two characters. The authors only considered candidates to be PNs if their F part is stored in the family name list (which contains 373 entries). The next step, the authors recognized the PNs G based on language model (bigram model) (statistic based approach).

Also in this way, the authors Wu [9] *et al* also used the hybrid algorithm which combines a class based statistical model with many different linguistic knowledge to recognize NE. For the authors Chen *et al* [7], they only focused on identifying the OG based on morphological analysis.

In another aspect, Liu *et al* [10] assert that PN recognition based on the keywords in the Chinese family name list is not effective for the wrong-rule PN. Typically, PN does not have family name, such as    (bn cng lp tn i Hi: classmate Dai Hai), a respected name  (L tiln sinh, ng L: sir Li), pen name, stage name, one character PN, etc. To solve this problem, the authors propose PN recognition based on the context around it. For example, the PN often appears in the context having words, such as: (president), (journalist), etc.

NE is a popular UKW type in SMT in general and Chinese-Vietnamese SMT in particular. Currently there are many studies with different approaches to re-translate UKW and improve MT performance. Based on orthographic cues given by words, Joao *et al* [4] have proposed two methods to overcome the UKW, namely such as cognates' detection and logical analogy. This approach has been successfully implemented for inflectional language pair English  Portugal.

Another handling UKW approach for re-translating UKW is conducted by Eck *et al* [5]. The authors looked for the definitions of UKW in the source language and translate its definitions (instead of translating the UKW). The definitions of UKW will be automatically extracted from online dictionaries and encyclopedias, and then they are translated through the SMT system. The translation result will replace the UKW in previous translation. The approach has been tested on language pair English  Spanish.

On the other hand, Zhang *et al* [6] translated a Chinese UKW by re-splitting the UKW into sub-words and translating the sub-words (sub-word based translation). Sub-word is a unit in the middle of character and word. In addition, the authors also found that the quality of translation will increase significantly if applying NER to translate UKW before using the sub-word based translation. Our approach is also similar to the approach. However, instead of re-splitting UKW into sub-words (greater than character), we re-split UKW into single characters, and then find their Sino-Vietnamese and filter out suitable meaning.

## 3   Background of Chinese NE

A Chinese NE includes four categories: PN, LC, OG and Factoid. PN is formed
by the following structure: ¡Family name¿ (F) ¡Given name¿ (G), both F and G part
have a length of one to two characters. For example, the PN , is F part and is G part.
Besides, the names of older people are usually prefixed with (lo: old). For instance,
means Lo Trang (Old Zhang). Younger people, on the other hand, are often prefixed
with (Tiu: small), so means Tiu Vng (Small Wang).

Normally, the Chinese LC has maximum of 10 characters which has structure as
follows: ¡name part¿ ¡keyword¿. In particular, ¡name part¿ is an item from the list
of Chinese LCs (approximately 30,000 name parts), the LC is usually terminated by
a ¡keyword¿ (approximately 120 keywords), in some cases there is no keyword at
the end of LC, for example: or (Thnh Ph Bc Kinh: Beijing city) where is a ¡name
part¿ and is a ¡key word¿.

OG is more complicated than PN and LC because it usually includes many other
entities combining together, its maximum length is 15 characters. Its structure is
often: [PN] [OG] [LC] [kernel name] * [organization type] ¡key word¿. Symbol
* means selecting at least one items. For example: (Hc vin Ngn ng Bc Kinh:
Beijing Language Institute), is a LC and is ¡keyword¿.

## 4   Re-translating NE-UKW

Our Chinese NE-UKW re-translating model is represented in Fig. 1. We use the
Stanford Chinese Segmenter[1] to segment Chinese corpus and Vietnamese corpus
is segmented by our groups tool. The tool was implemented by Dinh Dien *at el*,
according to Maximum Entropy Approach [11]. After segmenting word for
Chinese-Vietnamese corpus, we use MOSES[2] tool to train these corpuses (creating translation model and language model) and translate Chinese sentences. The
translation result (Vietnamese translation) continues to be identified if it contains
UKWs or not. Chinese-Vietnamese alphabets are different, so we can easily identify
the Chinese UKWs in Vietnamese translation. We match the words in Vietnamese
translation sentence to Chinese source sentence, the words appear in both the source
and the destination sentence to be UKWs.

Next, we will identify NE-UKW candidates and classify them by Standford
NER[3] tool. This tool is installed by CRF method (Conditional Random Field), the
current version of this tool divides Chinese NE into five classes. In the scope of this
paper, we are only interested and use three classes as the following table 1.

By this stage, the NE-UKWs are identified and classified into tree classes as
above. Corresponding to each subclass, we will translate as follows:

---

[1] http://nlp.stanford.edu/software/segmenter.shtml

[2] http://www.statmt.org/moses/

[3] http://nlp.stanford.edu/software/CRF-NER.shtml

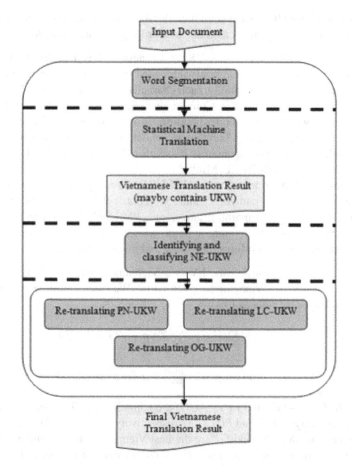

**Fig. 1** Identifying and re-translating NE-UKW model

**Table 1** Class used in Standford NER tool

| Class | Description | Example |
|---|---|---|
| GPE | Geographical and Political Entity | Ł (Vit Nam), (Thnh Ph  Bc Kinh) |
| PERSON | Person Name | (Trn T  Minh) |
| ORG | Organization Name | (Hc Vin Ngn Ng  Bc Kinh) |

## 4.1  Person Name UKW (PN-UKW)

A PN-UKW includes UKWs are labeled as PERSON. They consist of two parts: F (Family name) and G (Given name), in the form F + G. F and G have a length from one to two characters. We only consider candidates to be PN if their F part is stored in the Chinese Family list (our system includes 484 Family Names). The purpose of the filter is to avoid some error cases due to word segmentation and NER.

- Generating translations of PN-UKW: Chinese PN is translated into Sino-Vietnamese. Therefore, we are only interested in the Sino-Vietnamese meaning of Chinese characters. Normally, each character will have one Sino-Vietnamese respectively. In some cases, a Chinese character is translated into many Sino-Vietnamese words. Thus, corresponding to one PN-UKW, there will usually be a set of corresponding Vietnamese names. For example, Chinese name , there will be four Sino-Vietnamese translations respectively: Trn T Minh, Trn T Minh, Trn T Minh, Trn T Minh.
- Selecting a suitable translation: for PN-UKW has only one Vietnamese name, we will choose the Vietnamese name to replace PN-UKW in the final Vietnamese translation. Remaining cases, one PN-UKW has many Vietnamese names, we will select Vietnamese name with the highest probability. We calculate individual probabilities for family and given name, as follows:

Assuming FC is a Chinese family, FV is a Vietnamese family, GC is a Chinese given name and GV is a Vietnamese given name. We calculate these probabilities according to Bayes formula as follows:

$$F^* =_{FV}^{argmax} p(FV|FC) =_{FV}^{argmax} p(FV) * p(FC|FV) \tag{1}$$

$$G^* =_{GV}^{argmax} p(GV|GC) =_{GV}^{argmax} p(GV) * p(GC|GV) \tag{2}$$

In which, $P(FV-FC)$ and $P(GV-GC)$ are Chinese and Vietnamese translation models respectively, $P(FV)$ and $P(GV)$ are respectively Vietnamese family and given name language models, $F^*$ and $G^*$ are the best Vietnamese family and given name corresponding Chinese ones. Because $FC$ and $GC$ have only from one or two characters, specially, most of Chinese family names have only one character. Therefore, we choose unigram language model for Vietnamese language. Final translation $F^*G^*$ will replace PN-UKW in the final translation. As above example, Trn T Minh is the translation with the highest probability of , we chose the translation to be the final translation.

## 4.2   Location Name UKW (LC-UKW)

LC-UKW is structured: ¡name part¿ ¡keyword¿. In some cases, there is no ¡keyword¿ in LC. A ¡name part¿ is also translated into Vietnamese to be Sino-Vietnamese meaning, its translation method is similar to PN-UKW. The key point of this section is identifying and translating ¡keywords¿ in LC. We identify them based on a list of 120 ¡keywords¿, their meanings are Pure-Vietnamese. One difference between LC and PN is that Vietnamese translation of LC-UKW is reordered, the reordering is conducted as follows (fig. 2):

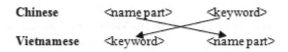

**Fig. 2** Reordering method of LC-UKW translation

For example, LC-UKW , ¡key word¿ will be identified and translated as follows (fig. 3):

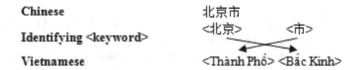

**Fig. 3** Example of reordering LC-UKW

## 4.3 Organization Name UKW (OG-UKW)

OG usually includes many other entities combining together, for example: OG in-cludes: is [OG] and is ¡keyword¿. Word order of OG is reversed in the opposite direction when it is translated into Vietnamese. For example, [] [] [] ¡¿ is translated into Vietnamese in order like ¡¿ [] [] [] ([Vin] [Nghiln Cu] [ngn ng] [Trung ng]).

A noticeable point is that the reordering in OG is not character based reordering but sub-word based reordering. Sub-word can be a character, a word consisting of many characters or maybe the OG. As above example , it is not reordered to be [] [] [] [] [] [] [] (character based) that the correct order is ¡¿ [] [] []. The OG-UKW re-translating process is conducted as follows:

- Identifying and tagging components in OG including NE, not NE (NNE) and keyword (KEY).
- Defining the boundary of sub-words in NE. We use the Maximum Matching algorithm based on Chinese-Vietnamese dictionary to define sub-words bound-ary.
- Reordering and re-translating the sub-words. The NE sub-words are translated as mentioned in part 4.1 and 4.2, the remaining sub-words will be translated based on the Chinese-Vietnamese dictionary.

This process is illustrated in Fig. 4

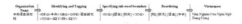

**Fig. 4** Re-translating OG-UKW

## 5   Experiments

Our bilingual corpus includes 12,000 Chinese-Vietnamese sentence pairs, which are collected from Chinese conversation textbooks and Chinese online forums. Documents in the corpus are mostly communication text, the length of the sentences is relatively short, an average of about 10 words in a sentence. The corpus quality is fairly clean, its content is uniform and spread over 12,000 sentences. We use 90% of the total number of sentences for training, 5% of sentences for testing and the remaining 5% of the sentences for developing. Training corpus (the sentences for training and developing) is trained by MOSES tool with the default parameters (SMT Baseline). We use these corpuses to perform three experiments such as word un-segmentation translation, word segmentation translation and re-translating NE-UKW for word segmentation case. Besides, we also translate sentences containing NE-UKWs through Google Translator. The experiment is not intended to assess quality of our system with Google, with a very large corpus, Google seems to translate all the input sentences, the UKW rarely arises. The purpose of this experiment is to prove our improvement in a narrow scope about translating Chinese text containing NE-UKW.

In the Baseline system, we consider the Chinese characters and the Vietnamese spelling words as the meaningful independent units. We insert one space between Chinese characters and insert one space between spelling words with the punctuations.

In the word segmentation system, we segment Chinese word by Stanford Chinese Segmenter. For Vietnamese, we segment word by our groups word segmentation tool. The segmenter was implemented by Dinh Dien *et al*, according to Maximum Entropy approach.

Basing on the translation result of MOSES in word segmentation case, we proceed to re-translate this result. Depending on selecting test sentence that the BLEU score has difference depending on the selection. The choice giving result with many NE-UKWs is sure that the BLEU score of word segmentation system is much lower than the re-translating NE-UKW system and vice versa. The fig. 5 shows the translation result of the test selected following the format: each 20 sentences in the corpus, the first 18 sentences for training, the 19th sentence for developing and the 20th for testing.

Besides, in order to clarify the improvement of the re-translating approach, we also have computed the Precision of re-translation for 30 sentences containing NE-UKWs. We have evaluated the Precision by the following formula:

$$Precision = \frac{\sum CorrectPairs}{\sum Total} \tag{3}$$

Total = 30 in this case. The 30 sentences were re-translated through our system. The system translated exactly 25 NE-UKWs, gaining 83%.

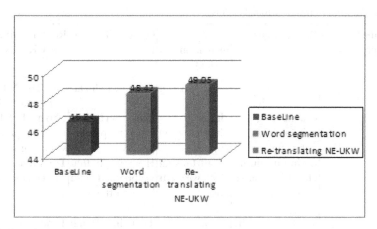

**Fig. 5** Experiment result

## 6 Discussion

Translation result in the word segmentation case usually gives the result better than the word un-segmentation case. This problem has been presented in work [3]. However, the word segmentation translation system gives result with more UKWs. The results of the word segmentation will be continued to re-translate NE-UKW by our system. In addition, the test corpus is also translated through Google Translator. With extremely large corpus, the Google Translator seems to translate all the Chinese words. However, also like the baseline system, although translation result of Google rarely arises UKW but it often gives wrong meaning. Here are five specific cases in the test corpus (table 2):

All systems have errors of meaning, word order, etc. However, we do not discuss about these errors. Here, we only focus on the errors of the systems when they translate sentences containing NEs. The baseline system gives some results (not contain UKW) in all five cases, but the results are mostly inaccurate except sentence 3 and 4 (translating correctly PN Trng Th Mu and LC i hc Thanh Hoa because of the PN and LC existed in training corpus with the highest probability). In the remaining cases, the characters in NEs also exist in the training corpus, so that the baseline system has chosen Vietnamese meanings having the highest probability to be the outcome. However, these meanings are not true for the NE. Typically, in case 2, (correct translation as Gim c Bch: Director Bach), however, the baseline system translated the word to be trng gim c (white director). The incorrect translation is due to /trng having highest probability. The similar errors are for case 1 and 5.

For Google Translator, because this system has to translate twice when translating from Chinese into Vietnam; errors in Vietnamese sentence including errors of Chinese-English translation and English-Vietnamese translation. In all five cases, the Google mostly recognizes NEs, but its translation result is wrong except for case 4. Some typical errors are: (1) lack of word error ( including three words but it only

**Table 2** Translation result of four systems

| ID | Chinese sentence | Baseline system | Word segmentation system | Google Translator | Re-translating NE-UKW system |
|---|---|---|---|---|---|
| 1 | | ti l Tp i nm, cng ty philn dch . | Ti l , ca cng ty philn dch . | Ti hc c dch Da-nian | Ti l Tp i Niln, ca cng ty philn dch. |
| 2 | | Cho ng, trng gim c. Ti n phng vn v bn cng vic. | cho ng , . Ti n phng vn v bn cng vic. | Qun l mu trng ca bn. Ti n p dng v phng vn. | Cho ng, Bch Kinh L. Ti n phng vn v bn cng vic. |
| 3 | | ti xin gii thiu mt cht, y l tt bn ca ti Trng Th Mu. | Ti gii thiu mt cht , y l ti tt bn . | Hy ti gii thiu vi bn, y l ngi bn tt ca ti Zhang WTO. | Ti gii thiu mt cht, y l tt bn ca ti Trng Th Mu. |
| 4 | | Ti hc i hc Thanh Hoa | Ti hc i hc Thanh Hoa | Ti i hc Thanh Hoa | Ti hc i hc Thanh Hoa |
| 5 | | Trn T Sng chng khon giao dch s lm vic | chng khon lm vic | Chen Ziming lm vic trong cc c phiu | Trn T Minh chng khon s giao dch lm vic |

generated two words Danian: Da Nian), (2) mistranslation because of unrecognizing PN ( translated as qun l mu trng (white management), word means trng (white) and Bch (Bach), the correct translation is Gim c Bch: Director Bach), (3) completely mistranslating (: Zhang WTO), (4) incorrectly translating Vietnamese PN (Google translated PN into Pinyin transcription, for example, in case 5, Google translated as Chen Ziming, while Vietnamese PN is Trn T Minh)

In the word segmentation translation system, the number of words in the corpus of this case will be less than a baseline system, so its alignment word dictionary as well as word recognition ability is also less than baseline system. As a result, this translation system arises many UKWs. On the other hand, with the abundance of the NE-UKW, although the corpus is extremely large, it is very difficult to cover all words in a language, so the system does not translate them is unavoidable. The word segmentation translation result is re-translated by our system. The translation performance significantly increased.

Besides, there are the obvious improvements, through experiments, we also have found some re-translating cases un-accurate. These incorrect cases fall into irregular NEs, typically the case 2 of Table 3 (: Bch Kinh L), the PN in this case has not ¡given name¿ part, has only ¡family name¿ part, the following ¡family name¿ is a title, this is a manner of respected naming in Chinese text. The correct translation of this case is to translate ¡family name¿ part into Sino-Vietnamese and the title is translated into Pure-Vietnamese (title : Gim c: Director). Our system detects UKW which has full condition of PN so that it re-translated the PN-UKW to be Sino-Vietnamese. For example, the Sino-Vietnamese of is kinh l. Clearly, this translation result is incorrect. Besides, the NE recognition quality of Stanford NER tool also affects the translation quality of our system. In some cases, the tool does not recognize or recognize NE incorrectly.

## 7 Conclusion

In this paper, we have proposed a method to handle the NE-UKW in Chinese-Vietnamese SMT based on their meaning relation. The experimental result shows that our system re-translated NE-UKW with good quality, contributing significantly to improve Chinese-Vietnamese SMT performance. Besides, we have also found that our system was ambiguous meaning when it encounters irregular NE-UKWs, the quality of NE recognition depends on the performance of the Stanford NER tool.

In the future, we continue to study and learn the other approaches to identify and translate NE-UKW more accurately, improving machine translation with the best quality.

**Acknowledgments.** This paper was performed under the sponsorship of Kim Tu Dien Multilingual Data Center.

## References

1. Tran, T.P., Dinh, D.: Dealing with affirmative-negative question in Chinese-Vietnamese statistical machine translation. Journal of Research, Development and Application on Information & Communication Technology 27, 140–150 (2012) (in Vietnamese)
2. Thanh Phuoc Tran, Dien Dinh: Identifying and reodering prepositions in Chinese-Vietnamese machine translation. First International Workshop on Vietnamese language and speech processing (VLSP), In conjunction with 9th IEEE-RIVF conference on Computing and Communication Technologies, Vietnam (2012).
3. Tran, T.P., Dinh, D.: The issue of word boundary in Chinese-Vietnamese statistical machine translation. The Thirteen Scientific Meeting of Ho Chi Minh City University of Science (2012) (in Vietnamese)
4. Joao Silva, Luisa Coheur, Angela Costa, Isabel Trancoso: Dealing with unknown words in statistical machine translation. In proceedings of the Eight International Conference on Language Resources and Evaluation (LREC'12) (2012).
5. Matthias Eck, Stephan Vogel, Alex Waibel: Communicating Unknown words in machine translation. In International Conference on Language Resources and Evaluation (2008).
6. Ruiqiang Zhang, Eiichiro Sumita: Chinese Unknown word Translation by Subword Resegmentation. In International Joint Conference on Natural Language Processing (2008).
7. Keh-Jiann, Chao-jan Chen: Knowledge Extraction for Indentification of Chinese Organization Names. In Second Chinese Language Processing Workshop, Hong Kong, (2000).
8. Jianfeng Gao, Mu Li, and Chang-Ning Huang: Improved Source-Channel Models for Chinese Word Segmentation. In ACL '03 Proceedings of the 41st Annual Meeting on Associa-tion for Computational Linguistics (2003).
9. Youzheng Wu, Jun Zhao and Bo Xu: Chinese Named Entity Recognition Combining a Statistical Model with Human Knowledge. In MultiNER '03 Proceedings of the ACL 2003 workshop on Multilingual and mixed-language named entity recognition - Volume 15 (2003).

10. Liu Hongjian, Guo Defang, Zhou Quan, Nagamatsu Kenji, Sun Qinghua: A pre-identification method for Chinese Named Entity Recognition (2010).
11. Dinh Dien, Vu Thuy: A maximum entropy approach for Vietnamese word segmentation. In Research, Innovation and Vision for the Future, 2006 International Conference on (2006).
12. Chinese names, http://www.chinesenames.org

# Towards Vietnamese Entity Disambiguation

Long M. Truong, Tru H. Cao, and Dien Dinh

**Abstract.** Entity Disambiguation (ED) is a fundamental task in Natural Language Processing (NLP). The term *Entity* is used to mean either a *Named Entity* or an *Abstract Concept*. Although there have been many works on the ED task for English and some for Vietnamese, this is the first time this paper tackles the general ED task for Vietnamese that deal with both named entities and abstract concepts. In this paper, we propose a method for linking named entities and abstract concepts in Vietnamese documents to the corresponding articles in the Vietnamese Wikipedia. In particular, it first has to recognize Vietnamese entity mentions, i.e., phrases that represent named entities or abstract concepts. Experimental evaluation is also presented to demonstrate the performance of the proposed method.

## 1 Introduction

In order to understand a document, we often have to read all the context of the document. However, the computers ability to understand a document is limited to only some specific cases ([2]). One machine approach is to understand some keywords in a document, based on a knowledge base, instead of all the words in the document. Among such important keywords are named entities (NE) and abstract

Long M. Truong
Ho Chi Minh City University of Science - VNUHCM,
Vietnam, and John von Neumann Institute - VNUHCM, Vietnam
e-mail: long.truong@jvn.edu.vn

Tru H. Cao
Ho Chi Minh City University of Technology - VNUHCM,
Vietnam, and John von Neumann Institute - VNUHCM, Vietnam
e-mail: tru@cse.hcmut.edu.vn

Dien Dinh
Ho Chi Minh City University of Science - VNUHCM, Vietnam
e-mail: diend@fit.hcmus.edu.vn

V.-N. Huynh et al. (eds.), *Knowledge and Systems Engineering, Volume 2,*
Advances in Intelligent Systems and Computing 245,
DOI: 10.1007/978-3-319-02821-7_26, © Springer International Publishing Switzerland 2014

concepts (AC). A named entity is one that is referred to by names, such as a person, an organization, or a location, while an abstract concept is not, e.g. *mouse* or *plant*.

Therefore, Named Entity Recognition (NER) and Entity Disambiguation (ED) are fundamental tasks in NLP. While NER is mainly to map an entity mention (i.e., a phrase used to refer to an entity) in a document to the right class of the entity referred to by that mention in ontology, ED is to link that mention to the right entity in a knowledge base (KB).

Figure 1 shows an example of Entity Disambiguation to the Vietnamese Wikipedia. Given the sentence óÀÌỎĂÕãy lÕà cuòệĒc chiòểln tranh thuòệĒc ỎÈòệĂa giÕành ỎÈòệĒc lòểắp ỎÈòểẩu tiÕận thÕành cÕểng trong lòệĂch sòệắóÀÍ, an ED system is expected to recognize all mentions óÀÌchiòểln tranhóÀÍ, óÀÌthuòệĒc ỎÈòệĂaóÀÍ, óÀÌỎÈòệĒc lòểắpóÀÍ, óÀÌlòệĂch sòệắóÀÍ and links them to matching Wikipedia articles. For instance, mention óÀÌỎÈòệĒc lòểắpóÀÍ in this sentence should be linked to Wikipedia article óÀÌỎĂòệĒc lòểắpóÀÍ, instead of óÀÌỎĂòệĒc lòểắp thòệÈng kỎãóÀÍ in mathematics.

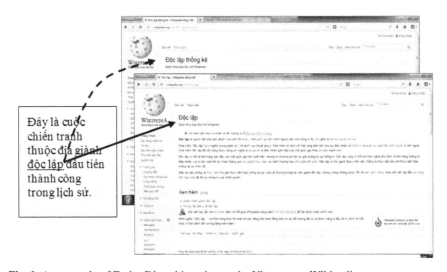

**Fig. 1** An example of Entity Disambiguation to the Vietnamese Wikipedia

In recent years, there have been many research works on ED for English documents, such as [5], [6], or [7] for linking only named entities to a KB, and [2], [3], or [8] for linking both named entities and abstract concepts. They used different approaches, namely, machine learning with diverse features (e.g. [2], [3], [5], [6], [7]), or combined heuristics-statistics one (e.g. [1]). For Vietnamese documents, [4] appears to be the first one, but dealt with only named entities.

In this paper, we first propose an ED method for Vietnamese language that considers both named entities and abstract concepts. We use the Vietnamese Wikipedia as the KB for linking entities to. Wikipedia, with various language versions, has become a reliable and widely used information resource for both humans and knowledge-based systems.

The rest of this paper is organized as follows. In Section 2, we review recent research works on English ED and Vietnamese ED. Section 3 presents in detail our proposed method. Sections 4 and 5 respectively show experimental results and our conclusions drawn from them.

## 2 Related Work

In an ED system, entity mentions in a document must be recognized before they are disambiguated. For mention recognition, [2] first proposed the idea of link probability that was used in [3] and [8] to recognize NE and AC mentions in a document. The process started by gathering all $n$-grams in the document and matched them to an anchor dictionary of Wikipedia. The purpose of using link probability is to discard non-sense phrases and stop words such as *is* and *an*. Meanwhile, also for English documents, [1] used GATE to recognize NEs and a Base Noun Phrase structure to recognize ACs.

For the disambiguation phase, the existing approaches can be classified into the Local, Global, and Collective ones. A local method disambiguates mentions independently and is based on local context compatibility between mentions and candidate entities. Global and collective methods assume that there is a relation among disambiguating decisions and there is coherence between co-occurrence entities in a document.

As a local method, for instance, [2] proposed and evaluated two different disambiguating ways. The first one was based on the overlapping between the local context of a given mention and the content of a Wikipedia candidate. The second one used some words to the left and the right of out-links in Wikipedia articles with their parts-of-speech, to train a Nave Bayes classifier for each mention. Meanwhile, [7] used classification algorithms that learn context compatibility for disambiguating mentions.

As a global method, [3] ranked Wikipedia candidates for a given mention based on three features, namely, Relatedness, Commonness and Context Quality. Relatedness is the semantic relatedness of a candidate entity to contextual entities. Commonness is the number of times a candidate entity used as a destination in Wikipedia. As a balanced measure of Relatedness and Commonness, Context Quality was defined as an unambiguous context of a given mention.

As a collective method, [8] proposed a disambiguating method by using graphs. First, a referent graph was constructed for a text based on local context compatibility and coherence between entities. A referent graph was defined as a weighted and undirected graph that contained all mentions in the text and all possible candidates of these mentions. Each node represents one mention or entity. So a referent graph had two types of edges: mention-entity and entity-entity. A mention-entity edge was established between a mention and an entity and its weight was calculated by cosine similarity implemented in a bag-of-words model. An entity-entity edge was established between two entities and its weight was calculated by using semantic relatedness (as in [3]) between these entities.

In [1], the authors proposed a method that combined heuristics and statistics for entity disambiguation in an incremental process, for English documents and for linking both NEs and ACs to the English Wikipedia. In each round of the iterative process, heuristics-based disambiguation was executed first to provide contextual information for the following statistics-based disambiguation based on similarity between mentions in a document and Wikipedia candidates. The newly disambiguated mentions were then exploited to disambiguate the remaining ambiguous mentions.

To our knowledge, so far the only existing system for Vietnamese entity disambiguation is the one reported in [4] and it only dealt with linking named entities. This paper is the first attempt to tackle the ED problem for Vietnamese with linking both NEs and ACs to the Vietnamese Wikipedia.

## 3 Proposed Method

### 3.1 Overview of the Proposed System

Figure 2 presents the three basic phases of our proposed Vietnamese ED system. Firstly, the Vietnamese Mention Recognition is to detect entity mentions in a document. Secondly, the Candidate Generation is to create a list of candidate entities for each mention in the document. Finally, the Entity Linking is to link each mention to the correct Vietnamese Wikipedia article.

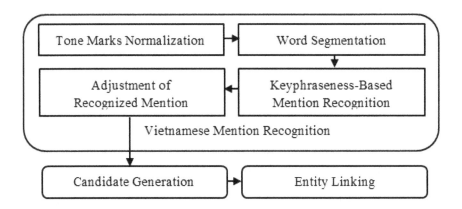

**Fig. 2** Overview of the proposed Vietnamese Entity Disambiguation system

Our purpose is to disambiguate Vietnamese entities in a document. Therefore the main problems to be solved are with the Mention Recognition and Candidate Generation.

Module Mention Recognition takes a raw document as input to the system and preprocesses that Vietnamese raw document by Tone Marks Normalization and Word Segmentation. Mentions in the preprocessed document are detected by using the Keyphraseness measure in [2]. Then the Adjustment of Recognized Mention removes all mentions that are Vietnamese stop-words or are not Wikipedia article titles. Output of module Mention Recognition is a list of recognized mentions, which are Wikipedia article titles.

Module Candidate Generation takes the recognized mentions in the previous phase as input and creates a list of Wikipedia candidate articles for each of the recognized mentions. This module is built by adapting the corresponding module in [1] but using coreference resolution heuristics for Vietnamese in [4].

The last module Entity Linking takes the recognized mentions as input and returns a linked Wikipedia article for each mention. It uses the same hybrid heuristics-statistics-based method in [1] as described in Section 2.

## 3.2 Mention Recognition

Due to the characteristics of Vietnamese regarding tone marks and word formation, a document needs to be preprocessed for recognizing mentions.

### 3.2.1 Tone Marks Normalization

Tone marks are distinguished features of Vietnamese. Vietnamese has six tone marks that are *unmarked*, *grave accent*, *acute accent*, *hook above*, *tilde*, and *dot below*[1]. There are two methods for marking in a syllable[2]. The first method, called *old style*, puts a tone mark in the middle character of a syllable. The second method, called *new style*, puts a tone mark in the main phoneme of a syllable. For example, *ha* and *ho* are two styles of tone marking for a syllable. Here, with the new style method, since character *h* is an onset, character *o* is a medial and character *a* is a nucleus, tone mark *grave accent* is put in character *a*.

In our system, all words are normalized by converting their tones into numbers. For example, *ha kh* and *ho kh* are converted into *hoa2 khi1*. Normalization of tone marks is applied to all tasks of Vietnamese words processing in the system. In particularly, it supports the keyphrasesness-based mention recognition for identifying mentions based on an anchor list, as described in Section 3.3.

### 3.2.2 Word Segmentation

Boundaries between Vietnamese words are different from English ones. They are not identified by white spaces and usually composed of special linguistic units called *morpho-sylable*. A morpho-syllable may be a morpheme or a word or none

---

[1] https://en.wikipedia.org/wiki/Vietnamese_alphabet
[2] http://vi.wikipedia.org/wiki/
   Quy_tac_dat_dau_thanh_trong_tieng_Viet

**Table 1** An example of ambiguous Vietnamese word segmentation

|    | 1 | 2 | 3 | 4 | 5 | 6 | 7 | 8 | 9 | 10 |
|----|---|---|---|---|---|---|---|---|---|----|
|    | MòệẼt | luòễắt | gia | còễấm | còệè | vòệỆi | tÕ̃ảnh | hÕ̃ảnh | hiòệẤn | nay |
|    | A, an | law | home increase | hold animal | resist | with | love state | image | appear | now |
| #1 | MòệẼt | luòễắt gia | | còễấm còệè | | vòệỆi | tÕ̃ảnh hÕ̃ảnh | | hiòệẤn nay | |
|    | A | lawyer | | contends | | with | situation | | present | |
| #2 | MòệẼt | luòễắt | gia còễấm | | còệè | vòệỆi | tÕ̃ảnh hÕ̃ảnh | | hiòệẤn nay | |
|    | A | law | poultry | | resists | with | situation | | present | |

of them[9]. As an example therein, the Vietnamese sentence óÀÌ*MòệẼt luòễắt gia còễấm còệè vòệỆi tÕ̃ảnh hÕ̃ảnh hiòệẤn nay*óÀÍ has 10 morpho-syllables, whose English literal translations are shown in Table 1.

These ten morpho-syllables have their own entries and meanings in a dictionary, but some of them are just morphemes in this sentence. Here, there are many different ways for word segmentation, but only two of them are grammatically correct and one of them is suitable in terms of its semantics. In this case, this sentence has six words óÀÌ*mòệẼt*óÀÍ, óÀÌ*luòễắt gia*óÀÍ, óÀÌ*còễấm còệè*óÀÍ, óÀÌ*vòệỆi*óÀÍ, óÀÌ*tÕ̃ảnh hÕ̃ảnh*óÀÍ, and óÀÌ*hiòệẤn nay*óÀÍ.

Word segmentation helps to choose semantically correct word phrases according to their context. In this work, we employ Left Right Maximum Matching (LRMM) [10] algorithm combined with the Vietnamese dictionary developed at the Language Lab of Kimtudien Company. We use 8-grams in LRMM algorithm as the maximum word length in the employed dictionary is 8 tokens. The algorithm takes in a list of tokens with corresponding offsets and a Vietnamese dictionary, and returns a list of words with corresponding offsets in a document. Input tokens are processed from the leftmost to the rightmost ones in a document, i.e., from the smallest offset to the largest offset. If the longest $n$-gram from chosen tokens exists in the dictionary, it is marked as a word and added into the result list. For example, successful cases using this method are óÀÌ*hòệắp tÕ̃ác xÕ̃a* | *mua bÕ̃án*óÀÍ and óÀÌ*thÕ̃ành lòễắp* | *nỖặòệỆc* | *ViòệẤt Nam* | *dÕ̃ān chòệẩ* | *còệỄng hÕ̃éa*óÀÍ.

### 3.2.3 Keyphraseness-Based Mention Recognition

For named entities, we choose all capital-based $n$-grams, except 1-gram, as named entity mentions. For abstract concepts, we identify their mentions in a document according to their keyphraseness. First, we build a list of anchors with corresponding keyphraseness based on Wikipedia articles. An anchor is the label of an out-link in Wikipedia. Then, we choose all $n$-grams that exist in that anchor list and exceed a keyphraseness threshold as abstract concept mentions. The keyphraseness threshold is set to 0.01 as in [2] and [3]. The order of choosing $n$-grams is from left to right.

The maximum length of $n$-grams is 7, which is the maximum length of Wikipedia article titles in tokens. The keyphraseness of a phrase is computed as the ratio between the number of Wikipedia articles that use the phrase as an anchor and the number of articles that mention the phrase. For example, if *tree* exists in 1,000 Wikipedia articles and 40 Wikipedia articles use it as an anchor, then its keyphraseness is 0.04. We use the Wikipedia version on 19/02/2012 to create a list of anchors with their keyphraseness.

### 3.2.4 Adjustment of Recognized Mention

The keyphraseness-based module may recognize mentions that are not Wikipedia article titles. For each of these mentions, we mark the longest $n$-grams that are Wikipedia article titles from left to right as mentions; if there are no such $n$-grams, then that mention is omitted.

For example, the keyphraseness-based module recognizes mention óÀÌ*tiòệẢu bang Hawaii*óÀÌ, but it is not a Wikipedia article title. In this case, it is adjusted into two mentions óÀÌ*tiòệẢu bang*óÀÌ and óÀÌ*Hawaii*óÀÌ, which are Wikipedia article titles.

Further, there are recognized mentions that are stop-words, such as óÀÌ*nòệÁn*óÀÌ in óÀÌ*nòệÁn ỎÈòệẼc lòệẳp*óÀÌ because there is Wikipedia article title *óÀÌNòệÁn (ỎÈòệẲa chòệẲt)*óÀÌ. Such stop-word mentions are also removed from the final list of recognized mentions. The used stop-word list is given by the Language Lab of Kimtudien Company.

## 4 Evaluation

### 4.1 Testing Dataset

Not as for English, there is not yet a testing dataset for Vietnamese entity disambiguation of both named entities and abstract concepts. So we have to build our own testing dataset based on the criteria described below. It is based on 13 Vietnamese Wikipedia articles and consists of 1,402 semi-manually marked mentions. The number of out-links in these 13 Wikipedia articles is a few because of Wikipedia rules that users have to follow when creating articles[3]. Therefore, we have to mark additional mentions in these testing articles so that:

- Each additionally marked mention must be a Wikipedia article title.
- In case of nested mentions, i.e., a long mention contains short mentions, the long mention is marked. For example, mention óÀÌ*ỎÈòệĒng bòệẳng sỎẻng HòệĒng*óÀÌ contains two mentions óÀÌ*ỎÈòệĒng bòệẳng*óÀÌ and óÀÌ*sỎẻng HòệĒng*óÀÌ. In this case, mention óÀÌ*ỎÈòệĒng bòệẳng sỎẻng HòệĒng*óÀÌ is marked.

---

[3] http://en.wikipedia.org/wiki/Wikipedia:Manual_of_Style Linking

All additionally marked mentions are then manually linked to their matching Wikipedia articles to be testing data. For example, the mention óÀÌchuòệẾtóÀÍ in óÀÌCon chuòệẾt nÕày bòệẮ hÔặóÀÍ is linked to the Wikipedia article óÀÌChuòệẾt (mÕáy tÕặnh)óÀÍ.

## 4.2  Evaluation of Module Mention Recognition

We evaluate our module Mention Recognition by comparing the offsets of recognized mentions and those of marked mentions in the testing dataset. A recognized mention is a correct mention when it has the same first offset and last offset as the corresponding mention in the testing dataset. Table 2 shows the performance of module Mention Recognition.

**Table 2** Performance of module Mention Recognition

| Precision | Recall | F1 |
|---|---|---|
| 0.88 | 0.76 | 0.82 |

There are four types of errors that affect the performance of our module Mention Recognition, as follows:

- First type of errors is caused by Word Segmentation. For example, our system recognizes mention óÀÌKinhóÀÍ in óÀÌKinh phÔặ hoòệắt_ỔÈòệẼng còệẮa LiÕận_HiòệẤp_QuòệÈcóÀÍ, because the module Word Segmentation did not group óÀÌKinhóÀÍ and óÀÌphÔặóÀÍ together due to the capital word óÀÌKinhóÀÍ and Wikipedia article title óÀÌKinhóÀÍ having keyphraseness greater than the set threshold. This type of error affects the precision of mention recognition.
- Second type of errors is caused by recognized mentions that are not included in the testing dataset. For example, for the text óÀÌnòệÁn kinh tòệẼ ỔÈòệẮng thòệẠ hai thòệẠ giòệẸióÀÍ, our method recognizes óÀÌthòệẠ haióÀÍ as a mention because there is the Wikipedia article title óÀÌThòệẠ haióÀÍ, which is one day of a week, whose keyphraseness is greater than the set threshold. However, we do not consider óÀÌthòệẠ haióÀÍ as a mention in the built testing dataset. This type of errors also affects the precision of mention recognition.
- Third type of errors is caused by Wikipedia article titles whose keyphraseness is 0, i.e., those that are not used as out-link labels in any Wikipedia article. As their keyphraseness is less than the set threshold, they are not recognized as mentions by our method, while they could be in the testing dataset. For example, óÀÌthòệẠ kòệÈ 196óÀÍ is a mention in the testing dataset, but it is not recognized by our method because its keyphraseness is 0. This type of error affects the recall of mention recognition.

- Fourth type of errors is caused by marked mentions in the testing dataset that are not Wikipedia article titles. For example, óÀÌỎÈòệĂa chòếât vỖễng biòệẢnóÀÍ is marked as a mention in the testing dataset as there is the Wikipedia article óÀÌỎÈòệĂa chòếât biòệẢnóÀÍ. Since there is no Wikipedia article title óÀÌỎÈòệĂa chòếât vỖễng biòệẢnóÀÍ, our method misses it. This type of errors also affects the recall of mention recognition.

## 4.3   Evaluation of Module Entity Linking

We evaluate our module Entity Linking directly on the 13 testing Wikipedia articles with all 1,402 correct mentions. That is, to evaluate this module alone, we assume that the previous results of module Mention Recognition are 100% correct. Table 3 shows its performance in Micro-Average Accuracy (MAA) measure. Our module uses the same hybrid heuristics-statistics-based method in [1] and it gives a similar performance figure as that for English entity disambiguation therein, whose MAA is 67.55%. That means the method works for Vietnamese as well.

**Table 3** Performance of module Entity Linking

| MAA |
| --- |
| 69.24 |

## 4.4   Evaluation of the Whole System

Table 4 shows the evaluation result of the whole system. It is affected by the errors of both the Mention Recognition and Entity Linking modules. Therefore, it is lower than that of entity linking alone. As analyzed, about 350 mentions in the testing dataset are not correctly recognized. Furthermore, about 280 out of 1,052 recognized mentions are not correctly linked to Wikipedia.

**Table 4** Performance of the whole system

| Precision | Recall | F1 |
| --- | --- | --- |
| 0.59 | 0.55 | 0.57 |

## 5   Conclusion

In this paper, we have developed the first system for Vietnamese Entity Disambiguation, which is to link both named entities and abstract concepts in a Vietnamese text to the correct articles in the Vietnamese Wikipedia. The main problem that we have solved is recognition of Vietnamese mentions, due to the special characteristics of Vietnamese tone marks and word formation. Our proposed Vietnamese mention recognition method has been then combined with an existing entity linking method to build up the Vietnamese ED system. The experimental results have showed that the proposed mention recognition method achieves high accuracy and the performance of the whole system is comparable to that of the respective one for English. Improvement of mention recognition and researching other entity linking methods for Vietnamese are among the topics of our current interests.

## References

1. Nguyen, H.T., Cao, T.H., Nguyen, T.T., Vo, L.T.T.: Heuristics and Statistics-based Wikification, In: Proc. of the 12th Pacific Rim International Conference on Artificial Intelligence, pp. 879-882 (2012)
2. Mihalcea, R., Csomai, A.: Wikify!: Linking Documents to Encyclopedic Knowledge. In: Proc. of the 16th ACM International Conference on Information and Knowledge Management, pp. 233-242 (2007)
3. Milne, D., Witten, I.H.: Learning to Link with Wikipedia. In: Proc. of the 17th ACM International Conference on Information and Knowledge Management, pp. 509-518 (2008)
4. Nguyen, H.T., Cao, T.H.: A Knowledge-based Method to Resolve Name Ambiguity in Vietnamese Texts. In: Addendum Contributions of the 5th International Conference on Research, Innovation and Vision for the Future, Studia Informatica Universalis, pp. 83-88 (2007)
5. Ji, H., Grishman, R., Dang, H.T.: An Overview of the TAC2011 Knowledge Base Population Track. In: Proc. of Text Analysis Conference (2011)
6. Ji, H., Grishman, R.: Knowledge Base Population Successful Approaches and Challenge. In: Proc. of the 49th Annual Meeting of the Association for Computational Linguistics: Human Language Technologies, pp. 1148-1158 (2011)
7. Zhang, W., Su, J., Tan, C.L., Wang, W.: Entity Linking Leveraging Automatically Genrated Annotation. In: Proc. of 23rd International Conference on Computational Linguistics, pp. 1290-1298 (2010)
8. Han, X., Sun, L., Zhao, J.: Collective Entity Linking in Web Text: A Graph Based Method. In: Proc. of the 34th Annual ACM Special Interest Group on Information Retrieval Conference, pp. 765-774 (2011)
9. Pham, T.X.T., Tran, T.Q., Dinh, D., Collier, N.: Named Entity Recognition in Vietnamese Using Classifier Voting. ACM Transactions on Asian Language Information Processing, 6(4) (2007)
10. Dinh, D.: Natural Language Processing. VNU-Ho Chi Minh Publisher (2006) (in Vietnamese)

# Maintenance of a Frequent-Itemset Lattice Based on Pre-large Concept

Bay Vo, Tuong Le, Tzung-Pei Hong, and Bac Le

**Abstract.** This paper proposes an effective approach for the maintenance of a frequent-itemset lattice in incremental mining based on the pre-large concept. First, the building process of a frequent-itemset lattice is improved using a proposed theorem regarding the paternity relation between two nodes in the lattice. Then, based on the pre-large concept, an approach for maintaining a frequent-itemset lattice with dynamically inserted data is proposed. The experimental results show that the proposed approach outperforms the batch approach for building the lattice in terms of execution time.

## 1 Introduction

Association rule mining is an important problems in data mining [1-2, 4-5, 8-12, 14]. For mining association rules, existing algorithms are often divided into two distinct phases [2, 4-5, 8-12, 14]: (i) mining frequent (closed) itemsets and (ii) generating the association rules from these frequent (closed) itemsets. Recently, an approach for mining association rules using a frequent (closed)-itemset lattice (FIL/FCIL)

Bay Vo
Ton Duc Thang University, Ho Chi Minh City, Vietnam
e-mail: bayvodinh@gmail.com

Tuong Le
University of Food Industry, Ho Chi Minh City, Vietnam
e-mail: tuonglecung@gmail.com

Tzung-Pei Hong
Department of Computer Science and Information Engineering,
National University of Kaohsiung, Taiwan, ROC
e-mail: tphong@nuk.edu.tw

Bac Le
Department of Computer Science, University of Science, Ho Chi Minh, Vietnam
e-mail: lhbac@fit.hcmus.edu.vn

V.-N. Huynh et al. (eds.), *Knowledge and Systems Engineering, Volume 2*,
Advances in Intelligent Systems and Computing 245,
DOI: 10.1007/978-3-319-02821-7_27, © Springer International Publishing Switzerland 2014

has been proposed [10-12]. The FIL/FCIL is then used for generating association rules. The building process of an FIL/FCIL is not better in terms of execution time and memory usage than the traditional approaches, which directly mine frequent (closed) itemsets. However, generating association rules from the FIL/FCIL is more efficient than from frequent (closed) itemsets [12] or from frequent itemsets using a hash table [10-11]. Therefore, this approach outperforms the traditional approaches when both phases of mining are considered.

In practical applications, databases are typically incremental, meaning that transactions could be removed or inserted. Therefore, mining association rules from incremental databases has attracted research interest. The first algorithm for mining association rules in incremental databases was Fast-UPdate (FUP) [3], an Apriori-based algorithm, which generates the candidates and repeatedly scans the databases. Hong et al. [6] proposed the pre-large concept (see Section 2.2) to reduce the need for rescanning the original database. With this concept, the original database does not need to be rescanned if the number of new transactions is equal to or less than a safety threshold. The maintenance cost of frequent itemsets is thus reduced with the pre-large concept. Although many algorithms have been developed to maintain the association rules in incremental databases, the maintenance of a FIL in incremental mining has received little attention. Compared to the maintenance of frequent itemsets, that of a FIL is more complex, with the algorithm having to consider the relations among the nodes in the lattice and update them. The present study proposes an effective approach for the maintenance of a FIL. The main contributions are:

1. A structure of FIL for quickly building a FIL is proposed.
2. A tidset-based maintenance of a pre-large FIL (TMP) algorithm is developed.

The rest of the paper is organized as follows. Section 2 presents the basic concepts. TMP algorithm is proposed in Section 3. Section 4 presents the results of experiments comparing the runtime of TMP algorithm with those of batch approach for building the lattice. Finally, the paper is concluded in Section 5 with a summary and some future research issues.

## 2 Basic Concepts

### 2.1 Frequent Itemsets Lattice

Consider a database $DB$ with $n$ transactions, with each transaction including a set of items belonging to $I$ where $I$ is the set of all items in $DB$. An example transaction database $D_1$ is presented in Table 1. The support of an itemset $X$, denoted by $\sigma(X)$ where $X \in I$, is the number of transactions in $DB$ which contain all the items in $X$. An itemset $X$ is called a frequent itemset if $\sigma(X) \geq minSup \times n$, where $minSup$ is a given threshold. An itemset with $k$ items is called a $k$-itemset.

Vo and Le [10] proposed an algorithm to build the FIL in which each node in the lattice has the form $\langle X, Tidset, Children \rangle$, where $X$ is a $k$-itemset, $Tidset$

**Table 1** Example transaction database $D_1$

| Transaction | Items |
|---|---|
| 1 | $A, C, T, W, Z$ |
| 2 | $C, D, W$ |
| 3 | $A, C, T, W$ |
| 4 | $A, C, D, W$ |
| 5 | $A, C, D, T, W$ |
| 6 | $C, D, T$ |

is the set of IDs associated with the transactions containing $X$, and *Children* = $\{Y \mid Y$ *is the* $(k+1) - itemset$ *and* $X \subset Y\}$.

When a node $XA$ in a FIL is created, the algorithm proposed in [10-11] has to find all nodes which are the children of $XA$ to update the lattice. This process has two loop and one if statements as follows: (i) let $Y \in X.Children$, (ii) let $YB \in Y.Children$ and (iii) if $XA \subset YB$ then $YB \in XA.Children$. For example, consider the lattice in Figure 2. When the algorithm creates the node $TC$, it has to consider all the child nodes associ-ated with $T$, which are $AT$ and $TW$. Then, the algorithm has to consider all the child nodes associated with $AT$ and $TW$, which are $\{ATW, ATC\}$ and $\{ATW\}$. However, the process of considering all child nodes of $TW$ does not find any node that is children of $TC$. The node $ATW$ is a duplicatand thus the process of considering all child nodes associated with $TW$ is meaningless. To overcome this weakness, this study proposes the following structure for a FIL:

**Definition 1.** Let $X$ be a $k$-itemset. The child nodes based on the equivalence class feature associated with $X$ are:

$$X.ChildrenEC = \{XA \mid \forall A \in I, A \notin X \text{ and } A \neq \emptyset\} \qquad (1)$$

**Definition 2.** Let $X$ be a $k$-itemset. The child nodes based on the lattice feature associated with $X$ are as follows:

$$X.ChildrenL = \{Y \mid Y \text{ is a } (k+1) - itemset, Y \notin X.ChildrenEC \text{ and } X \subset Y\} \qquad (2)$$

**Definition 3.** Each node in the lattice is a tuple as follows:

$$\langle Itemset, Tidset, ChildrenEC, ChildrenL \rangle \qquad (3)$$

where

- *Itemset* is a $k$-itemset.
- *Tidset* is the set of IDs associated with the transactions containing *Itemset*.
- *ChildrenEC* contains the child nodes based on the equivalence class feature asso-ciated with *Itemset*.
- *ChildrenL* contains the child nodes based on the lattice feature associated with *Itemset*.

**Theorem 1.** *Let XA be the node of k-itemset. $\forall XB \in X.ChildrenEC$ and A is before B in the order of frequent 1-itemsets (sorted in ascending order of frequency). Then $\nexists Y \in XB.ChildrenEC \cup XB.ChildrenL$ so that $Y \in XA.ChildrenL$.*

*Proof.* Assume that there exists the node $Y$ in the lattice so that (i) $Y \in XA.ChildrenL$ then $XA \subset Y$, (ii) $Y \in XB.ChildrenEC \cup XB.ChildrenL$ then $XB \subset Y$, (iii) $Y$ is a $(k+1)$-itemset. From (i), (ii) and (iii), $Y$ can only be $XAB$. However, the algorithm used the deep-first-search strategy; therefore $XAB$ has not yet been created. Therefore, Theorem 1 is proven.

With the new structure in Definition 3 and Theorem 1, the tidset-based FIL (TFIL) building algorithm can easily find the nodes which belong to $XA.ChildrenL$ in three two loop and one if statements as follows: (i) let $Y \in X.ChildrenL$, (ii) let $YB \in Y.ChildrenEC$ and (iii) if $XA \subset YB$ then $YB \in XA.ChildrenL$. Separating *Children* into *ChildrenEC* and *ChildrenL* makes the TFIL better than the algorithm in [10-11] because it eliminates a large number of candidates. The TFIL algorithm is presented in Figure 1.

The TFIL algorithm is applied to the example database in Table 1 with $minSup = 50\%$ to illustrate its use. First, TFIL finds all the frequent 1-itemsets and sorts them in ascending order of frequency. The result of this step is $I_1 = \{A, D, T, W, C\}$. The algorithm uses the depth-first-search strategy to generate the candidates associated with each equivalence class. The frequent $(k-1)$-itemsets in turn are combined with the remaining $(k-1)$-itemsets in this equivalence class to create the k-itemset candidates. The frequent itemsets from these candidates are used to create the frequent $(k+1)$-itemsets. When each node in the lattice is created, the algorithm calls the procedure **Update_Lattice** to update the childnodes. The results for the example database are shown in Figure 2. Note that the dashed and solid lines represent *ChildrenL* and *ChildrenEC*, respectively.

## 2.2   Pre-large Concept in Incremental Mining

The pre-large concept was proposed by Hong et al. [6]. It is based on a safety threshold $f$ to reduce the need of rescanning the original databases for efficiently maintaining association rules. The safety number $f$ of inserted transactions is derived as follows:

$$f = \left\lfloor \frac{(S_U - S_L) \times |D|}{1 - S_U} \right\rfloor \tag{4}$$

where $S_U$ is the upper threshold, $S_L$ is the lower threshold, and $|D|$ is the number of original database $D$'s transactions. When the number of new transactions is equal to or less than $f$, the algorithm does not need to rescan the original database.

When two thresholds are used, each itemset has three cases: frequent, pre-large, and infrequent. This divides itemsets in the original and updated databases into nine cases [6, 7] as presented in Table 2.

*Example.* Consider the database in Table 1 with $|D_1| = 6$. Assume that $S_U = 70\%$ and $S_L = 50\%$. Then if the number of new transactions is equal to or less than

```
Input: transaction database D with n transactions and frequent
threshold minSup
Output: lattice containing all frequent itemsets of D
Function FIL()
1.let L_root be the root of the lattice
2.let FI_1 be the frequent 1-itemsets
3.for each A_i ∈ FI_1 do
4.   add ⟨A_i.Itemset, A_i.Tidset, {}, {}⟩ as a node to L_root
5.Enumerate_Lattice(FI_1)
6.return L_root

Procedure Enumerate_Lattice(FI_k)
1.for each I_i ∈ FI_k do
2.   FI_{k+1} = {}
3.   for each I_j ∈ FI_k with j > i do
4.      X.Itemset = I_i.Itemset ∪ I_j.Itemset
5.      X.Tidset = I_i.Tidset ∩ I_j.Tidset
6.      X.support = X.Tidset.count
7.      if X.support ≥ minSup × n then
8.         add the address of X to I_i.ChildrenEC
9.         add the address of X to I_j.ChildrenL
10.        add X to FI_{k+1}
11.        Update_Lattice(I_i, X)
12.   Enumerate_Lattice(FI_{k+1})

Procedure Update_Lattice(Parent, X)
1.for each Child I_i in Parent.ChildrenL do
2.   for each Child I_j in I_i.ChildrenEC do
3.      if X.Itemset ⊂ I_j.Itemset then
4.         Add the address of I_j to X.ChildrenL
```

**Fig. 1** TFIL algorithm

$f = \lfloor ((0.7 - 0.5) \times 6)/(1 - 0.7) \rfloor = 4$, the algorithm does not need to rescan the original database to determine the support of infrequent itemsets.

## 3  Tidset-Based Maintenance of PFIL Algorithm

### 3.1  The Proposed Algorithm

First TMP uses the lower threshold $S_L$ to build a pre-large FIL (PFIL) and determine the safety threshold $f$. Then when the number of incremental transactions is equal to or less than $f$, TMP updates the PFIL without scanning database. Only when the number of transactions is larger than $f$, TMP scans the original database and all the incremental transactions. TMP is presented in Figure 3.

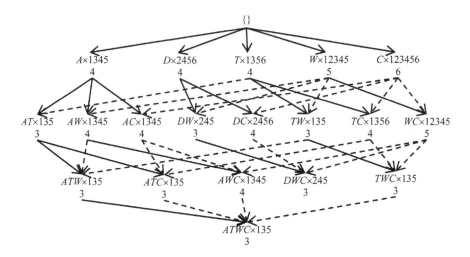

**Fig. 2** FIL for $D_1$ with $minSup = 50\%$ obtained using TFIL algorithm

**Table 2** Nine cases of itemsets

| Case | Original updated | Result |
|------|------------------|--------|
| 1 | frequent - frequent | frequent |
| 2 | frequent - pre-large | frequent/pre-large, determined from existing itemsets |
| 3 | frequent - infrequent | frequent/pre-large/infrequent, determined from existing itemsets |
| 4 | pre-large - frequent | pre-large/frequent, determined from existing itemsets |
| 5 | pre-large - pre-large | pre-large |
| 6 | pre-large - infrequent | pre-large/infrequent, determined from existing itemsets |
| 7 | infrequent - frequent | pre-large/infrequent, rescan the original if the number of inserted transactions is larger than $f$ |
| 8 | infrequent - pre-large | infrequent/pre-large, rescan the original if the number of inserted transactions is larger than $f$ |
| 9 | infrequent - infrequent | infrequent |

## 3.2   The Illustration of TMP Algorithm

In this section, an example is given to illustrate the process of TMP in three times of increment. Given the initial database $D = \emptyset$, $S_L = 50\%$ and $S_U = 65\%$.

### 3.2.1   First Increment of Database ($D_1$ with the Six Transactions in Table 1)

Because the condition in line 1 is true, the algorithm performs lines 2 and 3. The result of this increment is the PFIL shown in Figure 2. Then the algorithm will

```
Input:
- the original datbase D
- the safety threshold f determined from D
- incremental database D'
- upper threshold Su and the lower threshold SL
Output: PFIL
INCREMENTAL-FIL()
1.if |D| = 0 then
2.   call function FIL to build PFIL for D' using SL
```
3.   compute $f = \left\lfloor \frac{(S_U - S_L) \times |D|}{1 - S_U} \right\rfloor$
```
4.else if |D'| > f then
5.   call function FIL to build PFIL for D + D' using SL
```
6.   compute $f = \left\lfloor \frac{(S_U - S_L) \times (|D| + |D'|)}{1 - S_U} \right\rfloor$
```
7.else
8.   clear tidset information in each node in PFIL
9.   update node information in the first level of L1 and mark
the nodes which have changed information and their supports sat-
isfy SL.
10.  call procedure UPDATE-PFIL to update all nodes in PFIL with
L1 as parameter
```
11. $f = f - |D'|$
```
12.D = D + D'

UPDATE-PFIL(Lr)
1.for all li ∈ Lr.ChildrenEC do
2.   if li is marked then
3.      for all lj ∈ Lr.ChildrenEC, with j > i do
4.         if lj is marked then
5.            let O be the directly child node of li and lj
6.            if O exists then
7.               O.Tidset = li.Tidset ∩ lj.Tidset
8.               if O.Tidset.count > 0 then
9.                  O.support = O.support + O.Tidset.count
10.                 if O.support ≥ SL × (|D| + |D'|) then
11.                    mark O
12.   UPDATE-PFIL (li)
```

**Fig. 3** TMP algorithm

update $f = \lfloor ((S_U - S_L) \times |D|)/(1 - S_U) \rfloor = \lfloor ((0.65 - 0.5) \times 6)/(1 - 0.65) \rfloor = 2$ and $D = \emptyset + D_1$.

### 3.2.2 Second Increment of Database (D2 with Two Transactions in Table 3)

Because $|D_2| = 2 = f$, the algorithm performs lines 8, 9, 10 and 11 in the procedure **INCREMENTAL-FIL**.

1. TMP clears all the tidset information associated with all the nodes in the PFIL.

**Table 3** Example transaction database $D_2$

| Transaction | Items |
|---|---|
| 7 | $A, T, W, Z$ |
| 8 | $C, T, W, Z$ |

2. TMP inserts the tidset information (only for the transactions 7 and 8) associated with the frequent 1-itemsets in the PFIL and then marks the updated nodes (Figure 4).

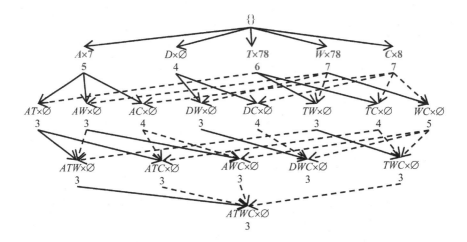

**Fig. 4** PFIL of step 2 in the second increment

3. TMP calls the procedure UPDATE-PFIL, which is called recursively in depth-first search, to update the tidset information of all nodes in the PFIL. The result of this step is shown in Figure 5.
4. TMP updates the safety threshold $f = f - |D_2| = 22 = 0$.
5. TMP updates the database $D = D_1 + D_2$.

Figure 5 shows that only a small number of nodes which have their tidset information updated in the PFIL are used to update the lattice.

### 3.2.3    Third Increment of Database ($D_3$ with One Transaction in Table 4)

Because $|D| > 0$ and $|D_3| = 1 > f = 0$, the algorithm performs lines 5 and 6 in the procedure **INCREMENTAL-FIL**.

1. TMP calls the function FIL to create the PFIL with D with nine transactions in Tables 1, 3 and 4. The result of this step is shown in Figure 5.

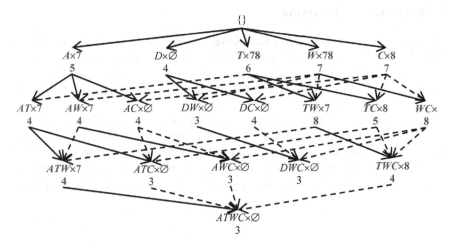

**Fig. 5** PFIL of step 3 in the second increment

**Table 4** Example transaction database $D_3$

| Transaction | Items |
|---|---|
| 9 | $A, D, T, W, Z$ |

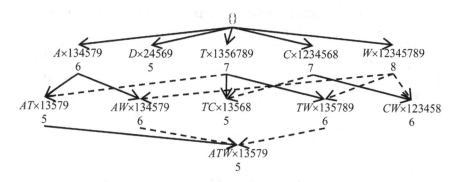

**Fig. 6** PFIL of step 1 in the third increment

2. TMP calculates the safety threshold $f = \lfloor((S_U - S_L) \times |D|)/(1 - S_U)\rfloor = \lfloor((0.65 - 0.5) \times 9)/(1 - 0.65)\rfloor = 3$. Therefore, for the fourth increment, if $|D_4| \leq 3$, the algorithm updates the PFIL without scanning the database ($D_1 + D_2 + D_3$).

## 4   Experimental Results

All experiments presented in this section were performed on an ASUS laptop with an Intel i3-3110M 2.4GHz CPU and 4GB of RAM. All the programs were coded in C# (version 4.5.50709). The experiments were conducted using the following UCI data-bases (http://fimi.cs.helsinki.fi/data/): Accidents (340,183 transactions and 468 items) and Chess (3,196 transactions and 76 items).

We compare runtime (total execution time) of TMP in ten times of increment with the runtime of the TFIL algorithm for the database in ten times. Note that TFIL used the upper threshold $S_U$ to build FIL. According to Figures 7, the runtime of TMP algorithm is smaller than the runtime of TFIL. Especially on the large database (Acci-dents database), the proposed approach is much more effective than TFIL algorithm.

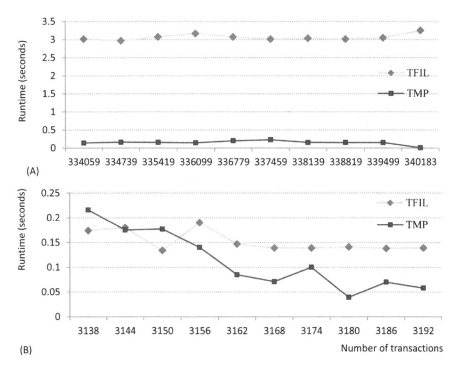

**Fig. 7** Runtimes of (A) TMP ($S_U = 50\%$ and $S_L = 47\%$), TFIL ($S_U = 50\%$) for Accidents and (B) TMP ($S_U = 70\%$, $S_L = 67\%$) and TFIL ($S_U = 70\%$) for Chess

## 5   Conclusion and Future Work

This paper proposed an effective algorithm for the maintenance of a FIL. The proposed approach has two phases: (i) building the FIL and (ii) maintaining the PFIL. In the first phase, a theorem is proposed for quickly building a FIL. In the second phase, we have proposed an effective approach, TMP, which uses the pre-large concept for the maintenance of PFIL. The experiment section shows that TMP is more effective than TFIL.

The disadvantage of our approach is that it must build FIL based on the whole da-tabase (integration from the original and inserted) when the number of inserted trans-actions is larger than the safety threshold. For future work we will study how to store the information to avoid rescanning the original database.

# References

1. Agrawal, R., Imielinski, T., Swami, A.N.: Mining association rules between sets of items in large databases. In: SIGMOD 1993, pp. 207–216 (1993)
2. Agrawal, R., Srikant, R.: Fast algorithms for mining association rules. In: VLDB 1994, pp. 487–499 (1994)
3. Cheung, D.W., Han, J., Ng, V.T., Wong, C.Y.: Maintenance of discovered association rules in large databases: An incremental updating approach. In: The 12th IEEE International Conference on Data Engineering, USA, pp. 106–114 (1996)
4. Grahne, G., Zhu, J.: Fast algorithms for frequent itemset mining using FP-trees. IEEE Transactions on Knowledge and Data Engineering 17, 1347–1362 (2005)
5. Han, J., Pei, J., Yin, Y.: Mining frequent patterns without candidate generation. In: SIGMODKDD 2000, pp. 1–12 (2000)
6. Hong, T.P., Wang, C.Y., Tao, Y.H.: A new incremental data mining algorithm using pre-large itemsets. Intelligent Data Analysis 5(2), 111–129 (2001)
7. Lin, C.W., Hong, T.P., Lu, W.H.: The Pre-FUFP algorithm for incremental mining. Expert Systems with Applications 36(5), 9498–9505 (2009)
8. Pasquier, N., Bastide, Y., Taouil, R., Lakhal, L.: Efficient mining of association rules using closed itemset lattices. Information Systems 24(1), 25–46 (1999)
9. Song, W., Yang, B., Xu, Z.: Index-BitTableFI: An improved algorithm for mining frequent itemsets. Knowledge-Based Systems 21, 507–513 (2008)
10. Vo, B., Le, B.: Mining traditional association rules using frequent itemsets lattice. In: 39th International Conference on CIE, Troyes, France, July 6-8, pp. 1401–1406 (2009)
11. Vo, B., Le, B.: Interestingness measures for association rules: Combination between lattice and hash tables. Expert Systems with Applications 38(9), 11630–11640 (2011)
12. Vo, B., Hong, T.P., Le, B.: A lattice-based approach for mining most generalization association rules. Knowledge-Based Systems 45, 20–30 (2013)
13. Zaki, M.J., Gouda, K.: Fast vertical mining using diffsets. In: KDD 2003, pp. 326–335 (2003)
14. Zaki, M.J., Hsiao, C.J.: Efficient algorithms for mining closed itemsets and their lattice structure. IEEE Transactions on Knowledge and Data Engineering 17(4), 462–478 (2005)

# Mining Class-Association Rules with Constraints

Dang Nguyen and Bay Vo

**Abstract.** Numerous fast algorithms for mining class-association rules (CARs) have been developed recently. However, in the real world, end-users are often interested in a subset of class-association rules. Particularly, they may consider only rules that contain a specific item or a specific set of items. The nave strategy is to apply such item constraints into the post-processing step. However, such approaches require much effort and time. This paper proposes an effective method for integrating constraints that express the presence of user-defined items (for example (Bread AND Milk)) into the class-association rule mining process. First, we design a tree structure in that each node contains the constrained itemset. Second, we develop a theorem and a proposition for quickly pruning infrequent nodes and weak rules. Final, an efficient algorithm for mining CARs with item constraints is proposed. Experiments show that the proposed algorithm outperforms the post-processing approach.

## 1 Introduction

The integration of classification and association rule mining was firstly introduced by Liu et al. in 1998 [1]. The problem is described as follows. First, the complete set of CARs that satisfy the user-specified minimum support and minimum confidence thresholds is mined from the training dataset. Second, a subset of CARs is then selected to form the classifier. Numerous approaches have been proposed to solve this problem. Examples include CBA [1], CMAR [2], CPAR [3], MCAR [4], ACME [5], ECR-CARM [6], LOCA [7], and CAR-Miner [8].

Dang Nguyen
University of Information Technology, Ho Chi Minh, Vietnam
e-mail: nguyenphamhaidang@yahoo.co.uk

Bay Vo
Information Technology Department, Ton Duc Thang University, Ho Chi Minh, Vietnam
e-mail: bayvodinh@gmail.com

V.-N. Huynh et al. (eds.), *Knowledge and Systems Engineering, Volume 2,*
Advances in Intelligent Systems and Computing 245,
DOI: 10.1007/978-3-319-02821-7_28, © Springer International Publishing Switzerland 2014

In practice, end-users often consider only a subset of CARs, for instance, those that contain a user-defined itemset. The item constraints reduce the number of CARs and decrease the search space so that the performance of the mining process can be improved. Additionally, constrained CARs help to discover interesting or useful rules particular to the end-user. For example, while classifying the risk of populations for HIV infection, epidemiologists often concentrate on rules that include demographic information such as sex, age, and marital status. Under this context, the present study considers constraints in the form of the presence of specific items in the rule antecedents. The main contributions of this paper are stated as follows. Firstly, we propose a tree structure named Single Constraint Rule-tree (SCR-tree) for efficiently mining CARs with item constraints. At the first level, the tree contains the constrained node which includes the constrained itemset and frequent nodes which include frequent 1-itemsets. At the following levels, the tree contains constrained nodes only. Secondly, we develop a theorem and a proposition for quickly pruning infrequent nodes and weak classification rules. Finally, we propose a fast algorithm for mining CARs with item constraints.

## 2 Preliminary Concepts

Let $D$ be a training dataset with $n$ attributes $\{A_1, A_2, ..., A_n\}$ and $|D|$ objects (cases). Let $C = \{c_1, c_2, ..., c_k\}$ be a list of class labels. A specific value of an attribute $A_i$ and class $C$ are denoted by lower-case letters $a_i$ and $c$, respectively.

**Definition 1.** An itemset is a set of pairs, each of which consists of an attribute and a specific value for that attribute, denoted by $\{(A_{i1}, a_{i1}), (A_{i2}, a_{i2}), ..., (A_{im}, a_{im})\}$.

**Definition 2.** Let Constraint_Itemset be a specific itemset considered by end-users.

**Definition 3.** A class-association rule $r$ has form $\{(A_{i1}, a_{i1}), ..., (A_{im}, a_{im})\} \rightarrow c_j$, where $\{(A_{i1}, a_{i1}), ..., (A_{im}, a_{im})\}$ is an itemset and $c_j \in C$ is a class label.

**Definition 4.** A strong rule is defined as a rule with the highest confidence among rules generated from a given node. Otherwise, that rule is called a weak rule.

**Definition 5.** The actual occurrence $ActOcc(r)$ of rule $r$ in $D$ is the number of objects in $D$ that match $r$s antecedent.

**Definition 6.** The support of rule $r$, denoted by $Sup(r)$, is the number of objects in $D$ that match $r$s antecedent and is labeled with $r$s class.

**Definition 7.** The confidence of rule $r$, denoted by $Conf(r)$, is defined as:

$$Conf(r) = \frac{Sup(r)}{ActOcc(r)}$$

A sample training dataset is shown in Table 1 where each OID is an object identifier. It contains eight objects, three attributes, and two classes (1 and 2). Considering the rule $r : \{(A, a1)\} \rightarrow 1$. We have $ActOcc(r) = 3$ and $Sup(r) = 2$ because there are three objects with $A = a1$, in that two objects have the same class 1. In addition, $Conf(r) = \frac{Sup(r)}{ActOcc(r)} = \frac{2}{3}$.

**Table 1** Example of a training dataset

| OID | A | B | C | Class |
|-----|-----|-----|-----|-------|
| 1 | a1 | b1 | c1 | 1 |
| 2 | a1 | b2 | c1 | 2 |
| 3 | a2 | b2 | c1 | 2 |
| 4 | a3 | b3 | c1 | 1 |
| 5 | a3 | b1 | c2 | 2 |
| 6 | a3 | b3 | c1 | 1 |
| 7 | a1 | b3 | c2 | 1 |
| 8 | a2 | b2 | c2 | 2 |

## 3 Related Work

### 3.1 Mining Association Rules with Item Constraints

Since the introduction of mining association rules with item constraints [9], various strategies have been proposed. The nave strategy, post-processing approaches, first mines frequent itemsets by using an algorithm such as Apriori [10], FP-Growth [11], or Eclat [12] and then filters out the ones that do not satisfy the item constraints in the post-processing step. Some examples are Apriori+ [13] and FP-Growth+ [14]. This kind of strategy is very inefficient because all frequent itemsets must be generated and often a huge number of candidate itemsets must be tested in the last step. Another strategy, constrained itemset filtering, tries to integrate the item constraints into the actual mining process in order to generate only the frequent itemsets that satisfy the constraints. Since this strategy can use the properties of the constraints much more effectively, its execution time is much lower than those of other strategies. CAP [13] and MFS-Contain-IC [15] belong to this group.

The two strategies for mining association rules with item constraints cannot be applied for mining CARs with item constraints because they do not generate constrained CARs directly. Moreover, to calculate the confidence of association rules, algorithms for mining constrained association rules have to scan the original database again to count the support of rule antecedents. Since frequent itemsets in rule antecedents do not contain constrained itemsets, their support cannot be known directly.

## 3.2 CAR-Miner-Post Algorithm

Liu et al. [1] proposed a method for mining CARs based on the Apriori algorithm. However, the method is time-consuming because it generates a lot of candidates and scans the dataset several times. Vo and Le proposed another method for mining CARs by using an Equivalence Class Rule tree (ECR-tree) [6]. An efficient algorithm, called ECR-CARM, was also proposed in their paper. ECR-CARM scans the dataset only once and uses the intersection of object identifiers to determine the support of itemsets quickly. However, it needs to generate and test a huge number of candidates because each node in the tree contains all values of one attribute. Nguyen et al. [8] modified the ECR-tree structure to speed up the mining time. In their enhanced tree, named MECR-tree, each node contains only one value of an attribute instead of the whole group. Moreover, they also provided some theorems to identify the support of child nodes and prune unnecessary nodes quickly. Based on MECR-tree and these theorems, they presented the CAR-Miner algorithm for effectively mining CARs. However, CAR-Miner cannot be applied directly for mining CARs with item constraints. To deal with item constraints, an extended version of CAR-Miner named CAR-Miner-Post is proposed here. Firstly, CAR-Miner is used to discover all CARs from the dataset. Secondly, the post-processing step filters out rules that do not satisfy the item constraints. The pseudo code of the CAR-Miner-Post algorithm is shown in Figure 1.

---

**Input:** Dataset $D$, *minSup*, *minConf*, and *Constraint_Itemset*
**Output:** All CARs satisfying *minSup*, *minConf*, and *Constraint_Itemset*

1.   CARs = CAR-Miner( $L_r$ , *minSup*, *minConf*)
2.   Constraint_CARs = filterRules(CARs, *Constraint_Itemset*)

**Procedure:**
**filterRules**(CARs, *Constraint_Itemset*)
3.   Constraint_CARs = $\varnothing$
4.   for each *rule* $\in$ *CARs* do
5.       if *Constraint_Itemset* $\in$ *rule.antecedent* then
6.           Constraint_CARs = Constraint_CARs $\cup$ *rule*

---

**Fig. 1** CAR-Miner-Post algorithm

For detail on the CAR-Miner algorithm, please refer to the study by Nguyen et al. [8]. CAR-Miner-Post is easily implemented with slight modification of the original CAR-Miner but it fails to exploit the properties of the constraints. The main drawback of this approach thus lies in its computational complexity. In the proposed method, we try to push the constraints as deep inside the computation as possible. The most noticeable is that rather than inducting all tree nodes whose computational cost is much high, we form only the tree nodes that can generate rules satisfying item constraints to speed up the process.

We use the example in Table 1 to illustrate the process of CAR-Miner-Post with $minSup = 20\%$, $minConf = 60\%$, and $Constraint\_Itemset = \{(A, a3), (B, b3)\}$. Figure 2 shows the result of this process. In total, there are 13 classification rules generated from the dataset in Table 1 that satisfy $minSup = 20\%$ and $minConf = 60\%$. However, only two rules also satisfy $Constraint\_Itemset = \{(A, a3), (B, b3)\}$, as shown in Table 2.

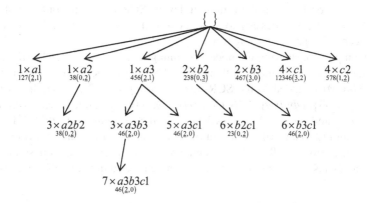

**Fig. 2** Tree generated by CAR-Miner-Post for the dataset in Table 1

**Table 2** Rules that satisfy $minSup = 20\%$, $minConf = 60\%$, and $Constraint\_Itemset = \{(A, a3), (B, b3)\}$

| ID | Node | CARs | Sup | Conf |
|----|------|------|-----|------|
| 1 | $3 \times a3b3$<br>46(2,0) | If A = a3 and B = b3 then Class = 1 | 2 | 2/2 |
| 2 | $7 \times a3b3c1$<br>46(2,0) | If A = a3 and B = b3 and C = c1 then Class = 1 | 2 | 2/2 |

## 4  Mining Class-Association Rules with Item Constraints

### 4.1  Tree Structure

This paper proposes the SCR-tree structure in that each node contains the following information:

1. *att*: a list of attributes.
2. *values*: a list of values, each of which is contained in one attribute in *att*.
3. $(Obidset_1, Obidset_2, ..., Obidset_k)$: each $Obidset_i$ is a set of object identifiers that contain an itemset and class $c_i$.

4. *pos*: stores the position of the class with the maximum cardinality of *Obidset_i*, i.e., $pos = \max\{|Obidset_i|\}$.

5. *total*: stores the sum of cardinality of all *Obidset_i*, i.e., $total = \sum_i^k |Obidset_i|$.

6. *const*: indicates whether the node contains the constrained itemset.

Unlike the MECR-tree, the SCR-tree stores not only the frequent nodes containing frequent 1-itemsets but also the constrained node containing the constrained itemset at the first level. At the following levels, SCR-tree stores only constrained nodes. Thus, it is not necessary to generate all rules, as done in CAR-Miner-Post. This noticeably improves mining time.

For example, considering the node containing itemset $X = \{(A, a3), (B, b3)\}$. $X$ is contained in objects 4 and 6, both of which belong to class 1. Therefore, the node $3 \times a3b3\,(\underline{46}, \emptyset)$ is added to the SCR-tree if *minSup* is 2. This node has $att = 3$, $values = \{a3, b3\}$, $Obidset_1 = 46$, $Obidset_2 = \emptyset$, $pos = 1$ (a line under position 1 of list $Obidset_i$), and $total = 2$. *pos* is 1 because the cardinality of *Obidset* for class 1 is maximum (2 versus 0). We use a bit representation for itemset attributes. For instance, the attributes $AB$ can be presented by 11 in bit representation, so the value of these attributes is 3. Bitwise operations can be used to quickly join itemsets.

## 4.2 Proposed Algorithm

In this section, we firstly introduce a theorem and a proposition as the basic concepts of the proposed method. Then, we present an effective and fast algorithm called Single Constraint CAR-Miner (SC-CAR-Miner) for mining CARs with item constraints based on the provided theorem and proposition.

**Proposition 1.** *To remove redundant rules, if multiple rules generated from a given node satisfy minSup and minConf, strong rule is selected (see Definition 4). This implies that rule has the form itemset* $\rightarrow c_{pos}$ *with* $Sup(r) = |Obidset_{pos}| \geq minSup$ *and* $Conf(r) = \frac{|Obidset_{pos}|}{total} \geq minConf$.

Assuming that *minSup* is 2 and *minConf* is 40%, the node $4 \times c1\,(\underline{146}, 23)$ has two rules, namely $r1 : c1 \rightarrow 1$ and $r2 : c1 \rightarrow 2$, that satisfy *minConf*, $r1$ is selected since $Conf(r1) = 3/5 > Conf(r2) = 2/5$.

**Theorem 1.** *Given two nodes* $att_1 \times values_1\,(Obidset_{1i})$ *and* $att_2 \times values_2\,(Obidset_{2i})$, *if* $att_1 = att_2$ *and* $values_1 \neq values_2$, *then* $Obidset_{1i} \cap Obidset_{2i} = \emptyset$.

*Proof.* Since $att_1 = att_2$ and $values_1 \neq values_2$, there exist $val_1 \in values_1$ and $val_2 \in values_2$ such that $val_1$ and $val_2$ have the same attributes but different values. Thus, if an object with $OID_i$ contained $val_1$, it could not include $val_2$. Thus, $\forall OID \in Obidset_{1i}$ and it can be inferred that $OID \notin Obidset_{2i}$. Consequently, $Obidset_{1i} \cap Obidset_{2i} = \emptyset$.

Theorem 1 implies that if two itemsets $X$ and $Y$ have the same attributes, it is not necessary to combine them as itemset $XY$ since $Sup(XY) = 0$. Considering two nodes $1 \times a1$ $(\underline{17}, 2)$ and $1 \times a2$ $(\emptyset, \underline{38})$ of which the attribute is $att = 1$, it can be seen that $Obidset_i(a1a2) = Obidset_i(a1) \cap Obidset_i(a2) = \emptyset$. Similarly, $3 \times a1b1$ $(\underline{1}, \emptyset) \cap$ $3 \times a1b2$ $(\emptyset, \underline{2}) = \emptyset$ since both $a1b1$ and $a1b2$ have the same attributes ($AB$) but different values.

The pseudo code of the proposed algorithm is shown in Figure 3.

---

**Input:** Dataset $D$, *minSup, minConf,* and *Constraint_Itemset*
**Output:** All CARs satisfying *minSup, minConf,* and *Constraint_Itemset*
**Procedure:**

**FIND-Lr**( $D$, *minSup, Constraint_Itemset*)
1.   Constraint_Node = findConstraint_Node( $D$, *minSup, Constraint_Itemset*);
2.   Frequent_Node = findFrequent_Node( $D$, *minSup*);
3.    $L_r$ = Constraint_Node $\cup$ Frequent_Node;
**SC-CAR-Miner**( $L_r$, *minSup, minConf*)
4.   CARs=$\varnothing$ ;
5.   for all $l_i \in L_r$.children  do
6.     if $l_i.const = false$ then
7.       break;
8.     GENERATE-RULE( $l_i$, *minConf*)
9.     $P_i = \varnothing$ ;
10.    for all $l_j \in L_r$.children,  with $j > i$  do
11.      if $l_j.att \neq l_i.att$   then // using Theorem 1
12.        $O.att = l_i.att \cup l_j.att$ ; // using bitwise operation
13.        $O.values = l_i.values \cup l_j.values$ ;
14.        $O.Obidset_i = l_i.Obidset_i \cap l_j.Obidset_i$ ;
15.        $O.pos = \max \{ |O.Obidset_i| \}$ ;
16.        $O.total = \sum_i^k |O.Obidset_i|$ ;
17.        $O.const = true$ ;
18.        if $|O.Obidset_{O.pos}| \geq minSup$ then // using Proposition 1
19.          $P_i = P_i \cup O$ ;
20.    SC-CAR-Miner( $P_i$, *minSup, minConf*);
**GENERATE-RULE**( $l$, *minConf*)
21.  conf = $|l.Obidset_{l.pos}| / l.total$ ;
22.  if conf $\geq minConf$ then // using Proposition 1
23.   CARs=CARs $\cup \{ l.itemset \rightarrow c_{pos} ( |l.Obidset_{l.pos}|, conf ) \}$ ;

---

**Fig. 3** SC-CAR-Miner algorithm for mining CARs with item constraints

Firstly, the root node of the SCR-tree $(L_r)$ is the union of the constrained node and the set of frequent nodes (Lines 1-3) at the first level of the tree. Note that infrequent nodes (based on Proposition 1) are excluded from $L_r$. Also, the nodes whose attributes belong to the attribute of the constrained node are not added to $L_r$ because they cannot combine with the constrained node to form frequent child nodes. Then, the procedure SC-CAR-Miner is called with the parameters $L_r$, $minSup$, and $minConf$ to mine all CARs with item constraints from dataset $D$.

The SC-CAR-Miner procedure considers each constrained node $l_i$ with all other nodes $l_j$ in $L_r$, with $j > i$ (Lines 5-7 and 10) to generate a candidate child node $O$. With each pair $(l_i, l_j)$, the algorithm checks whether $l_j.att \neq l_i.att$ (Line 11, using Theorem 1). If the condition holds, it computes the elements $att$, $values$, $Obidset_i$, $pos$, and $total$ for the new node $O$ (Lines 12-16) and node $O$ is a constrained node (Line 17). After computing all information of node $O$, the algorithm uses Proposition 1 to check whether this node can generate a rule satisfying $minSup$ (Line 18). Then, it adds node $O$ to $P_i$ ($P_i$ is initialized empty in Line 9) if the condition is true (Line 19). Finally, the procedure SC-CAR-Miner is called recursively with a new set $P_i$ as its input parameter (Line 20).

The function of procedure GENERATE-RULE($l$, $minConf$) is to generate a rule from node $l$. It firstly computes the confidence of the rule (Line 21), if the confidence of this rule satisfies $minConf$ by Proposition 1 (Line 22), then the rule is added to the set of CARs (Line 23).

### 4.3  Example

Considering the dataset in Table 1 with $minSup = 20\%$, $minConf = 60\%$, and $Constraint\_Itemset = \{(A, a3), (B, b3)\}$, the SCR-tree constructed by the proposed algorithm is shown in Figure 4.

The process of mining CARs with item constraints by using SC-CAR-Miner is explained as follows. The root node $(L_r = \{\})$ contains child nodes including both the constrained node $3 \times a3b3 (\underline{46}, \emptyset)$ and frequent nodes $\{4 \times c1 (\underline{146}, 23), \quad 4 \times c2 (7, \underline{58})\}$ at the first level. Nodes $\{1 \times a1 (\underline{17}, 2), 1 \times a2 (\emptyset, \underline{38}), 1 \times a3 (\underline{46}, 5), \quad 2 \times b2 (\emptyset, \underline{238}), 2 \times b3 (\underline{467}, \emptyset)\}$ are also frequent. However, their attributes belong to the attribute $AB$ (3 in bit representation) of the constrained node $3 \times a3b3 (\underline{46}, \emptyset)$, so they are removed from the root node $L_r$.

The procedure SC-CAR-Miner then generates nodes with the parameter $L_r$ at the second level and lower. Note that SC-CAR-Miner is executed only for the constrained node $3 \times a3b3 (\underline{46}, \emptyset)$. We use the node $l_i = 3 \times a3b3 (\underline{46}, \emptyset)$ as an example for illustrating the process of SC-CAR-Miner. $l_i$ joins with all nodes following it in $L_r$:

- With node $l_j = 4 \times c1 (\underline{146}, 23)$: since $l_j.att \neq l_i.att$, five elements are computed:

    1. $O.att = l_i.att \cup l_j.att = 3|4 = 7$ or 111 in bit representation
    2. $O.values = l_i.values \cup l_j.values = a3b3 \cup c1 = a3b3c1$
    3. $O.Obidset_i = l_i.Obidset_i \cap l_j.Obidset_i = (\underline{46}, \emptyset) \cap (\underline{146}, 23) = (\underline{46}, \emptyset)$

4. $O.pos = 1$
5. $O.total = 2$

Since $|O.Obidset_{pos}| = 2 \geq minSup$, $O$ is added to $P_i$ (by Proposition 1). There-fore, we have $P_i = \{7 \times a3b3c1\,(\underline{46}, \emptyset)\}$.

- With node $l_j = 4 \times c2\,(7, \underline{58})$: since $l_j.att \neq l_i.att$, five elements are computed:

1. $O.att = l_i.att \cup l_j.att = 3|4 = 7$ or 111 in bit representation
2. $O.values = l_i.values \cup l_j.values = a3b3 \cup c2 = a3b3c2$
3. $O.Obidset_i = l_i.Obidset_i \cap l_j.Obidset_i = (\underline{46}, \emptyset) \cap (7, \underline{58}) = \emptyset$
4. $O.pos = 0$
5. $O.total = 0$

Since $|O.Obidset_{pos}| = 0 < minSup$, $O$ is not added to $P_i$ (by Proposition 1).

After $P_i$ is created, SC-CAR-Miner is called recursively with parameters $P_i$, $minSup$, and $minConf$. Because $P_i$ has only one node, namely $7 \times a3b3c1\,(\underline{46}, \emptyset)$, the rule from this node is generated by the procedure GENERATE-RULE.

Rules with item constraints are easily generated in the same step of traversing node $l_i$ by calling the procedure GENERATE-RULE($l_i$, $minConf$) (Line 8). For instance, while traversing the node $l_i = 3 \times a3b3\,(\underline{46}, \emptyset)$, the procedure computes the confidence of the candidate rule (Line 21), conf $= |l_i.Obidset_{l_i.pos}| / l_i.total = 2/2 = 1$. The rule $\{(A, a3), (B, b3)\} \to 1\,(2, 1)$ is added to the rule set CARs be-cause conf $\geq minConf$. The meaning of this rule is If A = a3 and B = b3 then Class = 1 (support = 2 and confidence = 100%).

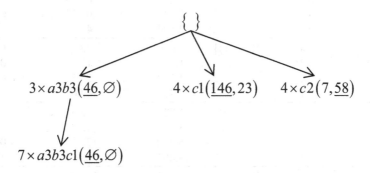

**Fig. 4** SCR-tree for the dataset in Table 1

It can be seen that the SC-CAR-Miner algorithm generates only CARs with item constraints instead of all CARs, as done in CAR-Miner-Post. Consequently, SC-CAR-Miner can lower the storage complexity while improve the mining process.

**Table 3** Characteristics of the experimental datasets

| Dataset | #attributes | #classes | #distinctive values | #objects |
|---------|-------------|----------|---------------------|----------|
| Breast | 12 | 2 | 737 | 699 |
| German | 21 | 2 | 1,077 | 1,000 |
| Lymph | 18 | 4 | 63 | 148 |
| Porker-hand | 11 | 10 | 95 | 1,000,000 |

## 5 Experiments

All experiments were conducted on a computer with an Intel Core i5 M 540 CPU at 2.53GHz and 4 GB of RAM, running Windows 7 Enterprise (32-bit) SP1. The experimental datasets were obtained from the University of California Irvine Machine Learning Repository (http://mlearn.ics.uci.edu). The algorithms were coded in C# using MS Visual Studio .NET 2010 Express. Characteristics of experimental datasets and experimental results are described in Table 3 and 4, respectively. $minConf = 50\%$ was used for all experiments.

**Table 4** Experimental results

| Dataset | minSup | Constraint_Itemset | #CARs | Time (s) CAR-Miner-Post | SC-CAR-Miner |
|---------|--------|--------------------|-------|------------------------|--------------|
| Breast | 1 | $\{(2,1)\}$ | 761 | 0.090 | 0.027 |
| | 0.5 | | 1,154 | 0.128 | 0.031 |
| | 0.3 | | 1,632 | 0.173 | 0.035 |
| | 0.1 | | 42,904 | 4.117 | 0.334 |
| German | 4 | $\{(1,0),(3,1)\}$ | 680 | 0.657 | 0.042 |
| | 3 | | 1,554 | 1.118 | 0.058 |
| | 2 | | 4,540 | 2.351 | 0.086 |
| | 1 | | 23,429 | 8.061 | 0.268 |
| Lymph | 4 | $\{(1,3),(2,2)\}$ | 1,720 | 3.452 | 0.046 |
| | 3 | | 3,624 | 4.794 | 0.061 |
| | 2 | | 25,220 | 17.527 | 0.271 |
| | 1 | | 118,884 | 52.361 | 1.223 |
| Porker-hand | 3 | $\{(1,4)\}$ | 5 | 22.104 | 2.798 |
| | 2 | | 5 | 22.290 | 2.960 |
| | 1 | | 5 | 22.365 | 2.979 |
| | 0.5 | | 110 | 55.853 | 7.027 |

The meaning of *Constraint_Itemset* = $\{(2,1)\}$ is that the obtained rules must include *Attribute2* along with *Value1* in the rule antecedent. Similarly, the final rules must contain *Attribute1* with *Value3* and *Attribute2* with *Value2* in the rule antecedent in case of *Constraint_Itemset* = $\{(1,3),(2,2)\}$.

The results show that SC-CAR-Miner is much more efficient than CAR-Miner-Post in all experiments. For example, considering the Lymph dataset with *Constraint_Itemset* = $\{(1,3),(2,2)\}$ and *minSup* = 1%, the mining time of SC-CAR-Miner is 1.223(s) while CAR-Miner-Post is 52.361(s). For this example, SC-CAR-Miner is 42.8 times faster than CAR-Miner-Post.

# 6 Conclusions and Future Work

This paper proposed an efficient method for mining CARs with item constraints. The constraints are in the form of a specific itemset. Unlike post-processing approaches, the proposed approach generates only rules that satisfy the item constraints. The framework of the proposed algorithm is based on the SCR-tree structure which includes only nodes containing the constrained itemset and the theorem and the proposition for quickly pruning infrequent nodes and weak classification rules. To validate the effectiveness and efficiency of the proposed method, a series of experiments was conducted on four datasets, namely Breast, German, Lymph, and Poker-hand. The experimental results show that the proposed method outperforms the post-processing method.

In the future, the SC-CAR-Miner algorithm will be extended for mining CARs with item constrains that are Boolean expressions over the presence of items (for example ((Shirts AND Shoes) OR Outerwear)) in rule antecedents.

**Acknowledgments.** This research is funded by Viet Nam National Foundation for Science and Technology Development (NAFOSTED).

# References

1. Liu, B., Hsu, W., Ma, Y.: Integrating classification and association rule mining. In: 4th International Conference on Knowledge Discovery in Databases and Data Mining, pp. 80–86 (1998)
2. Li, W., Han, J., Pei, J.: CMAR: Accurate and efficient classification based on multiple class-association rules. In: IEEE International Conference on Data Mining, pp. 369–376 (2001)
3. Yin, X., Han, J.: CPAR: Classification based on predictive association rules. In: 3rd SIAM International Conference on Data Mining, pp. 331–335 (2003)
4. Thabtah, F., Cowling, P., Peng, Y.: MCAR: multi-class classification based on association rule. In: 3rd ACS/IEEE International Conference on Computer Systems and Applications, pp. 33–39 (2005)
5. Thonangi, R., Pudi, V.: ACME: An associative classifier based on maximum entropy principle. In: Jain, S., Simon, H.U., Tomita, E. (eds.) ALT 2005. LNCS (LNAI), vol. 3734, pp. 122–134. Springer, Heidelberg (2005)
6. Vo, B., Le, B.: A novel classification algorithm based on association rules mining. In: Richards, D., Kang, B.-H. (eds.) PKAW 2008. LNCS (LNAI), vol. 5465, pp. 61–75. Springer, Heidelberg (2009)

7. Nguyen, L.T., Vo, B., Hong, T.P., Thanh, H.C.: Classification based on association rules: A lattice-based approach. Expert Systems with Applications, 11357–11366 (2012)
8. Nguyen, L.T., Vo, B., Hong, T.P., Thanh, H.C.: CAR-Miner: An efficient algorithm for mining class-association rules. Expert Systems with Applications, 2305–2311 (2013)
9. Srikant, R., Vu, Q., Agrawal, R.: Mining association rules with item constraints. In: 3rd International Conference on Knowledge Discovery in Databases and Data Mining, pp. 67–73 (1997)
10. Agrawal, R., Srikant, R.: Fast Algorithms for Mining Association Rules. In: 20th International Conference on Very Large Data Bases, pp. 487–499 (1994)
11. Han, J., Pei, J., Yin, Y.: Mining frequent patterns without candidate generation. In: ACM SIGMOD International Conference on Management of Data, pp. 1–12 (2000)
12. Zaki, M.J., Parthasarathy, S., Ogihara, M., Li, W.: New algorithms for fast discovery of association rules. In: 3rd International Conference on Knowledge Discovery in Databases and Data Mining, pp. 283–286 (1997)
13. Ng, R.T., Lakshmanan, L.V.S., Han, J., Pang, A.: Exploratory mining and pruning optimizations of constrained associations rules. In: ACM SIGMOD International Conference on Management of Data, pp. 13–24 (1998)
14. Lin, W.Y., Huang, K.W., Wu, C.A.: MCFPTree: An FP-tree-based algorithm for multi-constraint patterns discovery. International Journal of Business Intelligence and Data Mining, 231–246 (2010)
15. Duong, H., Truong, T., Le, B.: An Efficient Algorithm for Mining Frequent Itemsets with Single Constraint. In: Nguyen, N.T., van Do, T., Thi, H.A. (eds.) ICCSAMA 2013. SCI, vol. 479, pp. 367–378. Springer, Heidelberg (2013)

# Privacy Preserving Frequency-Based Learning Algorithms in Two-Part Partitioned Record Model

The Dung Luong and Dang Hung Tran

**Abstract.** In this paper, we consider a new scenario for privacy-preserving data mining called two-part partitioned record model (TPR) and find solutions for a family of frequency-based learning algorithms in TPR model. In TPR, the dataset is distributed across a large number of users in which each record is owned by two different users, one user only knows the values for a subset of attributes and the other knows the values for the remaining attributes. A miner aims to learn, for example, classification rules on their data, while preserving each user's privacy. In this work we develop a cryptographic solution for frequency-based learning methods in TPR. The crucial step in the proposed solution is the privacy-preserving computation of frequencies of a tuple of values in the users' data, which can ensure each user's privacy without loss of accuracy. We illustrate the applicability of the method by using it to build the privacy preserving protocol for the naive Bayes classifier learning, and briefly address the solution in other applications. Experimental results show that our protocol is efficient.

## 1 Introduction

Data mining have been used in various applications to support people discovering useful knowledge in large databases. However, there has been growing concern that the use of this technology is violating individual privacy [17] and many privacy preserving data mining approaches have been proposed for tackling the problem of privacy violation [3], [18], [23].

The Dung Luong
Academy of Cryotographic Techniques
e-mail: luongthedung@gmail.com

Dang Hung Tran
Hanoi National University of Education
e-mail: hungtd@hnue.edu.vn

V.-N. Huynh et al. (eds.), *Knowledge and Systems Engineering, Volume 2*, 319
Advances in Intelligent Systems and Computing 245,
DOI: 10.1007/978-3-319-02821-7_29, © Springer International Publishing Switzerland 2014

Privacy preserving data mining methods mainly divided into two groups: the perturbation-based methods and the cryptography-based methods. The methods based on perturbation (e.g., [1], [9], [11]) are very efficient, but have a tradeoff between privacy and accuracy. The methods based on cryptography (e.g., [18], [22], [26]) can safely preserve privacy without loss of accuracy, but have high complexity and communication cost. These privacy preserving data mining methods have been presented for various scenarios in which the general idea is to allow mining datasets distributed across multiple parties, without disclosing each party's private data [6].

In this paper, we study privacy preserving data mining in yet another scenario that exists in various practical applications but has not been investigated. In this scenario, the data set is distributed across a large number of users, and each record is owned by two different users, one user only knows the values for a subset of attributes while the other knows the values for the remaining attributes. We call this *two-part partitioned record model* (TPR, for short). A miner would like to learn the classification or description rules in the data set, without disclosing each user's private data.

Let us take some examples of TPR. Consider the scenario in which a sociologist wants to find out the depersonalization behavior of children depending on the parenting style of their parents [19]. The sociologist provides the sample survey to collect information about the parenting style from parents and behavior from their children. Clearly, the information is quite sensitive, parents do not want to objectively reveal their limitations in educating children, while it is also difficult to ask the children to answer honestly and truthfully about their depersonalization behavior. Therefore, in order to get accurate information, the researcher must ensure the confidentiality principle of information for each subject. In this case, each data record is privately owned by both the parents and their children.

Another example is the scenario where a medical researcher needs to study the relationship between living habits, clinical information and a certain disease [14], [15]. A hospital has a clinical dataset of the patients that can be used for research purpose and the information of living habits can be collected by a survey of patients, though, neither the hospital nor the patients are willing to share their data with the miner because of privacy. This scenario meets the TPR model, where each data object consists of two parts: one part consisting of living habits belongs to a patient, the remaining part consisting of clinical data of this patient is kept by the hospital. Furthermore, we can see that the TPR model is quite popular in practice, and that privacy preserving frequency mining protocols in TPR are significant and can be applied to many other similar distributed data scenarios.

In this paper we present a solution for a family of frequency-based learning algorithms in TPR. The contributions of the work include:

- The development of a cryptographic approach to privacy preserving frequency-based learning in TPR model. We proposed a protocol for privacy preserving frequency computation. The protocol ensures each user's privacy without loss of accuracy. In addition, it is efficient, requiring only 1 or 2 interactions between each user and the miner, while the users do not have to communicate with each other.

- The applicability of the approach. To illustrate it we present the design and analysis of the privacy preserving naive Bayes learning protocol in TPR as well as briefly show it on other methods.

## 2 Related Works

Randomization solutions used in [1], [4], [10], [11] can be applied to privacy preserving frequency computation in TPR model. The basic idea of these solutions is that every user perturbs its data before sending it to the miner. The miner then can reconstruct the original data to obtain the mining results with some bounded error. These solutions allow each user to operate independently, and the perturbed value of a data element does not depend on those of the other data elements but only on its initial value. Therefore, they can be used in various distributed data scenarios. Although these solutions are very efficient, their usage generally involves a tradeoff between privacy and accuracy, i.e. if we require the more privacy, the miner loses more accuracy in the data mining results, and vice-versa.

Our work is similar in spirit to [20], [26] that describe methods for doing some privacy preserving learning task in fully distributed setting. The essence of these methods is a private frequency computation method that allows the miner to compute frequencies of values or tuples in the data set, while preserving privacy of each user's data. These methods are based on cryptographic techniques, which provided strong privacy without loss of accuracy. The result of the private frequency computation is then used for various privacy preserving learning tasks such as naive Bayes learning, decision tree learning, association rule mining, etc. Here we aim at solving the same privacy preserving learning tasks but in TPR model. Note that in this setting, each user may only know some values of the tuple but not all, and therefore, the above mentioned cryptographic approaches cannot be used in TPR model.

In [8], [21] and [22], the authors developed a private frequency computation solution from the vertically distributed data based on secure scalar product protocols, where the final goal is to design privacy preserving protocols for learning naive Bayes classification, association rules and decision trees. In [16], private frequency computation was addressed for horizontally distributed data by computing the secure sum of all local frequencies of participating parties.

## 3 Privacy Preserving Frequency Mining in TPR Model

### 3.1 Problem Formulation

In TPR model, a data set (a data table) consists of $n$ records, and each record is described by values of nominal attributes. The data set is distributed across two sets of users $U = \{U_1, U_2, ..., U_n\}$ and $V = \{V_1, V_2, ..., V_n\}$. Each pair of users $(U_i, V_i)$ owns a record in which user $U_i$ knows values for a proper subset of attributes, and user $V_i$ knows the values for the remaining attributes. Note that in this setting, the set of attributes whose values known by each $U_i$ is equal, and so for each user $V_i$.

The miner aims to mine the frequency of a tuple of values in the data set without disclosing each user's private information. Assume that the tuple consists of values for some attributes belonging to $U_i$ and the remaining values for the remaining attributes belonging to $V_i$. In this case, each $U_i$ and $V_i$ outputs a boolean value, $u_i$ and $v_i$, respectively (either 1 or 0) to indicate whether the data it holds matched values (corresponding to its attributes) in the tuple. Therefore, the objective is to design a protocol that allows the miner to obtain the sum $f = \sum u_i v_i$ without revealing $u_i$ and $v_i$.

Our formula is still appropriate when the tuple consisting of values for some attributes only belongs to $U_i$ (or $V_i$). For example, when the tuple consists of values for some attributes only belonging to $U_i$, $U_i$ outputs a boolean value $u_i$ to indicate whether the data it holds matches all values in the tuple and $V_i$ outputs $v_i = 1$. Therefore, clearly the sum $f = \sum u_i = \sum u_i v_i$ is the frequency value which needs be computed.

To be applicable, we require that the protocol can ensure users' privacy in an environment that doesn't have any secure communication channel between the user and the miner, as well as it should not require any communication among the users. In addition, it should minimize the number of interactions between the user and the miner. Particularly, the user $U_i$ must not interact with the miner more than twice, and the user $V_i$ must interact with the miner exactly once. Those requirements make our protocol more applicable. For example, considering a real scenario when a miner uses a web-application to investigate a large number of users for his research, a user only needs to use his browser to communicate with the server one or two times, while he does not have to communicate with the others.

## 3.2 Definition of Privacy

The privacy preservation of the proposed protocol is based on the semi-honest security model. In this model, each party participating in the protocol has to follow rules using correct input, and cannot use what it sees during execution of the protocol to compromise security. A general definition of secure multi-party computation in the semi-honest model is stated in [12]. This definition was derived to make a simplified definition in the semi-honest model for privacy-preserving data mining in the fully distributed setting scenario [26], [20]. This scenario is similar to TPR model, so here we consider the possibility that some corrupted users share their data with the miner to derive the private data of the honest users. One requirement is that no other private information about the honest users be revealed, except a multivariate linear equation in which each variable presents a value of an honest user. In our model, information known by users is no more than information known by the miner, so we do not have to consider the problem in which users share information with each other.

**Definition:** *Assume that each user $U_i$ has a private set of keys $D_i^{(u)}$ and a public set of keys $E_i^{(u)}$, and each user $V_i$ has a private set of keys $D_i^{(v)}$ and a public set of* keys $E_i^{(v)}$. *A protocol for the above defined frequency mining problem protects each user's privacy against the miner along with $t_1$ corrupted users $U_i$ and $t_2$ corrupted users $V_i$ in the semi-honest model if, for all $I_1, I_2 \subseteq \{1, ..., n\}$ such that $|I_1| = t_1$ and $|I_2| = t_2$, there exists a polynomial-time algorithm $M$ such that*

$$\{M(f, [u_i, D_i^{(u)}]_{i \in I_1}, [E_j^{(u)}]_{j \notin I_1}, [v_k, D_k^{(v)}]_{k \in I_2}, [E_l^{(v)}]_{l \notin I_2})\}$$

$$\overset{c}{\equiv} \{View_{miner, \{U_i\}_{i \in I_1}, \{V_k\}_{k \in I_2}} [u_i, D_i^{(u)}, v_i, D_i^{(v)}]_{i=1}^n\}$$

*where $\overset{c}{\equiv}$ denotes computational indistinguishability.*

Basically, the definition states that the computation is secure if the joint view of the miner and the corrupted users (the $t_1$ users $U_i$ and the $t_2$ users $V_i$) during the execution of the protocol can be effectively simulated by a simulator, based on what the miner and the corrupted users have observed in the protocol using only the result $f$, the corrupted users' knowledge, and the public keys. Therefore, the miner and the corrupted users can not learn anything from $f$. By the definition, in order to prove the privacy of a protocol, it suffices to show that there exists a simulator that satisfies the above equation.

## 3.3 Privacy Preserving Protocol for Frequency Mining in TPR Model

In this section, we use ELGamal encryption scheme together with the joint decryption technique to build a privacy-preserving frequency mining protocol. This idea has been extensively used in previous works, e.g., [13], [25], [26]. Before describing our protocol, we briefly review ElGamal encryption scheme [20] as follows.

Let $G$ be a cyclic group of order $q$ in which the discrete logarithms are hard. Let $g$ be a generator of $G$, and $x$ be uniformly chosen from $\{0, 1, ..., q-1\}$. In ElGamal encryption schema, $x$ is a private key and the public key is $h = g^x$. Each user securely keeps their own private keys, otherwise public keys are publicly known.

To encrypt a message $M$ using the public key $h$, one randomly chooses $k$ from $\{0, ..., q-1\}$ and then computes the ciphertext $C = (C_1 = Mh^k, C_2 = g^k)$. The decryption of the ciphertext $C$ with the private key $x$ can be executed by computing $M = C_1(C_2^x)^{-1}$.

ElGamal encryption is semantically secure under the Decisional Diffie-Hellman (DDH) Assumption[5]. In ElGamal encryption scheme, one cleartext has many possible encryptions, since the random number $k$ can take many different values. ElGamal encryption has a randomization property in which it allows computing a different encryption of $M$ from a given encryption of $M$.

### 3.3.1 Protocol

In the proposed protocol, we assume that each user $U_i$ has private keys $x_i$, $y_i$ and public keys $X_i = g^{x_i}$, $Y_i = g^{y_i}$, and each user $V_i$ has private keys $p_i$, $q_i$ and public keys $P_i = g^{p_i}$, $Q_i = g^{q_i}$. Note that computations in this paper take place in the group G.

As presented in Subsection 3.1, our purpose is to allow the miner to privately obtain the sum $f = \sum_{i=1}^{n} u_i v_i$. The privacy preserving protocol for the miner to compute $f$ consists of the following phases:

- Phase 1. Each user $U_i$ and the miner work as follows:

  - Each $U_i$ randomly chooses $k_i$, $s_i$ from $\{0, 1, ..., q-1\}$. Next, it computes $C_1^{(i)} = g^{u_i} X_i^{s_i}$, $C_2^{(i)} = g^{s_i}$, $C_3^{(i)} = P_i X_i^{k_i}$ and $C_4^{(i)} = Q_i Y_i^{k_i}$, and sends them to the miner.

  - The miner computes $X = \prod_{i=1}^{n} C_3^{(i)}$ and $Y = \prod_{i=1}^{n} C_4^{(i)}$

- Phase 2. Each user $V_i$ does the following:

  - Get $C_1^{(i)}, C_2^{(i)}, X$ and $Y$ from the miner,

  - Choose randomly $r_i$ from $\{0, 1, ..., q-1\}$,

  - if $v_i = 0$ then compute $R_1^{(i)} = X^{q_i}$, $R_2^{(i)} = (C_2^{(i)})^{p_i r_i} Y^{p_i}$ and $R_3^{(i)} = P_i^{r_i}$, and send them to the miner

  - if $v_i = 1$ then compute $R_1^{(i)} = (C_1^{(i)})^{v_i} X^{q_i}$, $R_2^{(i)} = (C_2^{(i)})^{p_i r_i} Y^{p_i}$ and $R_3^{(i)} = (X_i)^{-1} P_i^{r_i}$, and send them to the miner.

- Phase 3. Each user $U_i$ and the miner work as follows:

  - Each $U_i$ gets $R_1^{(i)}, R_2^{(i)}$ and $R_3^{(i)}$ from the miner. Then, it computes $K_1^{(i)} = R_1^{(i)} (R_3^{(i)})^{s_i} X^{k_i y_i}$, $K_2^{(i)} = R_2^{(i)} Y^{k_i x_i}$, and sends them to the miner.

  - The miner computes $d = \prod_{i=1}^{n} \dfrac{K_1^{(i)}}{K_2^{(i)}}$. Next, it finds $f$ from $\{0, 1, ..., q-1\}$ that satisfies $g^f = d$, and outputs $f$.

### 3.3.2   Proof of Correctness

**Theorem 1.** *The above presented protocol correctly computes the frequency value $f = \sum_{i=1}^{n} u_i v_i$ as defined in Subsection 3.1.*

*Proof.* We show that the miner can compute the desired value $f$ by using the above protocol. Indeed,

$$d = \prod_{i=1}^{n} \frac{K_1^{(i)}}{K_2^{(i)}}$$

$$= \prod_{i=1}^{n} \frac{R_1^{(i)} (R_3^{(i)})^{s_i} X^{k_i y_i}}{R_2^{(i)} Y^{k_i x_i}}$$

If $v_i = 0$ then $g^{u_i v_i} = 1$, therefore

$$K_1^{(i)} = X^{q_i} g^{p_i r_i s_i} X^{k_i y_i}$$
$$= g^{u_i v_i} g^{p_i r_i s_i} X^{k_i y_i + q_i}$$

If $v_i = 1$, we have

$$K_1^{(i)} = (C_1^{(i)})^{v_i} X^{q_i} (X_i^{-1} P_i^{r_i})^{s_i} X^{k_i y_i}$$
$$= g^{u_i v_i} g^{x_i s_i v_i} X^{q_i} g^{-x_i s_i} g^{p_i r_i s_i} X^{k_i y_i}$$
$$= g^{u_i v_i} g^{p_i r_i s_i} X^{k_i y_i + q_i}$$

In both cases, we also have

$$K_2^{(i)} = (C_2^{(i)})^{p_i r_i} Y^{p_i} Y^{k_i x_i}$$
$$= g^{p_i s_i r_i} Y^{k_i x_i + p_i}$$

Finally, we obtain

$$d = \prod_{i=1}^{n} \frac{K_1^{(i)}}{K_2^{(i)}}$$

$$= \prod_{i=1}^{n} \frac{g^{u_i v_i} g^{p_i r_i s_i} X^{k_i y_i + q_i}}{g^{p_i r_i s_i} Y^{k_i x_i + p_i}}$$

$$= \prod_{i=1}^{n} g^{u_i v_i} \prod_{i=1}^{n} \frac{X^{k_i y_i + q_i}}{Y^{k_i x_i + p_i}}$$

$$= g^{\sum_{i=1}^{n} u_i v_i} \prod_{i=1}^{n} \frac{(g^{\sum_{j=1}^{n} (k_j x_j + p_j)})^{(k_i y_i + q_i)}}{(g^{\sum_{j=1}^{n} (k_j y_j + q_j)})^{(k_i x_i + p_i)}}$$

$$= g^{\sum_{i=1}^{n} u_i v_i} \frac{g^{\sum_{i=1}^{n} \sum_{j=1}^{n} (k_j x_j + p_j)(k_i y_i + q_i)}}{g^{\sum_{i=1}^{n} \sum_{j=1}^{n} (k_j y_j + q_j)(k_i x_i + p_i)}}$$

$$= g^{\sum_{i=1}^{n} u_i v_i}$$

Therefore, we can obtain $f$ from the equation $d = g^f = g^{\sum_{i=1}^{n} u_i v_i}$.

Note that, in practice, the value of $f$ is not too large, so that the discrete logarithms can be successfully taken (for example $f = 10^5$).

### 3.3.3  Proof of Privacy

In this section, we first show that under the DDH assumption, our protocol preserves each user's privacy in the semi-honest model. Then, we show that in the case of collusion of some corrupted users with the miner, the protocol still preserves the privacy of each honest user.

In our model, the communication only occurs between each user and the miner, thus the miner receives the messages of all users. Assume that each user can get the messages of the remaining users via the miner, then the information known by the miner and each user are the same during the execution of the protocol. Therefore, it is sufficient to only consider the view of the miner, as follow:

In Phase 1, the miner receives the messages $C_1^{(i)}$, $C_2^{(i)}$, $C_3^{(i)}$ and $C_4^{(i)}$ of each $U_i$. Here $(C_1^{(i)}, C_2^{(i)})$ is an ElGamal encryption of the value $g^{u_i}$ under the private/the public key pair $(x_i, X_i)$, and the value $s_i$ is randomly chosen from $\{1, 2, ..., q-1\}$. $C_3^{(i)}$ and $C_4^{(i)}$ are the first part of Elgammal encryptions of $P_i$ and $Q_i$ under the key pairs $(x_i, X_i)$ and $(y_i, Y_i)$.

In Phase 2, the messages $R_1^{(i)}$, $R_2^{(i)}$ and $R_3^{(i)}$ sent by each $V_i$ are equivalent to the first part of ElGamal encryptions $(\alpha X^{q_i}, g^{q_i})$, $(\beta Y^{p_i}, g^{p_i})$ and $(\gamma P_i^{r_i}, g^{r_i})$, respectively. Here $\alpha = 1$ or $(C_1^{(i)})^{v_i}$, $\beta = (C_2^{(i)})^{p_i r_i}$, $\gamma = 1$ or $(X_i)^{-1}$. Note that $q_i$, $p_i$ and $r_i$ are randomly chosen from $\{1, 2, ..., q-1\}$ and we have,

$$X = \prod_{i=1}^{n} X_i^{k_i} P_i = g^x$$

$$Y = \prod_{i=1}^{n} Y_i^{k_i} Q_i = g^y$$

where

$$x = \sum_{i=1}^{n} (k_i x_i + p_i)$$

$$y = \sum_{i=1}^{n} (k_i y_i + q_i)$$

and $k_i$ is randomly chosen from $\{0, ..., q-1\}$. In our protocol, each user uses $X$ and $Y$ as the public keys to encrypt its data. Thus, decrypting its encryptions requires the use of the private keys $x$ and $y$, where no individual user known these values.

Similarly, in Phase 3, the messages $K_1^{(i)}$ and $K_2^{(i)}$ sent by each $U_i$ can be represented as the first part of ElGamal encryptions $K_1 = (\alpha' X^{k_i y_i}, g^{y_i})$ and $K_2 = (\beta' Y^{k_i x_i}, g^{x_i})$.

As well known, the ElGamal encryption is semantically secure under the DDH assumption. So, the view of the miner can be efficiently simulated by a simulator for ElGamal encryptions.

Now, we show that the protocol preserves the privacy of the honest users against the collusion of the corrupted users with the miner, even up to $2n - 2$ corrupted users. We have the following theorem

**Theorem 2.** *The protocol in Subsection 3.3 preserves the privacy of the honest users against the miner and up to $2n - 2$ corrupted users. In cases with only two honest users, it remains correct as long as two honest users do not own the attribute values of the same record.*

*Proof.* In the proposed protocol, the information known by each user is the same, thus we need to only consider the case where a user $U_i$ and a user $V_j$ ($i \neq j$) are honest. The remaining cases can be proved similarly. Without loss of generality, we assume that $I = \{2, 3, 4, ..., n\}$ and $J = \{1, 3, 4, ..., n\}$.

Now we need to design a simulator $M$ that simulates the joint view of the miner and the corrupted users by a probabilistic polynomial-time algorithm, and then this simulator is combined with a simulator for the ElGamal ciphertexts to obtain a completed simulator. To do so, basically we show a polynomial-time algorithm for computing the joint view of the miner and the corrupted users. The computation of the algorithm is based on what the miner and the corrupted users have observed in the protocol using only the result $f$, the corrupted users' information, and the public keys. The algorithm outputs the simulated values for the encryptions generated by a simulator of ElGamal encryptions.

- $M$ simulates $C_1^{(1)}, C_2^{(1)}, C_3^{(1)}$ and $C_4^{(1)}$ using the random ElGamal ciphertexts.
- $M$ takes the following encryptions as its input

$$(a_1, a_1') = (\alpha g^{(k_1 x_1 + p_2)q_2}, g^{q_2})$$
$$(a_2, a_2') = (\beta g^{(k_1 y_1 + q_2)p_2}, g^{p_2})$$

where $\alpha = 1$ or $C_1^{(2)}$, $\beta = (C_2^{(2)})^{p_2 r_2}$, and it computes the following values

$$R_1^{'(2)} = a_1 Q_2^{\Sigma_{i \in I} k_i x_i + \Sigma_{j \in J} p_j} / g^{f - \Sigma_{l=3}^{n} u_l v_l - \lambda v_1 - \theta u_2}$$

$$R_2^{'(2)} = a_2 P_2^{\Sigma_{i \in I} k_i y_i + \Sigma_{j \in J} q_j}$$

where $\lambda, \theta \in \{0, 1\}$. Next, $M$ simulates $R_2^{(3)}$ using a random ElGamal ciphertext.
- $M$ takes the two following encryptions as its input

$$(b_1, b_1') = (R_1^{(1)}(R_3^{(1)})^{c_1} g^{(k_1 x_1 + p_2)y_1}, g^{y_1})$$
$$(b_2, b_2') = (R_2^{(1)} g^{(k_1 y_1 + q_2)x_1}, g^{x_1})$$

and computes

$$K_1^{'(1)} = b_1 . Y_1^{\Sigma_{i \in I} k_i x_i + \Sigma_{j \in J} p_j}$$
$$K_2^{'(1)} = b_2 . X_1^{\Sigma_{i \in I} k_i y_i + \Sigma_{j \in J} q_j}$$

This finishes the simulation algorithm.

### 3.4 Efficiency Evaluation

In this section, we show results of the complexity estimation of the protocol and the efficiency measurement of the protocol in practice

In the proposed protocol, the computational cost of each user $U_i$ in the first phase and in the third phase are 4 and 3 modular exponentiations, respectively. The computational cost of each user $V_i$ in the second phase is at most 4 modular exponentiations. The miner uses $2n$ modular multiplications in the first phase, and $2n$ modular multiplications and at most $n$ comparisons in the third phase.

For evaluating the efficiency of the protocol in practice, we build an experiment on the privacy preserving frequency mining in C environment, which runs on a laptop with CPU Pentium M 1.8 GHz and 1GB memory. The used cryptographic functions are derived from Open SSL Library.

We measure the computation cost of the frequency mining protocol for different numbers of users, from 1000 to 5000. Before executing the protocol, we generate three pairs of keys for each user, with the size of public keys set at 512 bits. Note that generating these keys belongs to the preparation period of the mining process, so it does not affect the computation time of the protocol. The results show that the average time used by each $U_i$ for computing the first-phase messages and the third-phase messages are about 27ms and 13ms, respectively. Each $V_i$ needs about an average 24ms to compute her messages. For the miner, Fig 1 shows that the

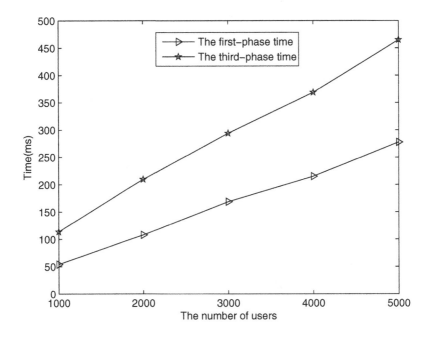

**Fig. 1** The time used by the miner for computing the key values $X, Y$ in the first phase and the frequency $f$ in third phrase

computation time for values $X$ and $Y$ in the first phase, and the computation time of the frequency value $f$ in the third phase. These times are very efficient and nearly linearly related to the number of users such as when the number of users is 5000, the miner uses only about 280 ms and 460 ms for phrase 1 and phrase 3, respectively.

# 4 Privacy Preserving Frequency-Based Learning in TPR Model

## 4.1 Privacy Preserving Protocol for Naive Bayes Classifier Learning

### 4.1.1 Privacy Preserving Problem for Naive Bayes Learning

The TPR model scenario is described in the following notation: There are $m$ attributes, $A_1, A_2, ..., A_m$ and one class attribute $C$. Each attribute $A_j$ ($1 \le j \le m$) has $d_j$ values $a_j^{(1)}, a_j^{(2)}, ..., a_j^{(d_j)}$ in its domain and the class attribute $C$ has $p$ values $c^{(1)}$, $c^{(2)}, ..., c^{(p)}$. Assume that there are $2n$ users $\{U_1, U_2, ...., U_n\}$ and $\{V_1, V_2, ..., V_n\}$. Each pair $(U_i, V_i)$ owns a vector $(a_{i1}, a_{i2}, ..., a_{im}, c_i)$, where $(a_{i1}, a_{i2}, ..., a_{im})$ denote an instance of the attribute vector $(A_1, A_2, ..., A_m)$ and $c_i$ is its class label. Here $U_i$ holds $r$ first values, $V_i$ holds the remaining values ($1 < r < m$) and the class label $c_i$. Our purpose is to allows the miner to use data from all users for naive Bayes classifier learning while still protecting each user's privacy.

For normal naive Bayes learning method, the miner can collects all users' data and place on a center server, and then trains his classifier at the center server. Following the method described in [7], the task of learning is to construct a classifier that can predict the correct class for a new instance by assigning it to the most probable class label $c$, given the attribute values $a_1, a_2, ..., a_m$ that describe the instance.

$$c = \arg\max_{c^{(l)} \in C} P(c^{(l)}) \prod_{j=1}^{m} P(a_j | c^{(l)})$$

$$= \arg\max_{c^{(l)} \in C} f(c^{(l)}) \prod_{j=1}^{m} \frac{f(a_j, c^{(l)})}{f(c^{(l)})}$$

where $f(a_j, c^{(l)})$ is the frequency of the pair $(a_j, c^{(l)})$ in all users' data and $f(c^{(l)})$ is the frequency of the attribute value $c^{(l)}$ in all users' data. Thus, the learning step in naive Bayes classifier consists of estimating $P(a_j^{(k)} | c^{(l)})$ and $P(c^{(l)})$ based on the frequencies $f(a_j^{(k)}, c^{(l)})$ and $f(c^{(l)})$ in the training data ($1 \le j \le m$, $1 \le k \le d_j$, $1 \le l \le p$).

In our problem, we assume that each user's data include the sensitive attribute values (without loss of generality, assume that all attribute values of each user is sensitive). Thus, each user is not willing to submit its data to the miner without protecting privacy. To allow the miner to train classifier while protecting each user's privacy, we design a privacy preserving protocol for naive Bayes learning. Our protocol allows the miner to obtain the classifier by privately computing the frequencies $f(a_j^{(k)}, c^{(l)})$, $f(c^{(l)})$ based on the primitive presented in section 3. Therefore, this protocol does not reveal any each user's privacy information to the miner beyond the frequencies in all user's data.

### 4.1.2 Protocol

We assume that each user has private keys and a public key as presented in Section 3.3. Our protocol is given as below

- Phase 1. Each user $U_i$ and the miner work as follows:

  - Each $U_i$ does the following:

    For $1 \leq j \leq m$, $1 \leq k \leq d_j$, $1 \leq l \leq p$

    · Set $u_i = 1$ if $((a_{ij} = a_j^{(k)}) \wedge (j \leq r)) \vee (j > r)$ else $u_i = 0$,

    · Randomly choose $k_i$, $s_i$ from $\{0, 1, ..., q - 1\}$,

    · Compute $C_{ij1}^{(k,l)} = g^{u_i} X_i^{s_i}$, $C_{ij2}^{(k,l)} = g^{s_i}$, $C_{ij3}^{(k,l)} = X_i^{k_i} P_i$ and $C_{ij4}^{(k,l)} = Y_i^{k_i} Q_i$,

    and send them to the miner.

  - The miner computes $X_j^{(k,l)} = \prod_{i=1}^{n} C_{ij3}^{(k,l)}$ and $Y_j^{(k,l)} = \prod_{i=1}^{n} C_{ij4}^{(k,l)}$

- Phase 2. Each user $V_i$ does the following:

  For $1 \leq j \leq m$, $1 \leq k \leq d_j$, $1 \leq l \leq p$

  - Set $v_i = 1$ if $((c_i = c^{(l)}) \wedge (j \leq r)) \vee ((a_{ij}, c_i) = (a_j^{(k)}, c^{(l)}) \wedge (j > r))$ else $v_i = 0$,

  - Get $C_{ij1}^{(k,l)}$, $C_{ij2}^{(k,l)}$, $X_j^{(k,l)}$ and $Y_j^{(k,l)}$ from the miner,

  - Choose randomly $r_i$ from $\{0, 1, ..., q - 1\}$,

  - if $v_i = 0$ then compute $R_{ij1}^{(k,l)} = (X_j^{(k,l)})^{q_i}$, $R_{ij2}^{(k,l)} = (C_{ij2}^{(k,l)})^{p_i r_i} (Y_j^{(k,l)})^{p_i}$ and $R_{ij3}^{(k,l)} = P_i^{r_i}$, and send them to the miner

  - if $v_i = 1$ then compute $R_{ij1}^{(k,l)} = (C_{ij1}^{(k,l)})^{v_i} (X_j^{(k,l)})^{q_i}$, $R_{ij2}^{(k,l)} = (C_{ij2}^{(k,l)})^{p_i r_i} (Y_j^{(k,l)})^{p_i}$ and $R_{ij3}^{(k,l)} = (X_i)^{-1} P_i^{r_i}$, and send them to the miner.

- Phase 3. Each user $U_i$ and the miner work as follows:

  - Each $U_i$ does the following:

    For $1 \leq j \leq m, 1 \leq k \leq d_j, 1 \leq l \leq p$

    · Gets $R_{ij1}^{(k,l)}, R_{ij2}^{(k,l)}$ and $R_{ij3}^{(k,l)}$ from the miner.

    · Computes $\quad K_{ij1}^{(k,l)} \quad = \quad R_{ij1}^{(k,l)}(R_{ij3}^{(k,l)})^{s_i}(X_j^{(k,l)})^{k_i y_i}$ and $\quad K_{ij2}^{(k,l)} =$ $R_{ij2}^{(k,l)}(Y_j^{(k,l)})^{k_i x_i}$, and sends them to the miner.

  - The miner does the following:

    For $1 \leq j \leq m, 1 \leq k \leq d_j, 1 \leq l \leq p$

    · Computes $d_j^{(k,l)} = \prod_{i=1}^{n} \dfrac{K_{ij1}^{(k,l)}}{K_{ij2}^{(k,l)}}.$

    · Finds $f_j(k,l)$ from $\{0,1,...,q-1\}$ that satisfies $g^{f_j(k,l)} = d_j^{(k,l)}$,

  - For $1 \leq l \leq p$ the miner computes $f(c^{(l)})$

  - The miner outputs classifier.

### 4.1.3 Correctness and Privacy Analysis

Basically, the correctness and privacy of our privacy preserving naive Bayes learning protocol can be derived from Theorem 1 and Theorem 2.

**Corollary 1.** *The protocol presented in subsection 4.1 allows the miner to obtain naive Bayes classifier correctly.*

*Proof.* By Theorem 1, this protocol correctly computes each $f(a_j^{(k)}, c^{(l)})$. Moreover, each $f(c^{(l)})$ can be directly obtained from frequencies $f(a_j^{(k)}, c^{(l)})$ by the formula $f(c^{(l)}) = \sum_{k=1}^{d_j} f(a_j^{(k)}, c^{(l)})$ . Therefore, the protocol outputs naive Bayes classifier correctly.

**Corollary 2.** *The protocol in Subsection 4.1 preserves the privacy of the honest users against the miner and up to $2n - 2$ corrupted users. In cases with only two honest users, it remains correct as long as two honest users do not own the attribute values of the same record.*

*Proof.* Note that in the protocol, the values $k_i$, $s_i$ and $r_i$ are independently and randomly chosen for every frequency value, so the computation is independently done for every frequency, and therefore this corollary follows immediately from Theorem 2.

## 4.2  Efficiency Evaluation of Privacy Preserving Protocol for Naive Bayes Classier Learning

We implemented an experiment to measure the computation times of each user and miner in the protocol. This experiment uses the environment and the keys size as described in Section 3.4. For each user's data, assume that each pair $(U_i, V_i)$ has a value vector of ten attributes in which $U_i$ has five values while $V_i$ has five remaining values and one class attribute with two values, each non-class attribute has five nominal values.

The results show that each user $U_i$ needs an average 1.38 seconds and 0.71 seconds for phase 1 and phase 3, respectively, and each $V_i$ needs only an average 1.12 seconds. Fig 2 shows the computation times of the miner in phase 1 and phase 3 for different numbers of users such as when the number of users is 5000, the miner uses only 27 seconds and 45 seconds for phase 1 and phase 3, respectively.

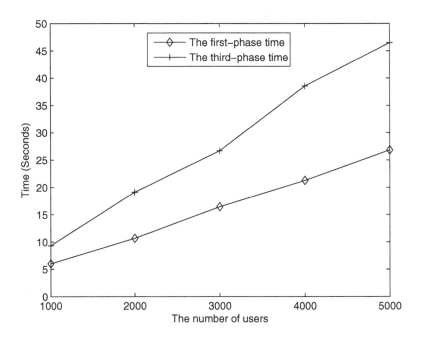

**Fig. 2** The time for computing the values $X_j^{(k,l)}$, $Y_j^{(k,l)}$ in the first phase and the frequencies in third phase

## 4.3   Privacy Preserving Association Rule Mining in TPR

We note that our method can also apply to other problems that require computation of frequencies in data sets such as association rules, decision tree learning, etc. Indeed, we here show another application in privacy preserving association rule mining.

The association rule mining problem can be formally stated in [2]. Given a database D, the problem is to find the association rules that have an implication of the form $X \Rightarrow Y$, where $X$ and $Y$ are the subsets of the set of items of D, and $X \cap Y = \phi$. Each rule $X \Rightarrow Y$ has a support that is the percentage of records containing both X and Y in D. The rule confidence is defined as the percentage of records containing both X and Y with regard to the total number of records containing X.

Association rules are required to meet a minimum support and a minimum confidence values defined by the miner. To this end, the Apriori algorithm in [2] consists of two steps. First is to find all frequent itemsets satisfied the minimum support in the data set. Second is from these frequent itemsets to find rules that meet the minimum confidence constraint. Clearly, association rules mining can be reduced to computing frequencies of tuples of values in data set.

Assume that the data set has the TPR model as described in 4.1.1. In general, for mining association rules in users' data while preserving each user's privacy, the miner can work as follows. Firstly, the miner computes frequencies of values or tuples of values by using the privacy preserving frequency mining method. Next, the miner can apply specified minimum support and confidence to form association rules.

## 5   Conclusion

In this paper, we proposed a method for privacy-preserving frequency-based learning in TPR model, which has not been investigated previously. Basically, the proposed method is based on ElGamal encryption scheme and it ensures strong privacy without loss of accuracy. We illustrated the applicability of the method by applying it to designing the privacy preserving protocol for some learning methods in TPR model such as naive Bayes learning, decision tree learning, association rules mining. We conducted experiments to evaluate the complexity of the protocols, and the results showed that the protocols are efficient and practical. Many other tasks of privacy preserving data mining in TPR model, such as regression analysis, that would be of interest for future work.

Also, although the proposed method is technically mature enough to be used in the privacy preserving frequency-based learning scenario with TPR model, there are still some issues we need to tackle to enhance the efficiency of the method. For example, in the proposed method, half of the users need two interactions with the miner, so a natural question is whether we can design a method in which each user needs only one interaction with the miner.

**Acknowledgements.** The authors wish to express their sincere thanks to the support given by Vietnam National Foundation for Science and Technology Development (NAFOSTED) under Project 102.01-2011.05.

# References

1. Agrawal, D., Aggarwal, C.C.: On the design and quantification of privacy preserving data mining algorithms. In: Proceedings of the Twentieth ACM SIGMOD-SIGACT-SIGART Symposium on Principles of Database Systems, pp. 247–255. ACM (2001)
2. Agrawal, R., Imieliński, T., Swami, A.: Mining association rules between sets of items in large databases. In: Proceedings of the 1993 ACM SIGMOD International Conference on Management of Data, pp. 207–216. ACM (1993)
3. Agrawal, R., Srikant, R.: Privacy-preserving data mining. In: Proceedings of the ACM SIGMOD Conference on Management of Data, pp. 439–450. ACM Press (2000)
4. Agrawal, R., Srikant, R., Thomas, D.: Privacy preserving olap. In: Proceedings of the 2005 ACM SIGMOD International Conference on Management of Data, pp. 251–262. ACM (2005)
5. Boneh, D.: The decision diffie-hellman problem. In: Buhler, J.P. (ed.) ANTS 1998. LNCS, vol. 1423, pp. 48–63. Springer, Heidelberg (1998)
6. Charu, A.C., Yu, P.S.: Privacy-Preserving Data Mining: Models and Algorithms. AS-PVU, Boston (2008)
7. Domingos, P., Pazzani, M.J.: On the optimality of the simple bayesian classifier under zero-one loss. Machine Learning 29(2-3), 103–130 (1997)
8. Du, W., Zhan, Z.: Building decision tree classifier on private data. In: Proceedings of the IEEE International Conference on Privacy, Security and Data Mining, pp. 1–8. Australian Computer Society, Inc. (2002)
9. Du, W., Zhan, Z.: Using randomized response techniques for privacy-preserving data mining. In: Proceedings of the Ninth ACM SIGKDD International Conference on Knowledge Discovery and Data Mining, pp. 505–510. ACM (2003)
10. Evfimievski, A., Gehrke, J., Srikant, R.: Limiting privacy breaches in privacy preserving data mining. In: Proceedings of the Twenty-Second ACM SIGMOD-SIGACT-SIGART Symposium on Principles of Database Systems, pp. 211–222. ACM (2003)
11. Evfimievski, A., Srikant, R., Agrawal, R., Gehrke, J.: Privacy preserving mining of association rules. In: Proceedings of the Eighth ACM SIGKDD International Conference on Knowledge Discovery and Data Mining, pp. 217–228. ACM (2002)
12. Goldreich, O.: Foundations of Cryptography: Basic Tools, vol. 1. Cambridge University Press, New York (2001)
13. Hirt, M., Sako, K.: Efficient receipt-free voting based on homomorphic encryption. In: Preneel, B. (ed.) EUROCRYPT 2000. LNCS, vol. 1807, pp. 539–556. Springer, Heidelberg (2000)
14. Jacobsen, B.K., Thelle, D.S.: The tromson heart study: The relationship between food habits and the body mass index. Journal of Chronic Diseases 40(8), 795–800 (1987)
15. Joachim, G.: The relationship between habits of food consumption and reported reactions to food in people with inflammatory bowel disease–testing the limits. J. Nutrition and Health 13(2), 69–83 (1999)
16. Kantarcioglu, M., Vaidya, J.: Privacy preserving naive bayes classifier for horizontally partitioned data. In: IEEE ICDM Workshop on Privacy Preserving Data Mining (2003)

17. European Parliament. Eu directive 95/46/ec of the european parliament and of the council on the protection of individuals with regard to the processing of personal data and on the free movement of such data. Official J. European Communities, 31 (1995)

18. Pinkas, B.: Cryptographic techniques for privacy-preserving data mining. SIGKDD Explor. Newsl. 4(2), 12–19 (2002)

19. Terry, D.J.: Investigating the relationship between parenting styles and delinquent behavior. McNair Scholars Journal 8(1), Article 11 (2004)

20. Tsiounis, Y., Yung, M.: On the security of elgamal based encryption. In: Imai, H., Zheng, Y. (eds.) PKC 1998. LNCS, vol. 1431, pp. 117–134. Springer, Heidelberg (1998)

21. Vaidya, J., Clifton, C.: Privacy preserving association rule mining in vertically partitioned data. In: Proceedings of the Eighth ACM SIGKDD International Conference on Knowledge Discovery and Data Mining. ACM (2002)

22. Vaidya, J., Kantarcioglu, M., Clifton, C.: Privacy-preserving naive bayes classification. The VLDB Journal 17(4), 879–898 (2008)

23. Verykios, V.S., Bertino, E., Fovino, I.N., Provenza, L.P., Saygin, Y., Theodoridis, Y.: State-of-the-art in privacy preserving data mining. SIGMOD Rec. 33(1), 50–57 (2004)

24. Wu, F., Liu, J., Zhong, S.: An efficient protocol for private and accurate mining of support counts. Pattern Recogn. Lett. 30(1), 80–86 (2009)

25. Yang, Z., Zhong, S., Wright, R.N.: Anonymity-preserving data collection. In: Proceedings of the Eleventh ACM SIGKDD International Conference on Knowledge Discovery in Data Mining, pp. 334–343. ACM (2005)

26. Yang, Z., Zhong, S., Wright, R.N.: Privacy-preserving classification of customer data without loss of accuracy. In: SIAM SDM, pp. 21–23 (2005)

# Mining Jumping Emerging Patterns by Streaming Feature Selection

Fatemeh Alavi and Sattar Hashemi

**Abstract.** One of the main challenges for making an accurate classifier based on mining emerging pattern is extraction of a minimal set of strong emerging patterns from a high-dimensional dataset. This problem is harder when features are generated dynamically and so the entire feature space is unavailable. In this scheme, features are obtained one by one instead of having all features available before learning starts.

In this paper, we propose mining Jumping Emerging Patterns by Streaming Feature selection (JEPSF for short) using a dynamic border-differential algorithm where builds a new border of the jumping emerging pattern space based on an old border of the jumping emerging pattern space for new coming features. This framework completely avoids going back to the most initial step to build a new border of the jumping emerging pattern space. This algorithm helps reducing number of irrelevant emerging patterns, what in turns brings significant advantages to the presented algorithm. Thus the number of jumping emerging patterns is reduced and strong emerging pattern is also extracted. We experimentally represent effectiveness of the proposed approach against other state-of-the-art methods, in terms of predictive accuracy.

## 1 Introduction

An emerging pattern is a combination of attributes values that often occurs in one class but rarely appears in other classes. An Emerging Pattern (EP for short) is defined as an itemset whose support increases significantly from one class of data to another [1]. A specific type of emerging patterns is named a Jumping Emerging Pattern (JEP for short) that is defined as an itemset whose support increases abruptly from zero in one dataset to non-zero in another dataset; the ratio of support-increase

Fatemeh Alavi · Sattar Hashemi
Department of Electrical and Computer Engineering, Shiraz University, Shiraz, Iran
e-mail: alavi@cse.shirazu.ac.ir, s_hashemi@shirazu.ac.ir

V.-N. Huynh et al. (eds.), *Knowledge and Systems Engineering, Volume 2,*
Advances in Intelligent Systems and Computing 245,
DOI: 10.1007/978-3-319-02821-7_30, © Springer International Publishing Switzerland 2014

being infinite [2]. JEPs represent contrast between different classes more strongly than any other types of EP. EPs are used in many real-world applications, such as failure detection [3], predicting disease [4] and discovering knowledge in gene expression data [5, 6, 7, 8].

The notion of streaming features has proposed to conduct feature selection in an uncertain feature space over time [9, 10]. On the contrary data streams, with streaming features, feature dimensions are constructed as a feature stream. Features are obtained one by one and each feature is processed upon its arrival.

Strong emerging patterns mining in a high-dimensional dataset is one of the main challenges specifically if the total feature space is unavailable in advance. The number of EPs is exponentially increasing in massive datasets that include millions or billions of features, such as image processing, gene expression data, text data, and so on. Therefore, strong emerging pattern extracting is an important problem. On the other hand, as features arrive one by one, a new EP space should be built by the border of the old EP space. Only new candidate EPs should be added to the old EP space without going back to the most initial step of mining of emerging patterns.

To effectively mine EPs from a high-dimensional dataset, we face two challenging research issues: (1) how to efficiently capture a minimal set of strong EPs from a high-dimensional dataset; and (2) how to mine strong EPs if the full feature space is unavailable before learning? Each new feature that arrives, extraction process of EPs is very time-consuming or infeasible if this process returns to the first phase of EPs mining.

Border-based approaches [1, 11, 12], inspired by the Max-Miner algorithm [13], provide a good structure and reduce size of mining results. A border is used to describe a large collection of EPs concisely, namely a border consists of minimal and maximal EPs. Since features arrive one by one, a border-based method is unable to mine EPs without going back to the first step of patterns mining. Inspired by the FP-tree [14], a CP-Tree miner was presented to better performance of EP mining [15]. Because previous methods were unable to handle more than seventy five dimensions, later a ZBDD EP-miner [16] is proposed, using Zero-Suppressed Binary Decision Diagrams (ZBDDs), for mining EPs from a high-dimensional dataset. But ZBDD EP-miner still mines an explosive number of EPs, even with a high support threshold. So it is yet a challenging research issue to mine a minimal set of strongly predictive EPs from a huge number of candidate EPs.

In a recent study [17], EPs mining integrates with Streaming Feature selection (EPSF in short) to get an accurate EP classifier. Integrating streaming feature selection into EP mining can efficiently extract a minimal set of predictive patterns from a high-dimensional dataset. But EPSF can mine 1-itemsets only.

In comparison to [17], we propose a mining Jumping Emerging Pattern by Streaming Feature selection (JEPSF for short) method to overcome mentioned problems. Our proposed method has four novel contributions:

- JEPSF removes irrelevant features, so the pattern space of EPs is reduced and the candidate EPs are more effective in the classification of a test data.
- Since jumping emerging patterns are stronger than other emerging patterns, JEPs are extracted from a high-dimensional dataset and the most expressive

JEPs [2] are used for a classification purpose. The most expressive JEPs have a support higher than other JEPs, thus they are noise-tolerant.

- In response to new features that arrive, JEPSF builds a new border of the JEP space based on a previous border of the JEP space. An efficient strategy is presented for mining of JEPs without going back to the first extraction stage, named a dynamic border-differential algorithm. The dynamic border-differential algorithm adjusts borders of the JEP space for new coming features dynamically.
- JEPSF dose not need to store all features in the memory, so it can mine JEPs without knowing the whole feature set in advance.

To manage a process of JEPs extracting online, this strategy consists of two stages. In the first stage, irrelevant EPs are removed. Each irrelevant feature generates irrelevant EPs, hence irrelevant features are removed before EPs mining process. In the second stage, a dynamic border-differential algorithm extracts JEPs when new features arrive.

The reminder of the paper is organized as follows. Section 2 presents the backgrounds on emerging patterns and borders. Section 3 presents our approach, and experimental results are reported in Section 4. Finally, Section 5 provides our conclusion and future work.

## 2 Preliminary Knowledge

### 2.1 Emerging Patterns

Suppose we have a dataset D defined upon a set of N features $(f_1, f_2, ..., f_N)$ with a discrete domain $dom(f_i)$ for each feature $f_i$ and the class label C. Let $I$ be the set of all items, $I = \cup_{i=1}^{N} dom(f_i)$, N denotes the number of features and $i = 1, ..., N$. An itemset $X$ is a subset of $I$ and its $support_D(X) = \frac{count_D(X)}{|D|}$, that $count_D(X)$ is the number of instances in D comprising $X$ and $|D|$ is the number of instances in D. The class label C is two classes, positive and negative classes. The dataset D can be divided into $D_p$ and $D_n$, the positive and negative instances respectively. The Growth Rate (GR for short) of X from $D_n$ to $D_p$ is defined as follows.

**Definition 1. (GR: Growth Rate)** [1] Growth rate of an itemset X from $D_n$ to $D_p$,

$$GR_{D_n \to D_p}(X) = \begin{cases} 0 & \text{if } support_{D_p}(X) = 0 \text{ and } support_{D_n}(X) = 0 \\ \infty & \text{if } support_{D_p}(X) \neq 0 \text{ and } support_{D_n}(X) = 0 \\ \frac{support_{D_p}(X)}{support_{D_n}(X)} & \text{otherwise} \end{cases} \quad (1)$$

**Definition 2. (EP: Emerging Pattern)** [1] Given a growth rate threshold $\rho > 1$. An itemset X is said to be an emerging pattern from $D_n$ to $D_p$ if $GR_{D_n \to D_p}(X) \geq \rho$. If $GR_{D_n \to D_p}(X) = \infty$, an itemset X is called a Jumping EP from $D_n$ to $D_p$.

Since a JEP is stronger than other types of EPs, we use JEPs for a classification purpose exclusively. The most expressive JEP [2] is a type of JEPs with a high support that the number of these patterns is much less than the number of all JEPs. We obtain all the most expressive JEPs of each class $C_i$ in a training set that are subsets of a test instance t. Mined patterns contribute to be decided which the class should be assigned to a test instance t.

**Definition 3. (Aggregate Score)** [2] Given a test instance t, an EP e and a set of the most expressive JEPs (**MEJEP** for short) of class $C_i \in C$, the aggregate score of t for $C_i$ is defined as,

$$score(t, C_i) = \sum_{e \subseteq t,\, e \in MEJEP} support_{C_i}(e) \tag{2}$$

K scores are calculated for each test instance t, one score per class, that a class with the highest score is label of t, so $label(t) = argmax_{c_i \in C}\ score(t, C_i)$.

## 2.2  Border Representation for Emerging Pattern

A border provides an appropriate structure for concisely describing a large collection of EPs. The structure of a border contains only minimal and maximal EPs.

**Definition 4. (Border)** [1] An ordered pair $< L, R >$ is called a border if (a) both $L$ and $R$ are antichain ( a collection $S$ of sets is an antichain if X and Y are incomparable sets (i.e. $X \nsubseteq Y$ and $Y \nsubseteq X$) for all X,Y $\in S$ ) collections of sets, and (b) each element of $L$ is a subset of some element in $R$ and each element of $R$ is a superset of some element in $L$; $L$ is called the left bound of the border and $R$ the right bound. The collection of sets represented by a border $< L, R >$ is

$$[L, R] = \{Y \mid \exists X \in L, \exists Z \in R \quad such\, that \quad X \subseteq Y \subseteq Z\} \tag{3}$$

If each set contains above properties, (a) and (b), it is convex or interval closed.

**Example 1.** The collection represented by $< \{12\}, \{1234, 1256\} >$ is $\{12, 123, 124, 1234, 125, 126, 1256\}$. All sets are both supersets of 12 and subsets of 1234 or 1256. The horizontal-border [1] of a dataset is the border $[\{\emptyset\}, R]$ that represents all non-zero support itemsets in a dataset.

**Definition 5. (Border-Differential Operation)** [1] Given two horizontal borders $< \{\emptyset\}, R_p >$ of positive instances and $< \{\emptyset\}, R_n >$ of negative instances, their border differential operation is defined as follows:

$$[L, R] = [\{\emptyset\}, R_p] - [\{\emptyset\}, R_n] \tag{4}$$

**Example 2.** Let $R_p = \{1234\}$ and $R_n = \{23, 24, 34\}$. Then

$$[\{\emptyset\}, \{1234\}] - [\{\emptyset\}, \{23, 24, 34\}] = \{1, 12, 13, 14, 123, 124, 134, 234, 1234\}$$

The minimal itemset of the border difference operation is $L = \{1, 234\}$ and the maximal itemset is $R = \{1234\}$. Therefore, $< \{\emptyset\}, \{1234\} > - < \{23, 24, 34\} >=< \{1, 234\}, \{1234\} >$.

Note the difference and similarity between the two notations of $< L, R >$ and $[L, R]$. A border $< L, R >$ consists of the two bounds L and R, while the border $[L, R]$ consists of all items bounded by the left bound L and the right bound R.

# 3 A Framework of Mining JEPs with Streaming Feature Selection

## 3.1 Dynamic Border-Differential Algorithm

Since features arrive one by one, an efficient algorithm is necessary without fully going back to the first step of JEPs mining. In this strategy, JEPs are extracted in response to new arriving features efficiently. This strategy are named a dynamic border-differential algorithm. The dynamic border-differential algorithm adjusts a new border of the JEP space based on a previous border of the JEP space. Considering that the difference between two horizontal borders is performed continuously without going back to the first extraction step hence a time-complexity is reduced. The dynamic border-differential algorithm modifies boundary elements of the JEP space in response to data changes.

As mentioned before, the horizontal border represents all non-zero support itemsets of a dataset [1]. The difference between two horizontal borders gives the border of the JEP space. Whereas features are obtained one by one, adding features are stored in a window with size m of features; with having m features, a window is built . We need two windows, a current window and a new window. The current window contains m features and the new window is made when n new features arrive. If the new window overlaps with the current window, new combinations of items are mined. Therefore, n (n $<$ m) arriving features and k (k $<$ m) features of the current window together build the new window that the total n and k is m ( n+k = m).

When a new window is created, the border-differential algorithm is performed to modify a boundary elements of the JEP space. Thus, the new window is reconstructed with receiving of n new features and this process is repeated constantly. The new window is initialized by m arriving features ,and the next time, it is built by n new coming features and k previous features of the current window.

It should be noted that the maximum length of extracted patterns is m that should not be too small or too large. If m be too small, JEPSF cannot extract JEPs and, if m be very big, mining JEPs are less likely to be seen in the test instances.

About n and k should be considered that k should be chosen $k \geq m/2$ and n should be selected $n \leq m/2$ to get a new combination of items.

As shown in Fig. 1., the border $< L_{current}, R_{current} >$ denotes the border of the JEP space. After receiving n features, the border of the new window is denoted by $< L_{new}, R_{new} >$. The border of the JEP space is modified by calling the border differential algorithm. This algorithm uses two operations, union and difference, to adjust the border of the JEP space. The difference operation is performed for each new window that gives boundary elements of the new window, $L_{new}$ and $R_{new}$. The

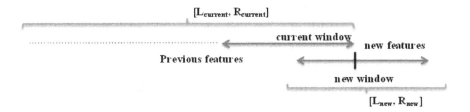

**Fig. 1** Dynamic border differential algorithm: The new window is created by k previous features of the current window and n new coming features. Both the current window and the new window consist of m features.

union operation integrates two borders of $< L_{current}, R_{current} >$ and $< L_{new}, R_{new} >$. The union operation changes the right bound by a union border operation on the current right bound and the new right bound, and the left bound by adding all items available in the new left bound and the current left bound. So the border-differential algorithm manipulates boundary elements of the JEP space, the left bound and the right bound, dynamically. We further state the union and difference border operations.

## 3.2 Border Operations: Difference and Union

There are two border operations for making a dynamic border-differential algorithm as discussed above. These operations include border union of two JEP spaces and border difference of two horizontal spaces that change boundary elements of the JEP space.

The operation of border union is used to integrate two JEP spaces. Suppose two borders $< L_{current}, R_{current} >$ and $< L_{new}, R_{new} >$ are given, the border union of two JEP spaces is shown in Algorithm 1.

---

**Algorithm 7.** Border Union of Two JEP Spaces

---

1 ;; inputs: two borders $< L_{current}, R_{current} >$ and $< L_{new}, R_{new} >$
2 ;; output: $[L_{dynamic}, R_{dynamic}] = [L_{current}, R_{current}] \cup [L_{new}, R_{new}]$
3 $R_{dynamic} \leftarrow R_{current} \cup R_{new}$ ;; union a new right bound and a current right bound gives the dynamic right bound
4 $L_{dynamic} \leftarrow$ all items available in a new left bound and a current left bound
5 ;; **return** $< L_{dynamic}, R_{dynamic} >$

---

Border union operation takes two borders $< L_{current}, R_{current} >$ and $< L_{new}, R_{new} >$ and returns new boundary elements of the JEP space. Union the new right bound and the current right bound gives a dynamic right bound, and all items available in a new left bound and a current left bound make a dynamic left bound.

**Example 3.** Let $R_{current} = \{1234\}$, $R_{new} = \{3467\}$, $L_{current} = \{23, 24\}$ and $L_{new} = \{7\}$ then

$$R_{dynamic} = \{1234\} \cup \{3467\} = \{123467\}, L_{dynamic} = \{23, 24, 7\}$$

The operation of border difference is fundamentally used to detect the border of the JEP space between two horizontal borders. Suppose the horizontal border of positive instances, $D_P$, is $< \{\emptyset\}, \{H_{P,1}, H_{P,2}, \ldots, H_{P,k_1}\} >$ and the horizontal border of negative instances, $D_N$ is $< \{\emptyset\}, \{H_{N,1}, H_{N,2}, \ldots, H_{N,k_2}\} >$, the border of the JEP space of $D_P$ is discovered by the border-differential algorithm in Algorithm 2.

We use a complete border-differential algorithm [11] for the subroutine Border-Diff that efficiently computes borders of the JEP space by manipulating the input borders.

---

**Algorithm 8.** DIFF

---

1  ;; input: horizontal borders $(< \{\emptyset\}, \{H_{P,1}, \ldots, H_{P,k_1}\} >, < \{\emptyset\}, \{H_{N,1}, \ldots,$
   $H_{N,k_2}\} >)$
2  ;; output: the border $< L_P; R_P >$
3  $L_P \leftarrow \{\}; R_P \leftarrow \{\}$;
4  **for** $j$ *from 1 to* $k_1$ **do**
5       if some $H_N$ is a superset of $H_{P,j}$ then **continue**;
6       border = Border-Diff($< \{\emptyset\}, \{H_{P,j}\} >, < \{\emptyset\}, \{H_{N,1}, \ldots, H_{N,k_2}\} >$)
7       $R_P = R_P \cup$ right bound of border
8       $L_P = L_P \cup$ left bound of border
9  10: **return** $< L_P, R_P >$

---

### 3.3  JEPSF: Mining Jumping Emerging Pattern with Streaming Feature Selection

A total framework of JEPSF is illustrated in Fig. 2. The key steps of JEPSF are (1) Step1: identifying irrelevant EPs, (2) Step 2: Performing the dynamic border-differential algorithm and (3) Step 3: Building a JEP-based classifier.

**Step 1: Identifying irrelevant EPs.** Because it is impractical to search the pattern space of JEP mining covering every possible combination of items for a large high-dimensional dataset, we propose that irrelevant EPs are removed to reduce the jumping emerging pattern space. Based on idea of Yu et al.[17], an EP can be divided into three categories, namely, irrelevant EPs (non-EPs), strongly EPs and redundant EPs. Irrelevant features generate irrelevant EPs, thus the pattern space is effectively reduced by elimination of irrelevant features.

In the following proposition, let F be a complete set of features, $f_i$ denotes the $i^{th}$ input feature, $a$ be a feature value in a discrete domain $f_i$ and C be a finite set of the different class labels that $C_i$ denotes the $i^{th}$ class label.

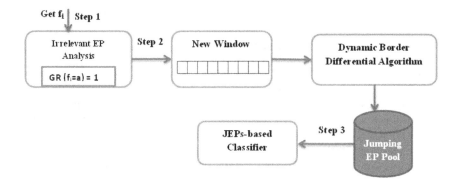

**Fig. 2** A framework of JEPSF: As a feature arrives, $f_i$, if it generates irrelevant EPs, JEPSF removes it. Otherwise, JEPSF adds it to a new window. Then the dynamic border differential algorithm receives m features from a new window and adds JEPs to a JEP pool.

**Proposition 1. (Irrelevant EP)** [17] For $\forall f_i \in F$, $\forall a \in dom(f_i)$, and $\forall C_i \in dom(C)$, $GR_{D-D_{C_i} \to D_{C_i}}(f_i = a) = 1$ holds iff $f_i$ is irrelevant to C.

Each irrelevant feature generates irrelevant EPs. If an attribute is irrelevant, GR for different values of an attribute is always 1. This means that the support change is the same for different classes and accordingly irrelevant EPs are not useful for describing a contrast between classes. Thus JEPSF can remove irrelevant features before EP extracting.

**Step 2: Performing the dynamic border differential algorithm.** If a coming feature is not an irrelevant feature, it is added to the new window. The dynamic border differential algorithm is performed when n new features are added to the new window. The new window consists of k previous features of the current window and n new coming features. After applying the dynamic border differential algorithm, the border of the JEP space is manipulated. Boundary elements, the left bound and the right bound, are adjusted by the union and difference operations.

**Step 3: Building JEP-based classifiers.** After all features arrived, with the extracted JEPs, the JEP-based classifier is constructed by the aggregate score function defined by Definition 3. The left bound consists of minimal itemsets which have a support higher than other JEPs. Therefore, the JEP-based classifier uses all items available in the left bound.

We illustrate the JEPSF algorithm in Algorithm 3. According to Proposition 1, our method can mine JEPs regardless of irrelevant features. Therefore, the JEPSF removes irrelevant features before the process of JEPs extracting to reduce the search space of JEPs and improve the predictive accuracy of classification.

As a new feature $f_i$ arrives, JEPSF checks whether it generates irrelevant EPs; and if so, it is ignored. Otherwise, an arriving feature is added to the new window (Stages 2 to 4).

---

**Algorithm 9.** The JEPSF algorithm

---

1  Initialize of the $L_{current} = \{\emptyset\}$ and setting $C$ as the class label.
2  Input a new feature $f_i$
3  ;; Identify irrelevant EPs by Proposition 1
4  **If** ;
5  $GR_{D_n \to D_p}(f_i = a) = 1$ **then** discard $f_i$ and go to step 2;
6  ;; Mine JEPs for each class by calling dynamic border-differential algorithm
7  ;; hold n arriving features and make a new window with m size
8  New_Window = n new coming features + k previous features
9  $R_p$ = instances of positive class ;; for a new window
10 $R_n$ = instances of negative class ;; for a new window
11 **for** $i=1:|C|$ **do**
12    |  $|C|$ denotes the number of classes
13    |  ;; compute new border of dataset of class $C_i$
14    |  New-border=
15    └  DIFF($< \{\emptyset\}, \{R_{C_i,1}, \ldots, R_{C_i,k_1}\} >,< \{\emptyset\}, \{R_{C-C_i,1}, \ldots, R_{C-C_i,k_2}\} >$)
16 Current-border = Border-Union(New-border, Current-border)
17 Current_Window = New_Window;
18 Repeat steps 2 to 16 until all features have reached
19 classify test instances by the score function defined in Definition 3

---

A new window is created by n new features with k previous features available in the current window. Then the dynamic border differential algorithm is called to adjust a new border of the JEP space. Instances of positive class are denoted by $R_p$ that the number of positive instances is $K_1$ and $R_n$ denotes instances of negative class that contains $K_2$ instances. For each class, the border differential algorithm (Algorithm 2) is performed separately; this means that if there are two classes, in the first $R_{C_i}$ and $R_{C-C_i}$ are instances of positive and negative classes respectively, and in the second time, $R_{C_i}$ and $R_{C-C_i}$ contain instances of negative and positive classes respectively. The border union operation takes two borders; the new border and the current border, returns a new border of the JEP space with adjusting boundary elements namely changing the left bound and the right bound. Then the current-border is gained by union border operation (Stages 5 to 16).

After all features arrive, test instances are classified by the function score defined in Definition 3. Since itemsets of the left bound are the most expressive JEPs, a test instance is classified based on itemsets available in the left bound.

## 4  Experimental Results

### 4.1  Experimental Setup

To complete evaluation of the proposed framework, we have selected fourteen (Table 1) datasets of the UCI machine learning repository and one dataset of the NIPS 2003 feature selection challenge (madelon).

Our proposed method is compared with two categories of classifiers, including the state-of-the-art EP classifiers and the well-known machine learning classifiers.

- Comparing JEPSF with two well-known EP classifiers, CAEP [18] and EPSF [17].
- Comparing JEPSF with the state-of-the-art machine learning classifiers, including Decision Tree J48 [19], SVM [20] and AdaBoost [21].

We gain the predictive accuracy using ten-fold cross-validation (CV-10). Results are reported as the mean classification accuracy over the ten folds. To discretize continuous datasets, we use the MDL discretization method in the WEKA software. In the experiments, we have initialized the minimum growth rate to 4 for CAEP and EPSF classifiers, and the minimum support to 0.05 for CAEP. We set a window size to m=10 (n=5 and k=5), in fact a window is made by 5 previous features and 5 new features. We have implemented EPSF and JEPFS in MATLAB. The accuracy of J48, SVM and AdaBoost are gained using the WEKA implementation.

**Table 1** Description of Datasets
#Features: number of features, #Instances: number of instances

| Dataset | #Features | #Instances | Dataset | #Features | #Instances |
|---------|-----------|------------|---------|-----------|------------|
| diabetes | 8 | 768 | promoters | 57 | 106 |
| german | 20 | 1000 | spect | 22 | 267 |
| hepatits | 19 | 155 | spectf | 44 | 267 |
| hypothyroid | 25 | 3163 | tictactoe | 9 | 958 |
| kr-vs-kp | 36 | 3196 | vote | 16 | 435 |
| labor-neg | 16 | 57 | wdbc | 30 | 569 |
| liver | 6 | 345 | madelon | 500 | 2000 |
| mushroom | 22 | 8124 | - | - | - |

## 4.2  Predictive Accuracy Comparison

In Tables 2 and 3, we have reported the predictive accuracy of the proposed JEPSF classifier and five classifiers, including two EP-based classifiers and three well-known machine learning classifiers on the fifteen benchmark datasets. The best result among all classifiers is specified by text bold for each dataset and the symbol "/" shows that a classifier runs out of memory due to a huge number of candidate Emerging Patterns.

Because JEP increases abruptly from zero in one dataset to non-zero in another dataset, JEP represents contrast between different classes more strongly than any other types. So JEPs are stronger than other types of EPs. As regards all items in a border are supersets of some items in the left bound, and are subsets of some items in the right bound, so all items in the left bound have the largest support among all items of the border. JEP-based classifier uses items of the left bound hence the number of mined JEPs is much smaller than the number of all JEPs. Thus JEPSF is stronger than the other classifiers generally.

In comparison with two EP-classifiers, CAEP and EPSF, JEPSF obtains significant predictive accuracy. CAEP mines all EPs without pruning the EP space hence it generates an explosive number of EPs from a high dimensional dataset. Because of the huge number of candidate EPs, it is difficult to store and retrieve them efficiently. So CAEP gives out of memory for a high dimensional dataset. On the other hand, JEPSF mines JEPs only and removes irrelevant EPs while CAEP mines all EPs, so the search space of JEPs in JEPSF is very samllar than the search space of EP mining in CAEP.

**Table 2** Comparison of predictive accuracy (%): JEPSF, EPSF and CAEP

| Dataset | JEPSF | EPSF | CAEP | Dataset | JEPSF | EPSF | CAEP |
|---|---|---|---|---|---|---|---|
| diabetes | **72.40** | 65.10 | 68.95 | promoters | **76.27** | 75.36 | / |
| german | **75.10** | 70.00 | 72.80 | spect | **82.09** | 79.46 | 66.92 |
| hepatitis | 82.54 | 76.67 | **86.00** | spectf | 79.05 | **82.04** | / |
| hypothyroid | **96.08** | 75.53 | 67.78 | tictactoe | **97.18** | 34.66 | 82.95 |
| kr-vs-kp | **98.56** | 74.91 | 83.49 | vote | **94.47** | **94.47** | 90.00 |
| labor-neg | 86.67 | 83.00 | **94.00** | wdbc | **94.20** | 81.38 | 81.96 |
| liver | **61.95** | 59.70 | 57.65 | madelon | **78.15** | 58.80 | / |
| mushroom | **100.00** | 97.30 | 96.18 | - | - | - | - |

**Table 3** Comparison of predictive accuracy (%): JEPSF, J48, SVM and AdaBoost

| Dataset | JEPSF | J48 | SVM | AdaBoost | Dataset | JEPSF | J48 | SVM | AdaBoost |
|---|---|---|---|---|---|---|---|---|---|
| diabetes | **72.40** | 64.59 | 64.72 | 65.18 | promoters | 76.27 | 73.18 | 82.81 | **88.55** |
| german | **75.10** | 71.00 | 72.70 | 70.20 | spect | **82.09** | 79.05 | 81.64 | 77.54 |
| hepatitis | **82.54** | 79.21 | 76.29 | 79.92 | spectf | 79.05 | 78.28 | 78.69 | **79.46** |
| hypothyroid | 96.08 | **96.52** | 95.57 | 95.38 | tictactoe | **97.18** | 89.66 | 67.84 | 77.34 |
| kr-vs-kp | 98.56 | **99.37** | 94.99 | 93.84 | vote | 94.47 | 94.94 | 94.93 | **96.07** |
| labor-neg | **86.67** | 84.33 | 85.96 | 86.00 | wdbc | **94.20** | 82.26 | 79.61 | 74.36 |
| liver | **61.95** | 59.15 | 61.20 | 60.87 | madelon | **78.15** | 69.30 | 56.35 | 62.70 |
| mushroom | **100.00** | **100.00** | 99.11 | 97.73 | - | - | - | - | - |

EPSF mines 1-itemsets of EPs only, there may not be enough EPs in some datasets and so EPSF achieves the low predictive accuracy in those datasets. Because JEPSF mines JEPs with a high support, it is superior to EPSF and CAEP.

In Table 2, compared to CAEP and ESPF, JEPSF gets the highest accuracy on twelve datasets. On three datasets, promoters, spectf and madelon, CAEP fails to deal with a high dimensional dataset. On eleven datasets, EPSF obtains the low predictive accuracy because there are not enough EPs in those datasets for classification.

In comparison with the well-known machine learning classifiers, JEPSF is significantly superior to SVM and also very competitive with J48 and AdaBoost as shown in Table 3. Predictive accuracy of JEPSF is improved by removing irrelevant features. So JEPSF is better than the well-known machine learning classifiers.

## 4.3 Reduction the Pattern Space of EPs

Given that irrelevant features generate irrelevant EPs, JEPSF removes irrelevant features before mining JEPs. Therefore the pattern space of EPs is reduced by removing irrelevant features. In addition, $\alpha$ constraint [16] can reduce the pattern space of EPs. $\alpha$ constraint is based on the a-priori principle. Any item whose support is less than $\alpha$ can be removed its supersets from $D_p$ and $D_n$ [16]. Thus the pattern space of EPs is reduced efficiently and also the EP mining is done faster.

## 5 Conclusion and Future Work

In this study, we proposed JEPSF to extract the most discriminating knowledge between two classes. We have developed an efficient algorithm for JEPs mining that captures JEPs for new coming features based on previous extracting JEPs. We presented that the pattern space of EP mining is reduced by removing irrelevant features and hence the proposed algorithm provides the higher predictive accuracy.

The experiments on the benchmark datasets have presented the effectiveness of the proposed approach against other state-of-the-art methods in terms of predictive accuracy.

Given that the process of the pattern mining is hard upon feature arrival, we propose that coming features store in a tree-like structure, inspired by the frequent pattern tree, as a future work. This requires that the tree is reconstructed with the arrival of each feature and then the candidate EPs can be extracted from the tree.

In addition, JEPSF can be extended for multi class and also datasets with higher dimensional as a future work.

## References

1. Dong, G., Li, J.: Efficient Mining of Emerging Patterns: Discovering Trends and Differences. In: Proceedings of the Fifth ACM SIGKDD International Conference on Knowledge Discovery and Data Mining, pp. 43–52. ACM (1999)
2. Li, J., Dong, G., Ramamohanarao, K.: Making use of the most expressive jumping emerging patterns for classification. Knowledge and Information Systems 3, 131–145 (2001)
3. Lo, D., Cheng, H., Han, J., Khoo, S., Sun, C.: Classification of software behaviours for failure detection: a discriminative pattern mining approach. In: KDD 2009, pp. 557–566 (2009)
4. Li, J., Liu, H., Downing, J.R., Yeoh, A.E.J., Wong, L.: Simple rules underlying gene expression profiles of more than six subtypes of acute lymphoblastic leukemia (all) patients. Bioinformatics 19(1), 71–78 (2003)
5. Fang, G., Pandey, G., Wang, W., Gupta, M., Steinbach, M., Kumar, V.: Mining low-support discriminative patterns from dense and high-dimensional data. IEEE Transactions on Knowledge and Data Engineering 24(2), 279–294 (2012)
6. Mao, S., Dong, G.: Discovery of highly differentiative gene groups from microarray gene expression data using the gene club approach. J. Bioinformatics and Computational Biology 3, 1263–1280 (2005)

7. Boulesteix, A.-L., Tutz, G., Strimmer., K.: A CART-based approach to discover emerging patterns in microarray data. Bioinformatics 19(18), 2465–2472 (2003)
8. Li, J., Wong, L.: Identifying good diagnostic gene groups from gene expression profiles using the concept of emerging patterns. Bioinformatics 18(5), 725–734 (2002)
9. Wu, X., Yu, K., Wang, H., Ding, W.: Online streaming feature selection. In: ICML 2010, pp. 1159–1166 (2010)
10. Zhou, J., Foster, D., Stine, R.A., Ungar, L.H.: Streamwise feature selection. J. of Machine Learning Research 7, 1861–1885 (2006)
11. Dong, G., Li, J.: Mining border descriptions of emerging patterns from dataset pairs. Knowledge and Information Systems 8, 178–202 (2005)
12. Li, J., Ramamohanarao, K., Dong, G.: The Space of Jumping Emerging Patterns and Its Incremental Maintenance Algorithms. In: Proceedings of the Seventeenth International Conference on Machine Learning, ICML 2000, USA, pp. 551–558 (2000)
13. Bayardo, R.J.: Efficiently mining long patterns from databases. In: SIGMOD 1998, pp. 85–93 (1998)
14. Han, J., Pei, J., Yin, Y.: Mining frequent patterns without candidates generation. In: SIGMOD, pp. 1–12 (May 2000)
15. Fan, H., Ramamohanarao, K.: An efficient single-scan algorithm for mining essential jumping emerging patterns for classification. In: Chen, M.-S., Yu, P.S., Liu, B. (eds.) PAKDD 2002. LNCS (LNAI), vol. 2336, pp. 456–462. Springer, Heidelberg (2002)
16. Loekito, E., Bailey, J.: Fast mining of high dimensional expressive contrast patterns using zero suppressed binary decision diagrams. In: KDD 2006, pp. 307–316 (2006)
17. Yu, K., Ding, W., Simovici, D., Wu, X.: Mining Emerging Patterns by Streaming Feature Selection. In: Proceedings of the 18th ACM SIGKDD International Conference on Knowledge Discovery and Data Mining (KDD 2012), China, pp. 60–68 (2012)
18. Dong, G., Zhang, X., Wong, L., Li, J.: CAEP: Classification by Aggregating Emerging Patterns. In: Arikawa, S., Nakata, I. (eds.) DS 1999. LNCS (LNAI), vol. 1721, pp. 30–42. Springer, Heidelberg (1999)
19. Quinlan, J.R.: C4.5: programs for machine learning, vol. 1. Morgan Kaufmann (1993)
20. Cortes, C., Vapnik, V.: Support vector machine. Machine Learning 20(3), 273–297 (1995)
21. Freund, Y., Shapire, R.: A decision-theoretic generalization of on-line learning and an application to boosting. In: Vitányi, P.M.B. (ed.) EuroCOLT 1995. LNCS, vol. 904, pp. 23–37. Springer, Heidelberg (1995)

# An Approach for Mining Association Rules Intersected with Constraint Itemsets

Anh Tran, Tin Truong, and Bac Le

**Abstract.** When the number of association rules extracted from datasets is very large, using them becomes too complicated to the users. Thus, it is important to obtain a small set of association rules in direction to users. The paper investigates the problem of discovering the set of association rules intersected with constraint itemsets. Since the constraints usually change, we start the mining from the lattice of closed itemsets and their generators, mined only one time, instead of from the dataset. We first partition the rule set with constraint into disjoint classes of the rules having the same closures. Then, each class is mined independently. Using the set operators on the closed itemsets and their generators, we show the explicit representations of the rules intersected with constraints in two shapes: rules with confidence of equal to 1 and those with confidence of less than 1. Due to those representations, the algorithm *IntARS-OurApp* is proposed for mining quickly the rules without checking rules directly with constraints. The experiments proved its efficiency.

## 1 Introduction

Mining association rules from datasets [1] usually outputs a large number of association rules. However, the users only take care of a small one of them which contains the rules satisfying given constraints. Some models and types of constraints have been considered in [5, 13, 10]. In the recent results [2, 4, 7], we concentrate on the mining frequent itemsets with constraints involved directly items. The paper focuses the problem, stated as follows: *For $\mathcal{T}$ – a given dataset of transactions, $\mathcal{A}$ – the*

Anh Tran · Tin Truong
Department of Mathematics and Computer Science, University of Dalat, Vietnam
e-mail: {anhtn,tintc}@dlu.edu.vn

Bac Le
University of Natural Science, Vietnam National University, Ho Chi Minh, Vietnam
e-mail: lhbac@fit.hcmus.edu.vn

V.-N. Huynh et al. (eds.), *Knowledge and Systems Engineering, Volume 2*,     351
Advances in Intelligent Systems and Computing 245,
DOI: 10.1007/978-3-319-02821-7_31, © Springer International Publishing Switzerland 2014

*set of all items appeared in $\mathscr{T}$, the minimum support $1 \leq s_0 \leq |\mathscr{T}|$ and the minimum confidence $0 < c_0 \leq 1$ and two constraints $\emptyset \neq \mathscr{G}\mathscr{T}, \mathscr{K}\mathscr{L} \subseteq \mathscr{A}$, let us find the constraint-based association rule set*

$$\mathscr{A}\mathscr{R}\mathscr{S}_{\cap\mathscr{G}\mathscr{T},\mathscr{K}\mathscr{L}}(s_0, c_0) \equiv$$
$$\{L' \to R' \in \mathscr{A}\mathscr{R}\mathscr{S}(s_0, c_0) : L' \cap \mathscr{G}\mathscr{T} \neq \emptyset, R' \cap \mathscr{K}\mathscr{L} \neq \emptyset\} \quad (1)$$

*where: $\mathscr{A}\mathscr{R}\mathscr{S}(s_0, c_0)$ is the set of association rules in the normal meaning according to the thresolds $s_0$ and $c_0$.*

The post-processing approach solves the problem in three phases as follows. The first is to mine all frequent itemsets. The second is to generate association rules from those itemsets. Selecting the ones satisfying two constraints is done in the last phase. We can see that the approach is naive because of several reasons as follows. First, the numbers of frequent itemsets and association rules can grow exponentially. Mining them spends much time since there could be a large number of redundant candidates. Further, the post-processing step contains many intersection operations on itemsets. Second, whenever the constraints change, we need to solve the problem from the beginning.

The authors of the recent papers [2, 3, 5, 6, 9, 11] concentrate on the discovering for the condensed, lossless representations of frequent itemsets as well association rules, e.g., the lattice of closed itemsets and their generators. Only based on them, we can obtain all frequent itemsets and corresponding association rules (see [2, 3]). Hence, using the lattice for constrained based mining is a natural approach. In fact, in [2], we presented the model of mining frequent itemsets based on it. The model had been applied successfully for mining frequent itemsets with some types of constraint such as [2, 3, 7].

In the paper, based on it we mine association rules intersected with constraints. First, the lattice of frequent closed itemsets and their generators is mined. The task is quickly finished since the number of closed itemsets is usually small compared to the one of all frequent itemsets (see [12]). Then, using the results of partitioning the association rule set and the structure of each rule class [3], the constraint-based association rule set is partitioned into disjoint equivalence rule classes. Without loss of the generality, we just consider the mining each class independently. Each class $\mathscr{A}\mathscr{R}\mathscr{S}_{\cap\mathscr{G}\mathscr{T},\mathscr{K}\mathscr{L}}(L, S)$ is represented by a pair of two frequent closed itemsets $(L, S)$ for $L \subseteq S, L \cap \mathscr{G}\mathscr{T} \neq \emptyset, S \cap \mathscr{K}\mathscr{L} \neq \emptyset, c_0 \leq (support(S)/support(L))$. The left side $L'$ of a rule is a frequent itemset such that: (1) its closure is identical to $L$ and (2) the intersection of it and $\mathscr{G}\mathscr{T}$ is not empty. In the case $L = S$, the corresponding right one, namely $R'$, is contained in the difference of $L'$ from $L$ having the non-empty intersection with $\mathscr{K}\mathscr{L}$. Otherwise, for $L \subset S, R'$ is contained in the difference of $L'$ from $S$ satisfied two following conditions: (1) the closure of the two-side union $L' + R'^1$ is equal to $S$ and (2) the intersection of $R'$ and $\mathscr{K}\mathscr{L}$ is non-empty. The paper proposes and applies explicit representations for $L'$ and $R'$ to mine quickly them without testing the constraints.

---

[1] The notation "+" represents the union of two disjoint sets.

The rest of the paper is organized as follows. Section 2 recalls some preliminaries of association rule mining. Section 3 proposes our mining approach. Experimental results and the conclusion are shown in Sect. 4 and Sect. 5.

## 2 Preliminaries

### 2.1 Basic Concepts of Association Rule Mining

Let $\mathcal{O}$ be a dataset of transactions, $\mathcal{A}$ the set of items related to objects $o \in \mathcal{O}$ and $\mathcal{R}$ a binary relation on $\mathcal{O} \times \mathcal{A}$. A triple $(\mathcal{O}, \mathcal{A}, \mathcal{R})$ is called a data mining context. We consider two set functions: $\lambda : 2^{\mathcal{O}} \to 2^{\mathcal{A}}, \rho : 2^{\mathcal{A}} \to 2^{\mathcal{O}}$ as follows: $\forall \emptyset \neq A \subseteq \mathcal{A}, \emptyset \neq O \subseteq \mathcal{O} : \lambda(O) \equiv \{a \in \mathcal{A} : (o,a) \in \mathcal{R}, \forall o \in O\}, \rho(A) \equiv \{o \in \mathcal{O} : (o,a) \in \mathcal{R}, \forall a \in A\}$, where $2^{\mathcal{O}}, 2^{\mathcal{A}}$ are the classes of all subsets of $\mathcal{O}$ and $\mathcal{A}$. We denote $h(A) \equiv \lambda(\rho(A))$ as the closure of $A$. An itemset $A \subseteq \mathcal{A}$ is called closed iff [2] it is equal to its closure [13], i.e. $h(A) = A$. For $G, A : \emptyset \neq G \subseteq A \subseteq \mathcal{A}, G$ is called a generator [9] of $A$ if $h(G) = h(A)$ and $(\forall G' : \emptyset \neq G' \subset G \Rightarrow h(G') \subset h(G))$. The class of all generators of $A$ is named by $\mathcal{G}(A)$.

Let $s_0$ and $c_0$ be the minimum support and minimum confidence. For an itemset $S'$, the number $|\rho(S')|$ is called the support of $S'$, denoted by $supp(S')$. $S'$ is called frequent iff $supp(S') \geq s_0$ [1]. Let $\mathcal{CS}, \mathcal{FS}$ be the classes of all closed itemsets and of all frequent itemsets and $\mathcal{FCS} \equiv \mathcal{CS} \cap \mathcal{FS}$ the class of all frequent closed itemsets. For $\emptyset \neq L' \subset S'$ and $R' = S' \setminus L'$, we call $r : L' \to R'$ the rule determined by $L', R'$. The confidence of $r$ is defined by $conf(r) \equiv supp(S')/supp(L')$ and it is called an association rule iff $conf(r) \geq c_0$ and $supp(r) \equiv supp(S') \geq s_0$ [1]. The set of all association rules is denoted by $\mathcal{ARS}(s_0, c_0)$. The set of the ones with constraints is defined in (1).

### 2.2 Explicit Structures of Frequent Itemsets

For frequent closed itemset $L$, the equivalence class of the subsets of $L$ having the same closure $L$ is written by $[L]$. Formally, $[L] \equiv \{\emptyset \neq L' \subseteq L : h(L') = L\}$. Following theorems of 2 and 3 in [2], we have propositions 1, 2 for the structures of frequent itemsets having the same closures and the functions for deriving them non-repeatedly.

**Proposition 1 ([2]).** *For $L \in \mathcal{FCS}$:*

*1. $L' \in [L] \iff \exists L_i \in \mathcal{G}(L), L'' \subseteq L \setminus L_i : L' = L_i + L''$.*
*2. Let us call $L_U := \bigcup_{L_i \in \mathcal{G}(L)} L_i, L_{U,i} := L_U \setminus L_i, L_- := L \setminus L_U$ (A),*

$$\mathcal{FS}(L) := \{L_i + L_i' + L^{\sim} \mid L_i \in \mathcal{G}(L), L^{\sim} \subseteq L_-, L_i' \subseteq L_{U,i},$$
$$\nexists 1 \leq k < i : L_k \subset L_i + L_i'\}. \quad (2)$$

---

[2] We write "if and only if" simply as "iff".

*We have $[L] = \mathscr{FS}(L)$ and all itemsets of $\mathscr{FS}(L)$ are derived non-repeatedly.*

For $L, S \in \mathscr{FCS}, L \subseteq S, L' \in [L], Y := S \setminus L'$, let $\mathscr{M}(Y, L')$ be the class of the minimal elements of the class $\{S_k \setminus L' : S_k \in \mathscr{G}(S)\}$ [(B)]. The class of the sub-sets $R'$ of $Y$ in which the union of each of them with $L'$ has the closure $S$, is defined by $\lfloor Y \rfloor_{L'} \equiv \{R' \subseteq Y : h(L' + R') = S\}$.

**Proposition 2.** *For $L, S \in \mathscr{FCS}, L \subseteq S, L' \in [L], Y := S \setminus L':$*

1. $R' \in \lfloor Y \rfloor_{L'} \iff \exists S_0 \in \mathscr{G}(S), S_0' \subseteq S \setminus S_0 : R' = (S_0 + S_0') \setminus L'$.
2. *Assign that $R_U := \bigcup_{R \in \mathscr{M}(Y, L')} R$, $R_{U,k} := R_U \setminus R_k$, $R_- := Y \setminus R_U$* [(C)],

$$\mathscr{FS}(Y)_{L'} := \{R_k + R_k' + R^\sim \mid R_k \in \mathscr{M}(Y, L'), R^\sim \subseteq R_-, R_k' \subseteq R_{U,k},$$
$$\nexists 1 \leq j < k : R_j \subset R_k + R_k'\}. \quad (3)$$

*Thus, $\lfloor Y \rfloor_{L'} = \mathscr{FS}(Y)_{L'}$ and all itemsets of $\mathscr{FS}(Y)_{L'}$ are distinctly obtained.*

*Proof*

1. "$\Longrightarrow$": $R' \in \lfloor Y \rfloor_{L'}$ implies that $R' \subseteq Y, h(L' + R') = (S)$. There exists $S_0 \in \mathscr{G}(S) :$ $S_0 \subseteq L' + R'$. Thus, $S_0' := (L' + R') \setminus S_0 \subseteq S \setminus S_0$. Therefore, $R' = (S_0 + S_0') \setminus L'$.
   "$\Longleftarrow$": If $R' = (S_0 + S_0') \setminus L' = (S_0 \setminus L') + (S_0' \setminus L')$, where $S_0 \in \mathscr{G}(S), S_0' \subseteq S \setminus S_0$. Then, $R' \subseteq Y$, because $L' \cap Y = \emptyset$. Further, $S_0 = (S_0 \cap L') + (S_0 \setminus L') \subseteq L' + R'$. Thus, $h(S) = h(S_0) \subseteq h(L' + R') \subseteq h(S)$. Hence, $h(L' + R') = h(S)$.
2. – "$\subseteq$": If $R' \in \lfloor Y \rfloor_{L'}$, by statement 1, assume that $k$ is the minimum index such that $S_k \in \mathscr{G}(L' + Y)$ and $R_k = S_k \setminus L'$ is a minimal set, $R_k'' \subseteq (L' + Y) \setminus S_k : R' = (S_k + R_k'') \setminus L' = R_k + (R_k'' \setminus L')$. Let $R_k' = (R_k'' \setminus L') \cap R_U, R^\sim = (R_k'' \setminus L') \setminus R_U$. Then $R_k' \subseteq R_{U,k}, R^\sim \subseteq R_-$ and $R' = R_k + R_k' + R^\sim$. Assume that there exists $j$ such that $1 \leq j < k, R_j \in \mathscr{M}(Y, L')$ and $R_j \subseteq R_k + R_k'$. Then, $R' = R_j + R_j''$, where $R_j'' = R_j' + R^\sim$ and $R_j' = (R_k + R_k') \setminus R_j = (R_k \setminus R_j) + (R_k' \setminus R_j) \subseteq (L' + Y) \setminus S_j$. Therefore, $R_j'' \subseteq (L' + Y) \setminus S_j$: It contradicts to the selection of the index $k$! Thus, $R' \in \mathscr{FS}(Y)_{L'}$.
   – "$\supseteq$": If $Y' \in \mathscr{FS}(Y)_{L'}$, there exists $R_k = S_k \setminus L' \in \mathscr{M}(Y, L'), S_k \in \mathscr{G}(L' + Y), R_k' \subseteq R_{U,k}, R^\sim \subseteq R_-$ such that $R' = R_k + R_k' + R^\sim \subseteq Y$. If $R' \cap L' = \emptyset$, we have $R_k' + R^\sim = R' \setminus R_k = R' \setminus S_k$ and $R' = (S_k \setminus L') + (R' \setminus S_k)$. Otherwise, $L' \cap Y = \emptyset$ implies that $R' \cap S_k = S_k \setminus L' \subseteq R', S_k = (S_k \cap L') + (S_k \setminus L') \subseteq L' + R' \subseteq L' + Y$ and $h(L' + Y) = h(S_k) \subseteq h(L' + R') \subseteq h(L' + Y)$. Hence, $h(L' + R') = h(L' + Y)$. Then, $R' \in \lfloor Y \rfloor_{L'}$.
   – To prove the left, we assume that there exist $k, j$ such that $1 \leq j < k$ and $R_k + R_k' + R^\sim_k \equiv R_j + R_j' + R^\sim_j, R_k, R_j \in \mathscr{M}(Y, L'), R^\sim_k, R^\sim_j \subseteq R_-, R_k' \subseteq R_{U,k}, R_j' \subseteq R_{U,j}$. Since $R_j \cap R^\sim_k = \emptyset, R_j \subset R_k + R_k'$: a contradiction! Therefore, all itemsets of $\mathscr{FS}(Y)_{L'}$ are distinctly derived. $\square$

## 3 Mining Association Rules with Constraints

### 3.1 Partitioning Association Rule Set with Constraints

It is known in [3] that the association rule set $\mathscr{ARS}(s_0, c_0)$ is partitioned into equivalence classes $\mathscr{AR}(L, S)$ for $(L, S) \in \mathscr{NFCS}(s_0, c_0) \equiv \{(L, S) \in \mathscr{FCS} \times \mathscr{FCS} \mid L \subseteq S, c_0 \leq (supp(S)/supp(L))\}$ where

$$\mathscr{AR}(L, S) \equiv \{r : L' \to R' \mid h(L') = L, h(L' + R') = S\}. \tag{4}$$

**Definition 1.** For $(L, S) \in \mathscr{NFCS}(s_0, c_0)$, the set of association rules in $\mathscr{AR}(L, S)$ intersected with constraints is defnided by:

$$\mathscr{AR}_{\cap \mathscr{GT}, \mathscr{KL}}(L, S) \equiv \{r : L' \to R' \in \mathscr{AR}(L, S) \mid L' \cap \mathscr{GT} \neq \emptyset, R' \cap \mathscr{KL} \neq \emptyset\}.$$

We have easily Theorem 3 that helps us to avoid the duplication in the mining rules intersected with constraints. Without loss of the generality, we consider only the mining independently each rule class with constraints $\mathscr{AR}_{\cap \mathscr{GT}, \mathscr{KL}}(L, S)$ of the same support $supp(S)$ and the same confidence $(supp(S)/supp(L))$.

**Theorem 1 (Partitioning Association Rule Set with Constraints)**

$$\mathscr{ARS}_{\cap \mathscr{GT}, \mathscr{KL}}(s_0, c_0) = \sum_{(L, S) \in \mathscr{NFCS}(s_0, c_0))} \mathscr{AR}_{\cap \mathscr{GT}, \mathscr{KL}}(L, S).$$

**Definition 2.** For $L \in \mathscr{FCS}$, we define

$$[L]_{\cap \mathscr{GT}} \equiv \{L' \in [L] : L' \cap \mathscr{GT} \neq \emptyset\}.$$

For $(L, S) \in \mathscr{NFCS}(s_0, c_0), L' \in [L], Y := S \setminus L'$ :

$$\lfloor Y \rfloor_{L', \cap \mathscr{KL}} \equiv \{R' \in \lfloor Y \rfloor_{L'} : R' \cap \mathscr{KL} \neq \emptyset\}.$$

Since $\mathscr{AR}(L, S) = \{r : L' \to R' \mid L' \in [L], R' \in \lfloor S \setminus L' \rfloor_{L'}\}$, we have:

$$\mathscr{AR}_{\cap \mathscr{GT}, \mathscr{KL}}(L, S) = \{r : L' \to R' \mid L' \in [L]_{\cap \mathscr{GT}}, R' \in \lfloor S \setminus L' \rfloor_{L', \cap \mathscr{KL}}\}. \tag{5}$$

### 3.2 Post-processing Mining Approach

Using the derivation functions of $\mathscr{FS}(L)$ and $\mathscr{FS}(S \setminus L')_{L'}$ (given in propositions of 1 and 2) we can generate distinctly the rules $r' : L' \to R'$ such that $L' \in [L]$ and $R' \in \lfloor S \setminus L' \rfloor_{L'}$. Then, we choose the ones satisfying the constraints.

The corresponding algorithm, namely *IntARS-PostPro*, is shown in Fig. 1. Though it does not make any duplication in the execution, it runs slowly since two reasons. First, we need to compute the intersections of two rule sides with $\mathscr{GT}$ and $\mathscr{KL}$. Second, there are many generated redundant rule candidates.

```
IntARS-PostPro (GT, KL, (L, S))
1. AR∩GT,KL(L, S) := ∅;
2. for each L' ∈ FS(L) do
3.    if L' ∩ GT ≠ ∅ then
4.       Y := S\L';
5.       for each R' ∈ FS(Y)L' do
6.          if R' ∩ KL ≠ ∅ then
7.             AR∩GT,KL(L, S) := AR∩GT,KL(L, S) + {L'→R'};
8. return AR∩GT,KL(L, S);
```

**Fig. 1** The algorithm *IntARS-PostPro*

## 3.3 Our Approach

To overcome those limitations, we propose the explicit representations for two sides of the rules in order to mine quickly them without testing the constraints. The one $\mathscr{FS}_{\cap \mathscr{GT}}(L)$, for generating non-repeatedly and directly the left sides $L'$ of the rules intersected with $\mathscr{GT}$ ($\mathscr{FS}_{\cap \mathscr{GT}}(L) = [L]_{\mathscr{GT}}$), was shown in [4]. Here, we give the function $\mathscr{FS}_{\cap \mathscr{KL}}(S \setminus L')_{L'}$ for generating the right ones.

We have immediately that $\mathscr{AR}_{\cap \mathscr{GT}, \mathscr{KL}}(L, S) = \emptyset$ when $L \cap \mathscr{GT} = \emptyset$ or $S \cap \mathscr{KL} = \emptyset$. Hence, we just mine the class $\mathscr{AR}_{\cap \mathscr{GT}, \mathscr{KL}}(L, S)$ for $(L, S) \in \mathscr{NFCS}_{\cap \mathscr{GT}, \mathscr{KL}}(s_0, c_0) \equiv \{(L, S) \in \mathscr{NFCS}(s_0, c_0) \mid L \in \mathscr{FCS}_{\cap \mathscr{GT}}, S \in \mathscr{FCS}_{\cap \mathscr{KL}}\}$ where: $\mathscr{FCS}_{\cap X} \equiv \{S \in \mathscr{FCS}, S \cap X \neq \emptyset\}$, with $X \subseteq \mathscr{A}$.

### 3.3.1 Mining the Right Sides of the Rules for $L = S$

Assign that $Y := L \setminus L'$, from definition 2, we have $\lfloor Y \rfloor_{L', \cap \mathscr{KL}} = \{R' \subseteq Y : R' \cap \mathscr{KL} \neq \emptyset\}$. Let us define the derivation function $\mathscr{FS}_{\cap \mathscr{KL}}(Y)_{L'}$ by:

$$\mathscr{FS}_{\cap \mathscr{KL}}(Y)_{L'} \equiv 2^{Y \setminus \mathscr{KL}} \oplus (2^{Y \cap \mathscr{KL}} \setminus \{\emptyset\}) \tag{6}$$

where $\mathscr{X} \oplus \mathscr{L} := \{A + B : \emptyset \neq A \in \mathscr{X}, \emptyset \neq B \in \mathscr{L}\}$ for $\mathscr{X}, \mathscr{L} \subseteq 2^{\mathscr{A}} \setminus \{\emptyset\}, \mathscr{X} \cap \mathscr{L} = \emptyset$.

**Theorem 2.** *For* $(L, L) \in \mathscr{NFCS}_{\cap \mathscr{GT}, \mathscr{KL}}(s_0, c_0), Y := L \setminus L'$, *we have:*
$$\lfloor Y \rfloor_{L', \cap \mathscr{KL}} = \mathscr{FS}_{\cap \mathscr{KL}}(Y)_{L'}$$
*and the fact that all itemsets of* $\mathscr{FS}_{\cap \mathscr{KL}}(Y)_{L'}$ *are generated non-repeatedly.*

*Proof.* It is obvious since $Y = (Y \cap \mathscr{KL}) + (Y \setminus \mathscr{KL})$. □

### 3.3.2 Mining the Right Sides of the Rules for $L \subset S$

For $Y := S \setminus L'$, we split $\lfloor Y \rfloor_{L', \cap \mathscr{KL}}$ into two disjoint parts of $\lfloor Y \rfloor^1_{L', \cap \mathscr{KL}}$ and $\lfloor Y \rfloor^2_{L', \cap \mathscr{KL}}$. The first one contains frequent itemsets created from the generators of

$S$ which do not contain any item of $\mathcal{KL}$ and non-empty subsets of the intersection of $Y$ with $\mathcal{KL}$. The remaining part is generated from the generators involved with $\mathcal{KL}$. It follows that we can avoid the intersection of itemsets with $\mathcal{KL}$. Two functions of $\mathcal{FS}^1_{\cap \mathcal{KL}}(Y)_{L'}$ and $\mathcal{FS}^2_{\cap \mathcal{KL}}(Y)_{L'}$ are obtained for deriving distinctly all itemsets of $\lfloor Y \rfloor^1_{L', \cap \mathcal{KL}}$ and $\lfloor Y \rfloor^2_{L', \cap \mathcal{KL}}$ accordingly.

We split the class $\mathcal{M}(Y, L')$ (see $^{(B)}$) into $\mathcal{M}_{\neg \mathcal{KL}}(Y, L') \equiv \{M_k \in \mathcal{M}(Y, L') : M_k \cap \mathcal{KL} = \emptyset\}$ and $\mathcal{M}_{\cap \mathcal{KL}}(Y, L') \equiv \mathcal{M}(Y, L') \setminus \mathcal{M}_{\neg \mathcal{KL}}(Y, L')$. All m elements in $\mathcal{M}_{\neg \mathcal{KL}}(Y, L')$ are numbered as $R_1, R_2, .., R_m$. The ones in $\mathcal{M}_{\cap \mathcal{KL}}(Y, L')$ are $R_{m+1}, R_{m+2}, .., R_M$ (where $M := |\mathcal{M}(Y, L')|$) $^{(D)}$.

Deriving $\lfloor Y \rfloor^1_{L', \cap \mathcal{KL}}$ by $\mathcal{FS}^1_{\cap \mathcal{KL}}(Y)_{L'}$.

We define: $\lfloor Y \rfloor_{L', \neg \mathcal{KL}} \equiv \{R' \in \lfloor Y \rfloor_{L'} : R' \cap \mathcal{KL} = \emptyset\}$ and

$$\lfloor Y \rfloor^1_{L', \cap \mathcal{KL}} \equiv \{R'' + V, R'' \in \lfloor Y \rfloor_{L', \neg \mathcal{KL}}, \emptyset \neq V \subseteq Y \cap \mathcal{KL}\}. \tag{7}$$

We can apply first $\mathcal{FS}(Y)_{L'}$ (in (3)) to derive the itemsets $R''$ in $\lfloor Y \rfloor_{L'}$ and then choose the ones that have empty intersections with $\mathcal{KL}$ in order to obtain $\lfloor Y \rfloor^1_{L', \cap \mathcal{KL}}$. However, the difficulties similar to the post-processing approach also come. Using the function $\mathcal{FS}_{\neg \mathcal{KL}}(Y)_{L'}$ in (8) we can generate them directly.

**Lemma 1.** *For* $R_k \in \mathcal{M}_{\neg \mathcal{KL}}(Y, L')$, $R_{U, \neg \mathcal{KL}} := \bigcup_{R_k \in \mathcal{M}_{\neg \mathcal{KL}}(Y, L')} R_k$, $R_{\_, \neg \mathcal{KL}} := (Y \setminus \mathcal{KL}) \setminus R_{U, \neg \mathcal{KL}}$, $R_{U, \neg \mathcal{KL}, k} := R_{U, \neg \mathcal{KL}} \setminus R_k$ $^{(E)}$,

$$\mathcal{FS}_{\neg \mathcal{KL}}(Y)_{L'} := \{R_k + R'_k + R^{\frown} : R_k \in \mathcal{M}_{\neg \mathcal{KL}}(Y, L'), R'_k \subseteq R_{U, \neg \mathcal{KL}, k}, R^{\frown} \subseteq R_{\_, \neg \mathcal{KL}}, \nexists 1 \leq j < k : R_j \subset R_k + R'_k, R_j \in \mathcal{M}_{\neg \mathcal{KL}}(Y, L')\}, \tag{8}$$

1. *$\lfloor Y \rfloor_{L', \neg \mathcal{KL}} = \mathcal{FS}_{\neg \mathcal{KL}}(Y)_{L'}$.*
2. *All itemsets of $\mathcal{FS}_{\neg \mathcal{KL}}(Y)_{L'}$ come non-repeatedly.*

*Proof*

1. – "$\subseteq$": Following from Proposition 2.1, for $R' \in \lfloor Y \rfloor_{L', \neg \mathcal{KL}}$, there exists $k$, such that $S_k \in \mathcal{G}(S), S''_k \subseteq S \setminus S_k : R' = (S_k + S''_k) \setminus L' = S_k \setminus L' + S''_k \setminus L'$ (because of $S_k \cap S''_k = \emptyset$). Hence, $R' = R_k + (S''_k \setminus L')$ for $R_k \in \mathcal{M}(Y, L')$. Since $R' \cap \mathcal{KL} = \emptyset$, $R_k \cap \mathcal{KL} = \emptyset$, i.e. $R_k \in \mathcal{M}_{\neg \mathcal{KL}}(Y, L')$. Let $R'_k = (S''_k \setminus L') \cap R_{U, \neg \mathcal{KL}}, R^{\frown} = (S''_k \setminus L') \setminus R_{U, \neg \mathcal{KL}}$. Then, $R'_k \subseteq R_{U, \neg \mathcal{KL}, k}, R^{\frown} \subseteq R' \setminus R_{U, \neg \mathcal{KL}} \subseteq R_{\_, \neg \mathcal{KL}}$. Thus, $R' = R_k + R'_k + R^{\frown}$. Assume that there exists the index $j$ such that $1 \leq j < k, R_j \in \mathcal{M}_{\neg \mathcal{KL}}(Y, L'), R_j \subset R_k + R'_k$. Therefore, $R' = R_j + R''_j$ for $R''_j = (R_k + R'_k) \setminus R_j + R^{\frown}$. Since $(R_k + R'_k) \setminus R_j = (R_k \setminus R_j) + (R'_k \setminus R_j) \subseteq (L' + Y) \setminus S_j$, we have $R''_j \subseteq (L' + Y) \setminus S_j$. This contradicts to how we select index $k$! Hence, $R' \in \mathcal{FS}_{\neg \mathcal{KL}}(Y)_{L'}$.
   – "$\supseteq$": For $R' = R_k + R'_k + R^{\frown} \in \mathcal{FS}_{\neg \mathcal{KL}}(Y)_{L'}$. Since $R' \cap L' = \emptyset, R'_k + R^{\frown} = (R'_k + R^{\frown}) \setminus L'$. Then, $R' = S_k \setminus L' + (R'_k + R^{\frown}) = S_k \setminus L' + (R'_k + R^{\frown}) \setminus L'$. Based on Proposition 2.1, $R' \in \lfloor Y \rfloor_{L'}$ because of $R'_k + R^{\frown} \subseteq S \setminus S_k$. Further, since $R' \cap \mathcal{KL} = \emptyset$ ($R_{U, \neg \mathcal{KL}} \cap \mathcal{KL} = \emptyset, R_{\_, \neg \mathcal{KL}} \cap \mathcal{KL} = \emptyset$), $R' \in \lfloor Y \rfloor_{L', \neg \mathcal{KL}}$.

2. Assume that there exists $k, j$ such that $1 \leq j < k$ and $R_k + R'_k + R\tilde{}_k \equiv R_j + R'_j + R\tilde{}_j$, where: $R_j, R_k \in \mathscr{M}_{\neg \mathscr{KL}}(Y, L')$, $R\tilde{}_k, R\tilde{}_j \subseteq R_{\neg, \neg \mathscr{KL}}$, $R'_k \subseteq R_{U, \neg \mathscr{KL}, k}, R'_j \subseteq R_{U, \neg \mathscr{KL}, j}$. Since $R_j \cap R\tilde{}_k = \emptyset$, $R_j \subset R_k + R'_k$. It is not how we select index $k$! □

**Proposition 3 (Deriving directly, distinctly all itemsets of $\lfloor Y \rfloor^1_{L', \cap \mathscr{KL}}$).** *For*

$$\mathscr{FS}^1_{\cap \mathscr{KL}}(Y)_{L'} \equiv \{R' = R'' + V, R'' \in \mathscr{FS}_{\neg \mathscr{KL}}(Y)_{L'}, \emptyset \neq V \subseteq Y \cap \mathscr{KL}\}, \quad (9)$$

*1. $\mathscr{FS}^1_{\cap \mathscr{KL}}(Y)_{L'} \subseteq \lfloor Y \rfloor_{L', \cap \mathscr{KL}}$.*
*2. All itemsets of $\mathscr{FS}^1_{\cap \mathscr{KL}}(Y)_{L'}$ are generated non-repeatedly.*

*Proof.* Based on Lemma 1, (7) and the fact that $\lfloor Y \rfloor^1_{L', \cap \mathscr{KL}} \subseteq \lfloor Y \rfloor_{L', \cap \mathscr{KL}}$. □

Deriving $\lfloor Y \rfloor^2_{L', \cap \mathscr{KL}}$ by $\mathscr{FS}^2_{\cap \mathscr{KL}}(Y)_{L'}$.

We describe the class of right sides coming from the generators of $S$ involved to $\mathscr{KL}$ by:

$$\lfloor Y \rfloor^2_{L', \cap \mathscr{KL}} \equiv \{R' \in \lfloor Y \rfloor_{L', \cap \mathscr{KL}} \mid \exists R_k \in \mathscr{M}_{\cap \mathscr{KL}}(Y, L') : R' \supseteq R_k\}.$$

**Proposition 4 (Deriving directly, distinctly all itemsets of $\lfloor Y \rfloor^2_{L', \cap \mathscr{KL}}$).** *Using the notations* [(C)], *for*

$$\mathscr{FS}^2_{\cap \mathscr{KL}}(Y)_{L'} \equiv \{R_k + R'_k + R\tilde{} : R_k \in \mathscr{M}_{\cap \mathscr{KL}}(Y, L'), R'_k \subseteq R_{U, k}, R\tilde{} \subseteq R_{\neg},$$
$$\nexists 1 \leq j < k : R_j \subset R_k + R'_k, R_j \in \mathscr{M}(Y, L') \ ^{[(F)]}\}, \quad (10)$$

*we hold:*

*1. $\mathscr{FS}^2_{\cap \mathscr{KL}}(Y)_{L'} \subseteq \lfloor Y \rfloor_{L', \cap \mathscr{KL}}$*
*2. All elements of $\mathscr{FS}^2_{\cap \mathscr{KL}}(Y)_{L'}$ are derived distinctly.*

*Proof.* Similar to the proof of Lemma 1. □

The derivation function $\mathscr{FS}_{\cap \mathscr{KL}}(Y)_{L'}$.

Defining
$$\mathscr{FS}_{\cap \mathscr{KL}}(Y)_{L'} \equiv \mathscr{FS}^1_{\cap \mathscr{KL}}(Y)_{L'} + \mathscr{FS}^2_{\cap \mathscr{KL}}(Y)_{L'}, \quad (11)$$

we have Theorem 3 for generating efficiently all rules intersected with constraints.

**Theorem 3.** *For $(L, S) \in \mathscr{NFCS}_{\cap \mathscr{GF}, \mathscr{KL}}(s_0, c_0), L \subset S, L' \in [L], Y := S \setminus L':$*

*1. $\mathscr{FS}^1_{\cap \mathscr{KL}}(Y)_{L'} \cap \mathscr{FS}^2_{\cap \mathscr{KL}}(Y)_{L'} = \emptyset$, and $\lfloor Y \rfloor_{L', \cap \mathscr{KL}} = \mathscr{FS}_{\cap \mathscr{KL}}(Y)_{L'}$.*
*2. All itemsets of $\mathscr{FS}_{\cap \mathscr{KL}}(Y)_{L'}$ come distinctly.*

*Proof*

1. – If there exist $R'_1 \in \mathscr{F}\mathscr{S}^1_{\cap\mathscr{K}\mathscr{L}}(Y)_{L'}, R'_2 \in \mathscr{F}\mathscr{S}^2_{\cap\mathscr{K}\mathscr{L}}(Y,L')$ such that $R'_1 \equiv R'_2$. Thus, $R'_1 = R_{j_1} + R'_{j_1} + R^{\sim}_1 + V \equiv R_{k_2} + R'_{k_2} + R^{\sim}_2 = R'_2$ where: $k_2 > j_1$ (since $^{(\mathbf{D})}$), $R_{j_1} \in \mathscr{M}_{\neg\mathscr{K}\mathscr{L}}(Y,L'), R'_{j_1} \subseteq R_{U,\neg\mathscr{K}\mathscr{L},j_1}, R^{\sim}_1 \subseteq R_{-,\neg\mathscr{K}\mathscr{L}}, \emptyset \neq V \subseteq Y \cap \mathscr{K}\mathscr{L}, R_{k_2} \in \mathscr{M}_{\cap\mathscr{K}\mathscr{L}}(Y,L'), R'_{k_2} \subseteq R_{U,k_2}, R^{\sim}_2 \subseteq R_-$ (used $^{(\mathbf{E})}$ and $^{(\mathbf{C})}$). Since $R_{j_1} \subseteq R_{U,\neg\mathscr{K}\mathscr{L}} \subseteq R_U$, we have immediately $R_{j_1} \cap R^{\sim}_2 = \emptyset$. Thus, $R_{j_1} \subset R_{k_2} + R'_{k_2}$. That contradicts to $^{(\mathbf{F})}$. Hence, $\mathscr{F}\mathscr{S}^1_{\cap\mathscr{K}\mathscr{L}}(Y)_{L'} \cap \mathscr{F}\mathscr{S}^2_{\cap\mathscr{K}\mathscr{L}}(Y)_{L'} = \emptyset$.
   – The fact that $\lfloor Y \rfloor_{L',\cap\mathscr{K}\mathscr{L}} \supseteq \mathscr{F}\mathscr{S}_{\cap\mathscr{K}\mathscr{L}}(Y)_{L'}$ follows from (11), Proposition 3.1 and Proposition 4.1. What is left is to show that $\lfloor Y \rfloor_{L',\cap\mathscr{K}\mathscr{L}} \subseteq \mathscr{F}\mathscr{S}^1_{\cap\mathscr{K}\mathscr{L}}(Y)_{L'} + \mathscr{F}\mathscr{S}^2_{\cap\mathscr{K}\mathscr{L}}(Y)_{L'}$. We use the notations given at $^{(\mathbf{E})}$ and $^{(\mathbf{C})}$. For every $R' \in \lfloor Y \rfloor_{L',\cap\mathscr{K}\mathscr{L}}$, from Proposition 2.2, $R' = R_k + R'_k + R^{\sim}, R_k \in \mathscr{M}(Y,L'), R'_k \subseteq R_{U,k}, R^{\sim} \subseteq R_-$ and $R' \cap \mathscr{K}\mathscr{L} \neq \emptyset$: [Case 1] If $R_k \in \mathscr{M}_{\neg\mathscr{K}\mathscr{L}}(Y,L')$, then $(R'_k + R^{\sim}) \cap \mathscr{K}\mathscr{L} \neq \emptyset$. Let us call $R''_k = R'_k \cap R_{U,\neg\mathscr{K}\mathscr{L}} \subseteq R_{U,\neg\mathscr{K}\mathscr{L},k}, R'''_k = (R'_k \setminus R_{U,\neg\mathscr{K}\mathscr{L}}) \setminus \mathscr{K}\mathscr{L} \subseteq R_{-,\neg\mathscr{K}\mathscr{L}}, R''''_k = (R'_k \setminus R_{U,\neg\mathscr{K}\mathscr{L}}) \cap \mathscr{K}\mathscr{L} \subseteq R \cap \mathscr{K}\mathscr{L}, R^{\sim}_{\neg\mathscr{K}\mathscr{L}} = R^{\sim} \setminus \mathscr{K}\mathscr{L} \subseteq R_{-,\neg\mathscr{K}\mathscr{L}}$, and $R^{\sim}_{\cap\mathscr{K}\mathscr{L}} = R^{\sim} \cap \mathscr{K}\mathscr{L} \subseteq Y \cap \mathscr{K}\mathscr{L}$. Clearly, $\emptyset \neq (R'_k + R^{\sim}) \cap \mathscr{K}\mathscr{L} = R''''_k + R^{\sim}_{\cap\mathscr{K}\mathscr{L}}$. Now, $R' = [R_k + R''_k + (R'''_k + R^{\sim}_{\neg\mathscr{K}\mathscr{L}})] + (R''''_k + R^{\sim}_{\cap\mathscr{K}\mathscr{L}})$ for $R'''_k + R^{\sim}_{\neg\mathscr{K}\mathscr{L}} \subseteq R_{-,\neg\mathscr{K}\mathscr{L}}$ and $\emptyset \neq (R''''_k + R^{\sim}_{\cap\mathscr{K}\mathscr{L}}) \subseteq Y \cap \mathscr{K}\mathscr{L}$. Therefore, $R' \in \mathscr{F}\mathscr{S}^1_{\cap\mathscr{K}\mathscr{L}}(Y)_{L'}$. [Case 2] Otherwise, $R' \in \mathscr{F}\mathscr{S}^2_{\cap\mathscr{K}\mathscr{L}}(Y)_{L'}$.
2. Following directly from (11) and propositions of 3.2 and 4.2. $\qquad\square$

**Theorem 4 (Mining directly, distinctly all rules with constraints for each class).**
*For* $(L,S) \in \mathscr{N}\mathscr{F}\mathscr{C}\mathscr{S}_{\cap\mathscr{G}\mathscr{T},\mathscr{K}\mathscr{L}}(s_0,c_0), \mathscr{A}\mathscr{R}^*_{\cap\mathscr{G}\mathscr{T},\mathscr{K}\mathscr{L}}(L,S) \equiv \{r : L' \to R' \mid L' \in \mathscr{F}\mathscr{S}_{\cap\mathscr{G}\mathscr{T}}(L), R' \in \mathscr{F}\mathscr{S}_{\cap\mathscr{K}\mathscr{L}}(S \setminus L')_{L'}\}$, *the following statements hold true:*

1. $\mathscr{A}\mathscr{R}^*_{\cap\mathscr{G}\mathscr{T},\mathscr{K}\mathscr{L}}(L,S) = \mathscr{A}\mathscr{R}_{\cap\mathscr{G}\mathscr{T},\mathscr{K}\mathscr{L}}(L,S)$.
2. *All rules of* $\mathscr{A}\mathscr{R}^*_{\cap\mathscr{G}\mathscr{T},\mathscr{K}\mathscr{L}}(L,S)$ *are derived non-repeatedly.*

*Proof.* Directly from (5), Theorem 3 in [4], and theorems 2, 3. $\qquad\square$

The algorithm for mining efficiently $\mathscr{A}\mathscr{R}_{\cap\mathscr{G}\mathscr{T},\mathscr{K}\mathscr{L}}(L,S)$ is posted in Fig. 2.

### 3.3.3  An Example

For illustrating the approach, we consider the mining association rules with constraints of $\mathscr{G}\mathscr{T} = 1, \mathscr{K}\mathscr{L} = 7$ on dataset $\mathscr{T}_1$ shown in Table 1 for $s_0 = 2, c_0 = 0.5$. The lattice of frequent closed itemsets (underlined) together their generators (italicized) and supports (superscripted) is shown in Fig. 3.

We observe the mining on the class $\mathscr{A}\mathscr{R}_{\cap 1,7}(12467,12467)$ containing the rules with the same support 2 and the same confidence 1. For $L' = 16 \in [12467]_{\cap 1}, Y = 12467 \setminus 16 = 247$. Then, $Y \cap \mathscr{K}\mathscr{L} = 7$ and $Y \setminus \mathscr{K}\mathscr{L} = 24$. Hence, we discovered constrained rules: $16 \to 7 + \emptyset, 16 \to 7 + 2, 16 \to 7 + 4, 16 \to 7 + 24$.

In the case of $L \subset S$, we consider the rule class $\mathscr{A}\mathscr{R}_{\cap 1,7}(12467,12)$ with the confidence 0.5. For $L' = 1 \in [12]_{\cap 1}, Y = 2467$. Therefore, $\mathscr{M}(Y,L') = Minimal\{16 \setminus 1, 17 \setminus 1, 247 \setminus 1, 26 \setminus 1\} = \{6,7\}$. Thus, $\mathscr{M}_{\neg 7}(Y,L') = \{6\}$. It follows from $R_{U,\neg 7} = 6$ that $R_{U,\neg 7,1} = \emptyset$ and $R_{-,\neg 7} = (2467 \setminus 7) \setminus 6 = 24$. Then, $\mathscr{F}\mathscr{S}_{\neg 7}(2467)_1 = \{6 +$

```
IntARS-OurApp (GT, KL, (L, S))
1. AR*∩GT,KL(L, S) := ∅;
2. for each L' ∈ FS∩GT(L) do
3.    Y := S\L';
4.    if (L = S) then
5.       for each R'' ⊆ Y∩KL and ∅ ≠ R''' ⊆ Y\KL do
6.          AR*∩GT,KL(L, S) := AR*∩GT,KL(L, S) + {L'→R''+R'''};
7.    else
8.       for each R' ∈ FS∩KL(Y)L' do
9.          AR*∩GT,KL(L, S) := AR*∩GT,KL(L, S) + {L'→R'};
10. return AR*∩GT,KL(L, S);
```

**Fig. 2** The algorithm *IntARS-OurApp*

**Table 1** Dataset $\mathcal{T}_1$

| Trans | Items |
|-------|-------|
| 1 | 0 1 2   4   6 7 |
| 2 | 0   2 3   5   7 |
| 3 | 0       3 4 5 6 7 |
| 4 |     1 2   4 5 6 7 |
| 5 |     1 2   4 |
| 6 |     1 2 |

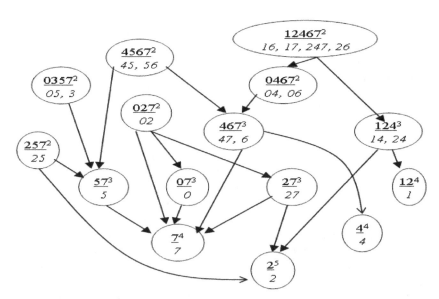

**Fig. 3** The lattice of frequent closed itemsets, their generators and supports from $\mathcal{T}_1$

$0+0, 6+0+2, 6+0+4, 6+0+24\}$. Since $Y \cap \mathcal{KL} = 7, \mathcal{FS}^1_{\cap 7}(2467)_1 = \{6 + 7, 62 + 7, 64 + 7, 624 + 7\}$, the rules of $r_1 : 1 \to 67, r_2 : 1 \to 627, r_3 : 1 \to 647$ and $r_4 : 1 \to 6247$ are discovered. We consider the right sides (in $\mathcal{FS}^2_{\cap 7}(2467)_1$) coming from $\mathcal{M}_{\cap 7} = \{7\}$. It is easy to know that $R_U = 67, R_{U,2} = 67 \setminus 7 = 6$ and $R_- = 2467 \setminus 67 = 24$. For $R'_2 = 0$, we have the rules of $1 \to 7 + 0 + 0, 1 \to 7 + 0 + 2, 1 \to 7 + 0 + 4$ and $1 \to 7 + 0 + 24$. For $R'_2 = 6$, since $R_1 = 6 \subset R_2 + R'_2 = 7 + 6$, we pass the repeatedly generation for the rules $r_1, r_2, r_3$ and $r_4$.

## 4 Experimental Results

The following experiments were performed on i5-2400 CPU, 3.10 GHz @ 3.09 GHz, 3.16 GB RAM, running Linux (Cygwin). The algorithms were coded in $C^{++}$. Two highly correlated datasets of *Mushroom* and *C20d10k*, coming from http://fimi.cs.helsinki.fi/data/, are used during these experiments. *Mushroom* describes the characteristics of the mushrooms. It includes 8124 transactions of 119 items. *C20d10k* is a census dataset from the PUMS sample file and includes 100000 transactions of 385 items.

Given minimum support $s_0$, *Charm-L* [12] and *MinimalGenerators* [11] are executed to mine from the dataset the lattice $\mathcal{FCS}$ of frequent closed itemsets (together their generators). For $s_0$, the constraints are selected from the set $\mathcal{A}^{\mathcal{F}}$ of all frequent items of $\mathcal{A}$ with the sizes of $K * |\mathcal{A}^{\mathcal{F}}|$ for $K = \frac{1}{8}, \frac{2}{8}, \frac{3}{8}$ and $\frac{1}{2}$. Since the

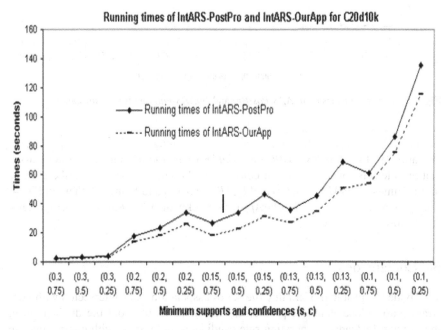

**Fig. 4** The running times of *IntARS-PostPro* and *IntARS-OurApp* for *C20d10k*

users are interested in the high-support items, we sort all items by the ascending order of their supports and fix a support threshold $H$ such that $|\{f \in \mathscr{A}^{\mathscr{F}} : supp(f) \geq H\}| \approx \frac{1}{2} * |\mathscr{A}^{\mathscr{F}}|$. The set $\mathscr{C}$ having the size $L$ of frequent items are constructed randomly by two subsets of $\mathscr{C}_1$ and $\mathscr{C}_2$ where $\mathscr{C}_1$ contains $P * L$ ($P := \frac{2}{3}$) high-support items (whose supports are greater than or equal to $H$) and $\mathscr{C}_2$ contains the remaining ones. Then $\mathscr{C}$ randomly splits into two disjoint subsets of $\mathscr{G}\mathscr{T}$ and $\mathscr{K}\mathscr{L}$. Thus, we consider eight pairs of ($\mathscr{G}\mathscr{T}$, $\mathscr{K}\mathscr{L}$) for each $s_0$.

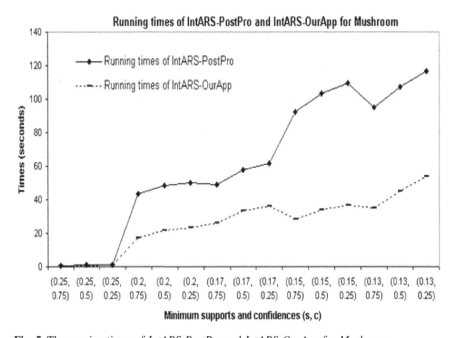

**Fig. 5** The running times of *IntARS-PostPro* and *IntARS-OurApp* for *Mushroom*

For each $(s_0, c_0)$, we traverse $\mathscr{F}\mathscr{C}\mathscr{S}$ for obtaining $\mathscr{N}\mathscr{F}\mathscr{C}\mathscr{S}_{\cap\mathscr{G}\mathscr{T},\mathscr{K}\mathscr{L}}(s_0, c_0)$ and apply in turn *IntARS-PostPro* and *IntARS-OurApp* to mine constrained association rule sets for eight pairs of constraints. We take in our account the average mining times of *IntARS-PostPro* and *IntARS-OurApp* and figure out them in Fig. 4, Fig. 5. They show that *IntARS-OurApp* runs quickly than *IntARS-PostPro*, especially for the low values of $s_0, c_0$.

## 5   Conclusions

We divide the problem of mining the set of association rules intersected with constraint into independent sub-problems. That helps us to avoid the duplication in the mining. Instead of generating rule candidates and testing with constraints, we find the explicit representations of them. Based on those representations, we design the algorithm *IntARS-OurApp* for mining rules. The tests on benchmark datasets showed its efficiency. The approach opens a direction to solve the tasks of mining association rules with the different types of constraint itemsets.

# References

1. Agrawal, R., Srikant, R.: Fast algorithms for mining association rules. In: Proceeding of the 20th International Conference on Very Large Data Bases, pp. 478–499 (1994)
2. Anh, T., Hai, D., Tin, T., Bac, L.: Efficient Algorithms for Mining Frequent Itemsets with Constraint. In: Proceedings of the Third International Conference on Knowledge and Systems Engineering, pp. 19–25 (2011)
3. Tran, A., Truong, T., Le, B.: Structures of Association Rule Set. In: Pan, J.-S., Chen, S.-M., Nguyen, N.T. (eds.) ACIIDS 2012, Part II. LNCS, vol. 7197, pp. 361–370. Springer, Heidelberg (2012)
4. Tran, A., Duong, H., Truong, T., Le, B.: Mining Frequent Itemsets with Dualistic Constraints. In: Anthony, P., Ishizuka, M., Lukose, D. (eds.) PRICAI 2012. LNCS (LNAI), vol. 7458, pp. 807–813. Springer, Heidelberg (2012)
5. Bonchi, F., Lucchese, C.: On closed constrained frequent pattern mining. In: Proc. IEEE ICDM 2004 (2004)
6. Boulicaut, J.F., Bykowski, A., Rigotti, C.: Free-sets: a condensed representation of boolean data for the approximation of frequency queries. Data Mining and Knowledge Discovery 7, 5–22 (2003)
7. Duong, H., Truong, T., Le, B.: An Efficient Algorithm for Mining Frequent Itemsets with Single Constraint. In: Nguyen, N.T., van Do, T., Thi, H.A. (eds.) ICCSAMA 2013. SCI, vol. 479, pp. 367–378. Springer, Heidelberg (2013)
8. Han, J., Pei, J., Yin, Y., Mao, R.: Mining frequent patterns without candidate generation: a frequent-pattern tree approach. Data Mining and Knowledge Discovery 8, 53–87 (2004)
9. Pasquier, N., Taouil, R., Bastide, Y., Stumme, G., Lakhal, L.: Generating a condensed representation for association rules. J. Intelligent Information Systems 24, 29–60 (2005)
10. Srikant, R., Vu, Q., Agrawal, R.: Mining association rules with item constraints. In: Proceeding KDD 1997, pp. 67–73 (1997)
11. Zaki, M.J.: Mining non-redundant association rules. Data Mining and Knowledge Discovery (9), 223–248 (2004)
12. Zaki, M.J., Hsiao, C.J.: Efficient algorithms for mining closed itemsets and their lattice structure. IEEE Trans. Knowledge and Data Engineering 17(4), 462–478 (2005)
13. Wille, R.: Concept lattices and conceptual knowledge systems. Computers and Math. with App. 23, 493–515 (1992)

# SE-Stream: Dimension Projection for Evolution-Based Clustering of High Dimensional Data Streams

Rattanapong Chairukwattana, Thanapat Kangkachit, Thanawin Rakthanmanon, and Kitsana Waiyamai

**Abstract.** Evaluation-based stream clustering method supports the monitoring and detection of the change in clustering structure. E-Stream is an evolution-based stream clustering method that supports different types of clustering structure evolution which are appearance, disappearance, self-evolution, merge and split. However, its runtime increases and its performance drops when face with high dimensional data. High dimensional data leads to more complexity in the clustering methods. In this paper, we present SE-Stream which extends E-Stream in order to support high dimensional data streams. A projected clustering technique to determine specific subset of dimensions for each cluster is proposed. The proposed technique reduces complexity of calculation. Each cluster describes itself by a set of selected dimensions. Experimental results show that SE-Stream gives better cluster quality compared with E-Stream and HP-Stream, a state of the art algorithm for projected clustering of high dimensional data streams. Further, it gives better execution time compared with E-Stream and comparable execution time compared with HP-Stream.

## 1 Introduction

In recent years, clustering data streams has become a research topic of growing interest. The streams clustering processes data in a single pass and summarizes it in real-time, while using limited resources. Streams clustering algorithms that support the monitoring and the change detection of clustering structures are called evolution-based stream clustering method. Many techniques have been proposed for clustering data streams [2, 6, 15]. However, very few have proposed to monitor and

Rattanapong Chairukwattana · Thanapat Kangkachit ·
Thanawin Rakthanmanon · Kitsana Waiyamai
Department of Computer Engineering, Kasetsart University, Bangkok, Thailand
e-mail: {rattanapong.chairukwattana,tkangkachit}@gmail.com,
        {fengtwr,fengknw}@ku.ac.th

V.-N. Huynh et al. (eds.), *Knowledge and Systems Engineering, Volume 2*,
Advances in Intelligent Systems and Computing 245,
DOI: 10.1007/978-3-319-02821-7_32, © Springer International Publishing Switzerland 2014

detect change of the evolving clustering structures [14, 8, 12]. E-Stream [14] is an evolution-based stream clustering method that support various types of clustering structure evolution which are appearance, disappearance, self-evolution, merge and split. However, its runtime increases and its performance drops when face with high dimensional data. High dimensional data leads to more complexity in the clustering methods. To deal with high-dimensional data streams, a clustering method needs to be designed to effectively adjust with the progression of the streams and must avoid multiple passes over the data.

Many solutions have proposed for clustering on high-dimensional datasets. In [2, 5, 4, 9, 10, 13, 1, 7], projected clustering techniques have been proposed to determine appropriated subset of dimensions for each cluster. However, almost of them are difficult to generalize to handle data streams because their clustering process cannot be done in a single pass. In [3], the concept of high dimensional projected clustering on data streams has been introduced. HP-Stream is considered as a state of the art algorithm for projected clustering of high dimensional data streams. For each cluster, HPStream performs a continuous refinement of the set of appropriate dimensions during the progression of the streams. Due to its adaptability to the nature of real datasets, HPStream is able to generate clustering with quality where each cluster is specific to a particular group of dimensions. In [11], the authors proposed a density-based projected clustering technique for high dimensional data streams. HDDStream works online (macro clusters with projected clustering) and in offline (micro clusters) modes. Compared with HPStream which requires the limit number of clusters, the number of clusters in HDDStream is variably adjusted over time, and the clusters can be of arbitrary shape.

This paper presents SE-Stream which extends E-Stream in order to support high dimensional data streams. SE-Stream supports many types of clustering structure evolution, reduces time complexity and improves efficiency from E-Stream. In order to handle high dimensional data streams, a projected clustering technique to determine specific subset of dimensions for each cluster is proposed. In terms of F-measure, experimental results show that SE-Stream outperforms E-Stream and HP-Stream. It gives comparable performance in terms of cluster purity. While execution-time of SE-Stream is much better than E-Stream, however it is a bit slower than HPStream. This can be explained by the fact that SE-Stream performs operations such as cluster split and cluster merge to maintain cluster quality. Performance of SE-Stream is compared on two real-world datasets: Network Intrusion and Forest Cover Type against E-Stream and HPStream.

To summarize, the contribution of our work is as follows:

topsep=0pt, partopsep=0pt SE-Stream is a new, high dimensional, projected data streams clustering.

topsep=0pt, pbrtopsep=0pt SE-Stream cluster representation is extended from E-Stream to keep set of projected dimensions.

topsep=0pt, pcrtopsep=0pt SE-Stream reduces time-complexity of E-Stream.

The remaining of the paper is organized as follows. Section 2 presents evolution-based stream clustering technique algorithm. Section 3 introduces our stream clustering algorithm called SE-Stream. Section 4 compares performance of SE-Stream with E-Stream and HPStream on two real-world data sets: Network Intrusion and Forest Cover Type. Section 5 concludes the paper.

## 2 Evolution-Based Stream Clustering with Dimension Projection

In this section, notations and definitions related to the evolution-based stream clustering with dimension projection are given. Assume that **data streams** consists of a set of multidimensional records $X_1 \ldots X_k$ arriving at time stamps $T_1 \ldots T_k \ldots$. Each **data point** $X_i$ is a multidimensional record containing d dimensions, denoted as $X_i = (x_i^1 \ldots x_i^d)$.

### 2.1 Cluster Representation Using Fading Cluster Structure with Histogram

A *Fading Cluster structure* has been proposed in [3] to use as cluster representation instead of storing all the data points in a cluster. [14] introduced a *Fading Cluster Structure with Histogram (FCH)*. For each cluster dimension, an $\alpha$-bin histogram is used to detect change of the clustering structure. In this paper, FCH is extended to support dimension projection and is defined as $FCH = (FC1(t), FC2(t), W(t), BS(t), H(t))$. Following is the description of FCH.

Let $N$ be the total number of data points of such cluster, $T_i$ be the time when data point $x_i$ is retrieved, and $t$ be the current time. The *fading weight* of data point $x_i$ is defined as $f(t - T_i)$ where $f(t) = 2^{-\alpha t}$ and $\alpha$ is the user-defined decay rate.

**FC1(t)** is a vector of weighted summation of each dimension at time $t$. The $j^{th}$ dimension is $FC1^j(t) = \sum_{i=1}^{N} f(t - T_i) \cdot (x_i^j)$.

**FC2(t)** is a vector of weighted sum of square of each dimension at time $t$. The $j^{th}$ dimension is $FC2^j(t) = \sum_{i=1}^{N} f(t - T_i) \cdot (x_i^j)^2$.

**W(t)** is a sum of all weights of data points in the cluster at time $t$, i.e., $W(t) = \sum_{i=1}^{N} f(t - T_i)$.

**BS(t)** is a bit vector of projected dimensions at time $t$. For the $j$-th dimension is

$$BS^j(t) = \begin{cases} 1 & \text{if dimension } j \text{ is within a set of relavant cluster dimensions} \\ 0 & \text{Otherwise} \end{cases} \quad \text{Note}$$

that number of projected dimensions are not the same for each cluster (see section 2.2 for more detail).

**H(t)** is a $\alpha$-bin histogram of data values with $\alpha$ equal width intervals. For the $l$-th bin histogram of $j$-th dimension at time $t$, the elements of $H^j$ are $H_l^j(t) = \sum_{i=1}^{N} f(t - T_i) \cdot (x_i^j) \cdot (y_{i,l}^j)$ where

$$y_{i,l}^j = \begin{cases} 1 & \text{if } l \cdot r + min(x^j) \leq x_i^j \leq (l+1) \cdot r + min(x^j); r = \frac{max(x^j) - min(x^j)}{\alpha} \\ 0 & \text{Otherwise} \end{cases}$$

## 2.2  Dimensions Projection

High dimensional data leads to more complexity in the clustering method and unsatisfied clusters quality. To reduce the effect of these problems, the set of dimensions associated to each cluster is selected instead of using all dimensions. To select only the relevant dimensions of each cluster, a small range of data spread based on radii is determined. To avoid error in computing radii of a cluster containing only one data point, the radii of each dimension of each cluster is determined with the help of the incoming data point $X_i$. Then, all the dimensions are ranked according to their radii in ascending order. Only top ($|FCH| \cdot l$) ranks are selected. Note that different clusters may have selected dimensions. This is due to the difference in relevance of each dimension.

Fig. 1 illustrates how our proposed dimensions projection method works to determine specific subset of dimensions for each cluster. Here the parameter $l$ is set to 1. Suppose that at timestamps $t$, the output clustering contains 3 clusters of data streams with 3 dimensions. For each cluster, radii of each dimension is computed by considering its fading cluster structure with histogram (FCH). Then, all the dimensions are ranked based their radii in ascending order. Only $|FCH| \cdot l$ ranks are selected i.e. dimension #1 and #2 of cluster #2 and dimension #2 of cluster #3. Notice that none of dimensions of cluster #3 is selected because of their high values of radii. At the bottom of Fig. 1, a bit vector of all the clusters is shown.

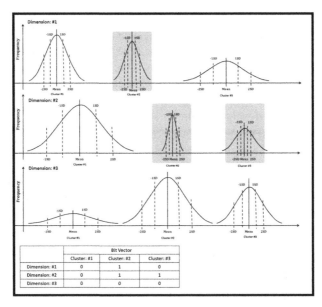

**Fig. 1** Dimension projection example

## 2.3   Distance Functions

To deal with the projected dimensions, distance functions are modified from E-stream to take into account only the relevant subset of dimensions for each cluster. Two types of distance are proposed: Cluster-Point distance and Cluster-Cluster distance. Each type of distance is described below. At timestamp $t$, a set of projected cluster dimensions is stored in a bit vector $BS(t)$.

- **Cluster-Point distance** is a distance from a data point to a cluster center. For each dimension, the distance is normalized by the radius (standard deviation) of the cluster data. This function is used to find the closet active cluster for an incoming data point. The Cluster-Point distance function $dist(C, X_i)$ of cluster $C$ and incoming data point $X_i$ at timestamp $t$ can be formulated as:

$$dist(C, X_i) = \frac{1}{n} \cdot \sum_{j \in BS(t)} \left| \frac{center_C^j - x^j}{radius_C^j} \right|$$

  where $n$ is the number of projected dimensions of cluster $C$ in the bit vector $BS(t)$.

- **Cluster-Cluster distance** is a distance between two cluster centers. Notice that the projected dimensions of the two clusters may differ. Thus, a set of the projected dimensions of the first cluster is determined. The cluster-cluster distance function $dist(C_a, C_b)$ of cluster $C_a$ and $C_b$ at timestamp $t$ can be formulated as:

$$dist(C_a, C_b) = \frac{1}{n} \cdot \sum_{j \in BS(t)} \left| center_{C_a}^j - center_{C_b}^j \right|$$

  where $n$ is the number of projected dimensions of cluster $C_a$ in the bit vector $BS(t)$.

## 3   SE-Stream Algorithm

In this section, we start by giving an overview of SE-Stream which extended from E-Stream. Projected clustering technique to determine specific subset of dimensions for each cluster is explained. Then, we present SE-Stream algorithm composing of a set of sub-algorithms. Finally, its time-complexity is discussed and analysed.

## 3.1   Overview of SE-Stream Algorithm

This section describes SE-Stream which is an extension of E-Stream to support high-dimensional data streams. Table 1 contains all the notations that are used in SE-Stream algorithm. SE-Stream main algorithm is given in Fig. 2.

SE-Stream is evolution-based algorithm that supports the monitoring and the change detection of clustering structure that can evolve over time. It is designed for high dimensional data streams. Various types of clustering structure evolution

| Algorithm *SE-Stream* | SE-Stream Runtime-complexity | E-Stream Runtime-Complexity |
|---|---|---|
| 1: Retrieve new data X$_i$ | | |
| 2: FadingAll | O(K) | O(K) |
| 3: CheckSplit | O(Kl) | O(Kd) |
| 4: If ((t - t$_{cd}$) mod T Or cluster split Or new FCH) | | |
| 5:    ProjectDimension | O(Kd) | n/a |
| 6: MergeOverlapCluster | O(K$^2$l) | O(K$^2$d) |
| 7: LimitMaximumCluster | O(K$^2$l) | O(K$^2$d) |
| 8: FlagActiveCluster | O(K) | O(K) |
| 9: (minDistance, index) ← FindClosestCluster | O(Kl) | O(Kd) |
| 10: If minDistance < radius_factor | | |
| 11:    Add X$_i$ to FCH$_{index}$ | | |
| 12: Else | | |
| 13:    Create new FCH from X$_i$ | | |
| 14: Waiting for new data | | |

**Fig. 2** Run-time complexity of E-Stream and SE-Stream algorithms

**Table 1** List of notations used in SE-Stream and its sub-procedures

| Notation | Definition |
|---|---|
| FCH | Cluster representation with fading cluster structure with histogram. |
| \|FCH\| | Current number of clusters. |
| FCH$_{temp}$ | Temporary of a cluster representation. |
| FCH$_i$ | i$^{th}$ Cluster representation of the i$^{th}$ cluster. |
| FCH$_i$.W | Weight of the i$^{th}$ cluster. |
| FCH$_i$.sd | Standard deviation of the i$^{th}$ cluster. |
| BS | Bit vector containing the relevant cluster dimensions (projected dimensions). |
| S | Set of pairs of split clusters. |
| $\lambda$ | Fading rate value. |
| l | Number of dimensions used in the projected dimension. |
| d | Total number of dimensions. |
| merge_threshold | Threshold value to determine which clusters should be merged. |
| K | Maximum number of clusters. |
| $\varepsilon$ | Threshold value to determine which cluster should be deleted. |
| X$_i$ | Data point. |
| t | Time stamp of data point. |
| t$_{cd}$ | Last time stamp of a computed dimension. |

are supported which are appearance, disappearance, self-evolution, merge and split. In line 1, the algorithm starts by retrieving a new data point. In line 2, it fades all clusters and deletes those have weight less than $\varepsilon$. In line 3, it splits a cluster when the behaviour inside is obviously separated. In line 4-5, dimension projection is performed if there is a cluster split or at appropriate checking period or detecting a new cluster($FCH$) created from $X_{i-1}$. Notice that it is not necessary to project dimensions in every round of the clustering process. Based on [15], an appropriate period for projecting dimension is based on time $T$ which can be defined as

$T = \left[\frac{1}{\lambda} \mid log \mid \frac{\varepsilon}{\varepsilon - 1} \mid \right) \mid \right]$. Details of the project dimension algorithm is given in Fig. 3 .In line 6, it checks for overlapping clusters and merges them. In line 7, when the number of cluster count exceeds the limit, it checks the closest pair of clusters and merges them until the number of cluster count does not exceed limit maximum number of clusters. In line 8, it checks all clusters whether their statuses are active. In line 9, it finds the closest cluster to the incoming data point. In line 10-13, if the distance between closest cluster and incoming point is less than *radius_factor* then the point is assigned to the cluster, otherwise it creates isolated data point. The flow of control returns to the top of algorithm and waits for a new data point. Following is the explanation of each sub-algorithm that has been extended from E-stream or added to SE-stream. Details of each sub-algorithm is given in Fig. 3.

| **Algorithm** *ProjectDimension* | **Algorithm** *FadingAll* |
|---|---|
| 1: Create $\|FCH_{temp}\|$ temp by adding $X_i$ with $\|FCH\|$ | 1: for i ← 1 to $\|FCH\|$ |
| 2: Compute the $\|FCH_{temp}\|$ * d radius along d dimensions | 2: fading $FCH_i$ |
| 3: Pick the $\|FCH\|$ * l dimensions with the least radius | 3: if ($FCH_i.W < \varepsilon$) |
| 4: Set timestamp of data to $t_{cd}$ | 4: delete $FCH_i$ |
| **Algorithm** *CCDistance(FCH$_i$, FCH$_j$)* | **Algorithm** *CPDistance(FCH i, X$_i$)* |
| 1: for k ← 1 to d | 1: for k ← 1 to d |
| 2: if (BS(FCH$_i$)$^k$ == 1) | 2: if (BS(FCH$_i$)$^k$ == 1) |
| 3: distance = abs(Center(FCH$_i$)$^k$ – Center(FCH$_j$)$^k$) | 3: distance = abs(Center(FCH$_i$)$^k$ –$X_i^k$ / Radius(FCH$_i$)$^k$) |
| 4: d' = the number of bits in BS(FCH$_i$) with value of 1 | 4: d' = the number of bits in BS(FCH$_i$) with value of 1 |
| 5: distance = distance / d' | 5: distance = distance / d' |
| 6: return distance | 6: return distance |
| **Algorithm** *CheckSplit* | **Algorithm** *MergeOverlapCluster* |
| 1: for i ← 1 to $\|FCH\|$ | 1: for i ← 1 to $\|FCH\|$ |
| 2: if (not isolate point) | 2: for j ← i + 1 to $\|FCH\|$ |
| 3: for j ← 1 to d | 3: overlap[i,j] ← CCDistance (FCHi, FCHj) |
| 4: if (BS(FCH$_i$)$^j$ == 1) | 4: m ← merge_threshold |
| 5: if (FCH/ have split point) | 5: if (overlap[i, j] > m * (FCH$_i$.sd + FCH$_j$.sd) |
| 6: split FCH$_i$ | 6: if (i, j) not in S |
| 7: S ← S U {(i, $\|FCH\|$)} | 7: merge (FCH$_i$, FCH$_j$) |
| **Algorithm** *LimitMaximumCluster* | **Algorithm** *FindClosestCluster* |
| 1: while $\|FCH\| > K$ | 1: for i ← 1 to $\|FCH\|$ |
| 2: for i ← 1 to $\|FCH\|$ | 2: if (FCHi) is active cluster |
| 3: for j ← i+1 to $\|FCH\|$ | 3: dist[i] ← CPDistance(FCHi, X$_i$) |
| 4: dist[i, j] ← CCDistance(FCH$_i$, FCH$_j$) | 4: (minDistance, i) ← min(dist[i]) |
| 5: (first, second) ← argmin$_{(i, j)}$(dist[i, j]) | 5: return (minDistance, i) |
| 6: merge (FCHfirst, FCHsecond) | |

**Fig. 3** Details of each sub-algorithm extended from E-Stream or added to SE-stream

**FadingAll.** SE-Stream performs fading of all clusters and deletes clusters with weight less than $\varepsilon$(an input parameter).

**CheckSplit.** SE-Stream finds the split point in the projected dimensions. If a splitting point is found in any cluster, it is slitted and dimensions are computed and stored with index pairs of split in $S$.

**MergeOverlapCluster.** SE-Stream finds the pairs of overlapping clusters. For each pair of clusters, cluster-cluster distance is calculated. If the distance is less than the *merge_threshold* and the merged pair is not already in $S$ then the two clusters are merged.

**LimitMaximumCluster.** SE-Stream checks weather the total number of clusters reaches its maximum *maximum_cluster*. If it exceeds the maximum, then the closest pair of clusters is merged until the number of the remaining clusters is less than or equal to the *maximum_cluster*.

**FindClosestCluster.** SE-Stream calculates cluster-point distance. Then, determine the closest active cluster to contain an incoming data point. In the distance calculation step, SE-Stream uses 2 types of distance: cluster-cluster distance (*CCDistance*) and cluster-point distance (*CPDistance*). To support projected clustering, these two distances have been extended from E-stream as shown in Fig. 3. In the MergeOverlapCluster and LimitMaximumCluster steps, projected dimensions are stored in a bit vector data structure. SE-Stream uses bitwise OR operation for merging the bit vectors of clusters. The result is new projected dimensions of the cluster that have been merged.

## 3.2 Time-Complexity of SE-Stream VS E-Stream

Evolution-based clustering methods are able to detect change of clustering structure evolution. As result, the clustering output is of high quality in terms of both F-measure and purity. However, with high-dimensional data streams, these clustering structure evolution detection operations, increases execution time of the algorithm. This section discuss the time-complexity of SE-Stream which is an evolution-based clustering method. Fig. 2 shows time-complexity of the different steps of SE-Stream compared with E-Stream.

In case of E-Stream, MergeOverlapCluster and LimitMaximumCluster procedures consume the most execution time. This is because distance between each pair of all the existing clusters need to be calculated, at least big-O $O(K^2)$ is required. Both of them perform distance calculation of every pair of active clusters to find the closet one. Notice also that its time-complexity depends on the number of dimensions for almost of the clustering operations. When face with high dimensional data, the run-time of E-Stream is extremely increased.

SE-Stream reduces time-complexity by performing dimension projection and thus decreases times for distance calculation. With its dimension projection, the number of dimension ($l$) is less than the total number of dimensions($d$). Instead of using all the dimensions, SE-Stream uses only selected dimensions for all of its clustering structure evolution detection operations then its run-time complexity is greatly reduced. For example,run-time complexity of MergeOverlapCluster and LimitMaximumCluster is reduced from $O(K^2d)$in E-Stream to $O(K^2l)$ in SE-Stream. From l value less than d value, it means SE-Stream use time less than E-Stream. For distance calculation, its time complexity is reduced from $O(K^2d)$ to $O(K^2l)$ is more than $O(Kd)$.

## 4  Evaluation and Experimental Results

In this section, experiments are performed in various aspects to evaluate the performance of SE-Stream. The clustering performance is measured by both execution time and cluster quality. All the experiments were conducted on a 2.6 GHz Intel®Core I5 computer with 2 GB memory and executed on Windows 7. To compare with our SE-Stream algorithm, both E-Stream and HPStream algorithms have been implemented in C++.

In the experiments, two standard UCI datasets are used i.e. Network Intrusion and Forest Cover Type. From both datasets, only numerical attributes of are selected. As result, the first dataset contains 494020 records with 34 attributes and 22 classes. The latter dataset contains 581012 records with 10 attributes and 7 classes. For the parameters setting, SE-Stream follows the same values as adopted in E-Stream i.e. stream speed, horizon, decay-rate $\lambda$, remove threshold $\varepsilon$ are set to 100, 2, 0.1 and 0.1 respectively. Otherwise the maximum number of clusters parameter is varied based-on the number of classes in a dataset. Therefore, it is set to 25 for Network Intrusion dataset and 15 for Forest Cover Type dataset.

To handle high dimensional data streams, both SE-Stream and HPStream use projected clustering technique to determine specific subset of dimensions for each cluster. To find the best-subset of dimensions, the number of projected dimensions ($l$) is varied from 100% to 10%. We observed that performance of SE-Stream continues to increase when number of projected dimensions is reduced to 40% for both datasets. When the number of dimensions is less than 40%, the performance of SE-Stream decreases dramatically due to lack of information. This is a reason why only performance of SE-stream with 40% and 50% of projected dimensions is reported in the following experimental results.

## 4.1   Cluster Quality

The cluster quality is measured by the average f-measure and purity in every 50000 incoming data points. Fig. 4 shows performance of SE-Stream, E-Stream and HP-Stream in terms of cluster quality on Network Intrusion dataset. The cluster quality is evaluated based on purity and f-measure. Compared to E-Stream, SE-Stream yields significantly higher f-measure in every progression of the data streams. In term of purity, SE-Stream and E-Stream offers comparable value. Compared to HP-Stream, SE-Stream produces higher f-measure in the overall progression of data streams. However, HPStream produces higher f-measure than SE-Stream in the first 200000 incoming data points. This is because HPStream performs an initial off-line process to find the initial clusters while SE-Stream and E-Stream perform clustering from scratch.

Another quality comparison is made on Forest Cover Type dataset and the result is shown in Fig.5. SE-Stream offers comparable cluster quality with E-Stream. Although SE-Stream outperforms HPStream in term of f-measure, their purity is considered equal.

In summary, we can say that SE-Stream produces better cluster quality than E-Stream and HPStream. The reason is that SE-Stream offers a higher f-measure value with equal purity value compared to the other techniques.

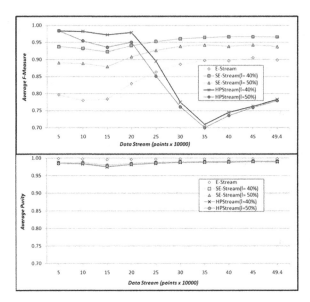

**Fig. 4** Performance comparison in terms of average purity and F-measure on network intrusion dataset

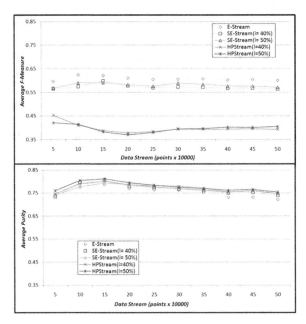

**Fig. 5** Performance comparison in terms of average purity and F-measure on forest cover type dataset

## 4.2    Execution Time

Fig. 6 compares performance of the different algorithms in terms of execution time. HPStream is more efficient than SE-Stream and E-Stream. This is due to the additional process to detect split and merge operations of SE-Stream and E-Stream which requires $O(K^2 l)$ time-complexity where $K$ is number of clusters and $l$ is number of projected dimensions. The gap between HPStream and SE-Stream is clearly evident for datasets with large number of clusters and dimensions such as Network Intrusion. In contrast, this gap is much reduced with Forest Cover Type dataset which contains less number of clusters and dimensions.

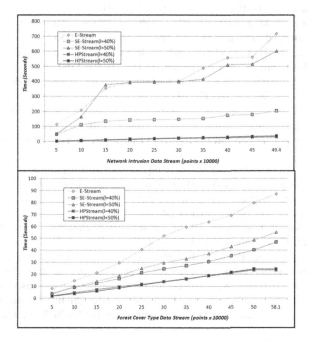

**Fig. 6** Performance comparison in terms of execution time

## 5    Conclusion and Future Work

In this paper, we present SE-Stream algorithm which is designed for clustering high dimensional data streams. SE-Stream inherits the ability to detect evolution of clustering structure from E-Stream. In term of quality and run-time, SE-Stream yields better results than E-Stream. In term of cluster description, SE-Stream provides explainable cluster description in terms of projected dimensions.

Notice that, there is no guarantee that the projected dimensions in SE-Stream are specific to each cluster. An interesting research direction is to propose a technique to ensure the best quality of the projected dimensions that are very cluster-specific. This will certainly improve quality of SE-stream to perform clustering on high-dimensional data streams more efficiently.

# References

1. Aggarwal, C.C.: On high dimensional projected clustering of uncertain data streams. In: Ioannidis, Y.E., Lee, D.L., Ng, R.T. (eds.) ICDE, pp. 1152–1154. IEEE (2009)
2. Aggarwal, C.C., Han, J., Wang, J., Yu, P.S.: A framework for clustering evolving data streams. In: Proceedings of the 29th International Conference on Very Large Data Bases, VLDB 2003, vol. 29, pp. 81–92. VLDB Endowment (2003)
3. Aggarwal, C.C., Han, J., Wang, J., Yu, P.S.: A framework for projected clustering of high dimensional data streams. In: Proceedings of the Thirtieth International Conference on Very Large Data Bases, VLDB 2004, vol. 30, pp. 852–863. VLDB Endowment (2004)
4. Aggarwal, C.C., Wolf, J.L., Yu, P.S., Procopiuc, C., Park, J.S.: Fast algorithms for projected clustering. SIGMOD Rec. 28(2), 61–72 (1999)
5. Agrawal, R., Gehrke, J., Gunopulos, D., Raghavan, P.: Automatic subspace clustering of high dimensional data for data mining applications. SIGMOD Rec. 27(2), 94–105 (1998)
6. Chen, K., Liu, L.: He-tree: a framework for detecting changes in clustering structure for categorical data streams. The VLDB Journal 18(6), 1241–1260 (2009)
7. Kim, M., Ramakrishna, R.S.: Projected clustering for categorical datasets. Pattern Recogn. Lett. 27(12), 1405–1417 (2006)
8. Kosonpothisakun, P., Kangkachit, T., Waiyamai, K.: E-stream++: Stream clustering technique for supporting numerical and categorical data. In: Proceedings of the 13th National Computer Science and Engineering Conference, NCSEC 2009, Bangkok, Thailand (2009)
9. Liu, W., Jia, O.: Clustering algorithm for high dimensional data stream over sliding windows. In: Proceedings of the 2011 IEEE 10th International Conference on Trust, Security and Privacy in Computing and Communications, TRUSTCOM 2011, pp. 1537–1542. IEEE Computer Society, Washington, DC (2011)
10. Moise, G., Sander, J., Ester, M.: P3c: A robust projected clustering algorithm. In: ICDM, pp. 414–425. IEEE Computer Society (2006)
11. Ntoutsi, I., Zimek, A., Palpanas, T., Kröger, P., Kriegel, H.-P.: Density-based projected clustering over high dimensional data streams. In: SDM, pp. 987–998 (2012)
12. Tossaporn, S., Thanapat, K., Kitsana, W.: Ce-stream: Evaluation-based technique for stream clustering with constraints. In: Proceedings of the 10th International Joint Conference on Computer Science and Software Engineering, JCSSE 2013, Khonkaen, Thailand (2013)
13. Sembiring, R.W., Zain, J.M., Embong, A.: Clustering high dimensional data using subspace and projected clustering algorithms. CoRR, abs/1009.0384 (2010)
14. Udommanetanakit, K., Rakthanmanon, T., Waiyamai, K.: E-stream: Evolution-based technique for stream clustering. In: Alhajj, R., Gao, H., Li, X., Li, J., Zaïane, O.R. (eds.) ADMA 2007. LNCS (LNAI), vol. 4632, pp. 605–615. Springer, Heidelberg (2007)
15. Wan, R., Wang, L.: Clustering over evolving data stream with mixed attributes. Journal of Computational Information Systems (June 2010)

# Mining Frequent Itemsets in Evidential Database

Ahmed Samet, Eric Lefèvre, and Sadok Ben Yahia

**Abstract.** Mining frequent patterns is widely used to discover knowledge from a database. It was originally applied on Market Basket Analysis (MBA) problem which represents the Boolean databases. In those databases, only the existence of an article (item) in a transaction is defined. However, in real-world application, the gathered information generally suffer from imperfections. In fact, a piece of information may contain two types of imperfection: imprecision and uncertainty. Recently, a new database representing and integrating those two types of imperfection were introduced: Evidential Database. Only few works have tackled those databases from a data mining point of view. In this work, we aim to discuss evidential itemset's support. We improve the complexity of state of art methods for support's estimation. We also introduce a new support measure gathering fastness and precision. The proposed methods are tested on several constructed evidential databases showing performance improvement.

## 1 Introduction

The majority of data mining algorithms were applied on precise and certain data constituting Boolean databases. This type of databases does only indicate if the considered item $I$ exists or not. However, in real life, gathered information are suffering from imperfection due to many factors such as acquisition reliability, human errors, information absence, etc. In [2], Lee detailed the two sides of imperfection that could manifest in a database. Indeed, we may encounter databases containing

Ahmed Samet · Sadok Ben Yahia
Laboratory of Research in Programming,
Algorithmic and Heuristic Faculty of Science of Tunis, Tunisia
e-mail: {ahmed.samet,sadok.benyahia}@fst.rnu.tn

Eric Lefèvre
Univ. Lille Nord de France UArtois, EA 3926 LGI2A, F-62400, Béthune, France
e-mail: eric.lefevre@univ-artois.fr

V.-N. Huynh et al. (eds.), *Knowledge and Systems Engineering, Volume 2*,
Advances in Intelligent Systems and Computing 245,
DOI: 10.1007/978-3-319-02821-7_33, © Springer International Publishing Switzerland 2014

*imprecise* and *uncertain* information. The imprecision is relevant to the content of an attribute value of a data object, while the concept of uncertainty is relevant to the degree of truth of its attribute value. Due to its adequate imprecision representation, the fuzzy theory [3] was largely used to extract fuzzy frequent patterns and association rules such that [4, 5, 6]. In [2], the author introduced a new type of databases that handle both imprecise and uncertain information. This database was modeled via the evidence theory [7, 8], which offers a certain level of flexibility in imperfect information representation. Those types of databases were denoted as the *Evidential database*. The Evidential database has brought flexibility in handling those imperfect knowledge but also added complexity in their treatment. Indeed, the number of patterns increases exponentially as far as the number of attributes arises in the database. Even the literature methodologies for estimating itemset's support are time consumer, since they rely on Cartesian product. In addition to the time limit, the proposed support functions such as those introduced in [9, 10], are not that precise in their estimation. The support computing do not explore all information that exist in the *Basic Belief Assignment*. This constraint makes from literature methods limited in their manner of support estimation and do not extract all existing frequent patterns within the evidential database. In this work, evidential data mining problem is tackled by putting our focus on the support estimation. Existing methods for support estimation are highlighted and we propose a ramification that considerably improves their original performance. We also introduce a new measure for evidential pattern's support estimation. This new measure improves the support computation where all pieces of information in a Basic Belief Assignment are considered. This method also presents an interesting performance comparatively to literature methods. We introduce the Evidential Data mining Algorithm (EDMA) that mines all frequent patterns in an evidential database. This paper is organized as follows: in section 2, the main principles of the evidence theory and Smets's TBM [11] interpretation are presented. In section 3, several state of art works are scrutinized and we highlight their limits. We present a ramification for their method that improves the performance. In section 4, we introduce a new method for evidential itemsets' support computing providing more precision in its estimation. Evidential Data Mining Apriori (EDMA) algorithm for mining frequent evidential patterns is introduced in section 5. The performance of this algorithm is studied in section 6. Finally, we conclude and we sketch issues of future work.

## 2 Evidence Database

In this section, evidential database concepts based on evidence theory formalism are presented. In the following, we define evidence theory main concepts based on a Transferable Belief Model interpretation [11].

## 2.1  Evidence Theory

The evidence theory or Dempster-Shafer theory proposes a robust formalism for modeling uncertainty. In the following, the evidence theory from a Smets's Transferable Belief Model (TBM) interpretation is presented. The TBM model represents quantified beliefs following two distinct levels: (i) a credal level where beliefs are entertained and quantified by belief functions; (ii) a pignistic level where beliefs can be used to make decisions and are quantified by probability functions. The evidence theory is based on several fundamentals such as the Basic Belief Assignment (BBA). A BBA $m$ is the mapping from elements of the power set $2^\theta$ onto [0, 1]:

$$m : 2^\theta \longrightarrow [0,1]$$

where $\theta$ is the *frame of discernment*. It is the set of possible answers for a treated problem and is composed of $N$ exhaustive and exclusive hypotheses:

$$\theta = \{H_1, H_2, ..., H_N\}.$$

A BBA $m$ do have some constraints such that:

$$\sum_{A \subseteq \theta} m(A) = 1 \tag{1}$$

Each subset X of $2^\theta$ fulfilling $m(X) > 0$ is called focal element. Constraining $m(\emptyset) = 0$ is the normalized form of a BBA and this corresponds to a closed-world assumption [12], while allowing $m(\emptyset) > 0$ corresponds to an open world assumption [13].

From a BBA another function is commonly defined from $2^\theta$ to [0, 1]: $Bel(A)$ is interpreted as the degree of justified support given to the proposition $A$ by the available evidence.

$$Bel(A) = \sum_{\emptyset \neq B \subseteq A} m(B) \tag{2}$$

Generally, in an information fusion problem, not all considered sources share the same domain (frame of discernment). This constraint prevents from using usual combination tools [7].

In this case, the Cartesian product allows the combination. Let it be two belief function $m_1$ and $m_2$ defined respectively in $\theta_1$ and $\theta_2$, the Cartesian product is expressed as follows:

$$m_{1 \times 2}^\theta(A \times B) = m_1^{\theta_1}(A) \times m_2^{\theta_2}(B). \tag{3}$$

After source's combination which integrates the credal stage of the TBM model, taking decision is necessarily. In [11], the pignistic probability is introduced allowing probabilistic decision from BBA following this formula:

$$BetP(H_n) = \sum_{A \subseteq \theta} \frac{|H_n \cap A|}{|A|} \times m(A) \qquad \forall H_n \in \theta \tag{4}$$

where $|\cdot|$ is the cardinality operator.

## 2.2 Evidence Database Concept

An evidential database stores data that could be perfect or imperfect. Uncertainty in such database is expressed via the evidence theory. An evidential database, denoted by $\mathscr{EDB}$, with $n$ columns and $d$ lines where each column $i$ ($1 \leq i \leq n$) has a domain $\theta_i$ of discrete values. Cell of line $j$ and column $i$ contains a normalized BBA as follows:

$$m_{ij} : 2^{\theta_i} \to [0,1] \quad with$$

$$\begin{cases} m_{ij}(\emptyset) = 0 \\ \sum_{A \subseteq \theta_i} m_{ij}(A) = 1. \end{cases} \tag{5}$$

**Table 1** Evidential transaction database $\mathscr{EDB}$

| Transaction | Attribute A | Attribute B |
|---|---|---|
| T1 | $m(A_1) = 0.7$ | $m(B_1) = 0.4$ |
|  | $m(\theta_A) = 0.3$ | $m(B_2) = 0.2$ |
|  |  | $m(\theta_B) = 0.4$ |
| T2 | $m(A_2) = 0.3$ | $m(B_1) = 1$ |
|  | $m(\theta_A) = 0.7$ |  |

In an evidential database, as shown in Table 1, an item corresponds to a focal element. An itemset corresponds to a conjunction of focal elements having different domains. Two different itemsets can be related via the inclusion or intersection operator. Indeed, the inclusion operator for evidential itemsets is defined as follows, let $X$ and $Y$ be two evidential itemsets:

$$X \subseteq Y \iff \forall x_i \in X, x_i \subseteq y_i.$$

where $x_i$ and $y_i$ are the $i^{th}$ element of $X$ and $Y$. For the same evidential itemsets $X$ and $Y$, the intersection operator is defined as follows:

$$X \cap Y = Z \iff \forall z_i \in Z, z_i \subseteq x_i \text{ and } z_i \subseteq y_i.$$

*Example 1.* In Table 1, $A_1$ is an item and $\{\theta_A \ B_1\}$ is an itemset such that $A_1 \subset \{\theta_A \ B_1\}$ and $A_1 \cap \{\theta_A \ B_1\} = A_1$.

## 3 Evidential Patterns' Cartesian Based Support

In this section, we present related works in support estimation for evidential databases. Afterwards, we sketch with an improvement that we propose to improve support's performance.

## 3.1 State of Art

Evidential data mining does not grasp so much attention. In [14], Hewawasam et al. proposed a methodology to estimate itemsets' support and modelize them in a tree representation: *Belief Itemset Tree* (BIT). The BIT representation brings easiness and rapidity for the estimation of the associative rule's confidence.

In [9], the authors introduced a new approach for itemset support computing and applied on a Frequent Itemset Maintenance (FIM) problem. All methods [9, 10] were based on Cartesian product between BBAs.

Let's study the support of an itemset $X = \prod_{i \in [1...n]} x_i$ such that $x_i$ is an evidential item belonging to the frame of discernment $\theta_i$. Since the items do not share the same discernment frame, any fusion rule cannot be applied. In the following, we study the belief support introduced by [9] computed by the following equation:

$$m_j(X) = \prod_{x_i \in X} m_{ij}(x_i) \tag{6}$$

where $m_j(X)$ is the Cartesian product of all BBA in the transaction $T_j$. Thus, the BBA of the itemset $X$ expressed in the entire $\mathcal{EDB}$ database becomes:

$$m_{\mathcal{EDB}}(X) = \frac{1}{d} \sum_{j=1}^{d} m_j(X). \tag{7}$$

Then, the support of $X$ in the $\mathcal{EDB}$ database becomes:

$$Support_{\mathcal{EDB}}(X) = Bel_{\mathcal{EDB}}(X). \tag{8}$$

The Cartesian product based support, presented above, fulfills several properties such that the *anti-monotony* property. A support measure satisfying the anti-monotony property consists in the fact that an itemset that contains an infrequent itemset is also infrequent. The opposite is true, all itemsets constituting a frequent one are also frequent. With this satisfied property, the construction of an Apriori based algorithm becomes straightforward [9].

## 3.2 Cartesian Support Ramification

The support measure, proposed by [9, 10] works (shown in equation 6), relies on Cartesian product. Indeed, the Cartesian product is the suited solution in case of combining BBAs with different frame of discernment. However, such solution waste execution time because of its exponential complexity. In this section, we focus our interest in simplifying the Cartesian based method for performance requirements.

Let us consider the evidential database $\mathcal{EDB}$ and the itemset $X = x_1 \times \cdots \times x_n$ constituted by the product of items (focal elements) $x_i$ ($1 \leq i \leq n$) of the exclusive frame of discernment $\theta_i$. For a transaction $T_j$, we have:

$$Support_{T_j}(X) = \prod_{i \in [1...n]} Support_{T_j}(x_i) = \prod_{i \in [1...n]} Bel(x_i) \qquad (9)$$

$$Support_{\mathscr{EDB}}(X) = \frac{1}{d} \sum_{j=1}^{d} Support_{T_j}(X) \qquad (10)$$

*Proof.* Let us consider two evidential items and focal elements $x_1$ and $x_2$ belonging respectively to $m_1$ and $m_2$ BBA such that $m = m_1 \times m_2$.

$$Bel\left( \prod_{x_i \in \theta_i, 1 \leq i \leq n} x_i \right) = \sum_{a \subseteq x_1 \times \cdots \times x_n} m_{1 \times \cdots \times n}(a)$$
$$Bel\left( \prod_{x_i \in \theta_i, 1 \leq i \leq n} x_i \right) = \sum_{y_1 \subseteq x_1, \ldots, y_2 \subseteq x_n} m_1(y_1) \times \cdots \times m_n(y_2)$$
$$Bel\left( \prod_{x_i \in \theta_i, 1 \leq i \leq n} x_i \right) = \sum_{y_1 \subseteq x_1} m_1(y_1) \times \cdots \times \sum_{y_n \subseteq x_n} m_n(y_n)$$
$$Bel\left( \prod_{x_i \in \theta_i, 1 \leq i \leq n} x_i \right) = Bel(x_1) \times \cdots \times Bel(x_n) = \prod_{i \in [1...n]} Bel(x_i)$$

In this section, the support is estimated with the $Bel(.)$ function which generates several limits. In the following, we highlight those limits and we propose a new support alternative: *The precise support.*

## 4   Precise Evidential Support Estimation

The evidential database relies on representing information's imperfection with BBAs. A BBA does not only represent belief accorded to a single hypothesis but also to their disjunction. As shown in section 2, from a piece of evidence (BBA), several functions exist allowing the pertinence's estimation of each hypothesis. The $Bel(.)$ (see equation 2), used for support definition in section 3, is not the only function that estimates the degree of veracity of each hypothesis in the superset $2^\theta$. In addition, $Bel(.)$ estimates the belief by referring only to a small subset of the superset. This limits make from belief based support measures imprecise. In the following, we propose a new alternative to the Cartesian support estimation allowing a precise support computing and a reasonable time scale performance.

### 4.1   Support Definition

Let us consider an evidential database $\mathscr{EDB}$ and the itemset $X = x_1 \times \cdots \times x_n$ constituted by the product of items (focal elements) $x_i$ $(1 \leq i \leq n)$ of the exclusive frame of discernment $\theta_i$. The degree of presence of an item $x_i$ in a transaction $T_j$ (BBA) can be measured as follow:

$$Pr : 2^\theta \rightarrow [0,1] \qquad (11)$$

$$Pr(x_i) = \sum_{x \subseteq \theta_i} \frac{|x_i \cap x|}{|x|} \times m(x) \qquad \forall x_i \in 2^{\theta_i}. \qquad (12)$$

As illustrated above, the $Pr(.)$ measure allows to compute $x_i$ presence in a single BBA. The $Pr$ measure is equal to the pignistic probability if $x_i \in \theta_i$. The evidential support of an itemset $X = \prod_{i \in [1...n]} x_i$ is then computed as follows:

$$Support_{T_j}^{Pr}(X) = \prod_{X_i \in \theta_i, i \in [1...n]} Pr(x_i) \tag{13}$$

$$Support_{\mathcal{EDB}}(X) = \frac{1}{d} \sum_{j=1}^{d} Support_{T_j}^{Pr}(X). \tag{14}$$

The presented approach for estimating itemset evidential support is similar to the support ramification given subsection 3.2. However, our approach presents several assets where our support inclusion is larger than those given in respectively [9, 10]. Indeed, in the previous cited works, the authors evaluate support of $X$ by considering only subsets included in it. The $Pr(.)$ function does not only consider all subsets of $X$ but also those having intersection with it. In addition, our support estimation provides an interesting performance since we get rid of the Cartesian product. The proposed support function sustains previous works on fuzzy [4] in case of dealing with consonant BBA[1]. It also sustains previous data mining works on binary databases [15] when BBA are certain[2]. Indeed, previous works have adopted a probabilistic orientation in support measure in those databases. Support generally represents the frequency of appearance of an itemset therefore it can be assimilated to an apriori probability. Interestingly enough, the precise support estimation function keeps the interesting anti-monotony property useful in infrequent itemsets removal. This fulfilled condition is proven in the proof given below.

*Proof*
Assuming an evidential database $\mathcal{EDB}$, let's consider two evidential itemsets $A$ and $A \times X$ where $A \subset A \times X$ such that $\forall x \in A$, $x \in A \times X$. We aim to prove that considering this condition $Support(A \times X) \leq Support(A)$.

$$Support_{T_j}(A \times X) = Pr(A) \times Pr(X)$$
$$Support_{T_j}(A \times X) \leq Support_{T_j}(A) \text{ Since } Pr(X) \in [0,1] \text{ then}$$
$$Support_{\mathcal{EDB}}(A \times X) \leq Support_{\mathcal{EDB}}(A)$$

### 4.2  Pr Table

Let us consider be the evidential database $\mathcal{EDB}$ containing $n$ attributes and $d$ transactions. The *Pr Table* is a table having $d$ rows ($j \in [1,d]$) where each one contains the $Pr(.)$ measure of all items (focal elements) found in the $j^{th}$ transaction of $\mathcal{EDB}$. Since the support function can be written as a simple product, the storage of item's $Pr$ measure in a table became a need. Table 2 shows the Pr Table extracted from the evidential database $\mathcal{EDB}$ presented in Table 1.

---

[1] A BBA is said consonant if focal elements are nested.
[2] A BBA with only one focal element $A$ is said to be certain and is denoted $m(A) = 1$.

**Table 2** Pr Table deduced from the evidential database $\mathscr{E}\mathscr{D}\mathscr{B}$ presented in Table 1

| Transaction | Transactional Support |
|---|---|
| T1 | $Pr^{\theta_A}(A_1) = 0.85$ |
|  | $Pr^{\theta_A}(A_2) = 0.15$ |
|  | $Pr^{\theta_A}(\theta_A) = 1.00$ |
|  | $Pr^{\theta_B}(B_1) = 0.60$ |
|  | $Pr^{\theta_B}(B_2) = 0.40$ |
|  | $Pr^{\theta_B}(\theta_B) = 1.00$ |
| T2 | $Pr^{\theta_A}(A_1) = 0.35$ |
|  | $Pr^{\theta_A}(A_2) = 0.65$ |
|  | $Pr^{\theta_A}(\theta_A) = 1.00$ |
|  | $Pr^{\theta_B}(B_1) = 1.00$ |
|  | $Pr^{\theta_B}(B_2) = 0.00$ |
|  | $Pr^{\theta_B}(\theta_B) = 1.00$ |

## 5 Evidential Data Mining Apriori: EDMA

In this section, we introduce the Evidential Data Mining Apriori (EDMA) algorithm that allows the extraction of all frequent itemsets. Each itemset having a support greater than a threshold *minsup* is considered as frequent and is retained. The proposed algorithm relies on Apriori algorithm basics [15].

Apriori exploits this assumption by generating frequents in a level-wise manner. First of all, it generates frequent items (level 1) by removing those candidates (items) that do not fulfill the *minsup* constraint. From the generated frequent, it seeks to find the frequent of the next level by composing those of the precedent level. The treatment comes to an end when no further frequent itemset can be generated.

Algorithm 1 sketches the EDMA algorithm where it generates $\mathscr{E}\mathscr{I}\mathscr{F}\mathscr{F}$ the set of all evidential frequent itemset. Their determination is based on support measure and *minsup* constraint fulfillment. This test is performed by *Frequent_itemset* function. The support is computed through the *Support_estimation* function. As it is shown, the support estimation does not rely anymore on calculating the Cartesian product of BBAs but on stored item's precise measure (Pr Table). Since the algorithm sweeps the search space in breadth first manner, EDMA generates candidates (i.e., *candidate_apriori_gen* function) from the frequent itemset of the previous level.

EDMA is an Apriori based algorithm that extract frequent evidential itemsets. It also generalizes several other known mining algorithms. In case of having only categorical items (categorical BBA), the database can be viewed as a binary transaction database. In this case, the proposed EDMA approach for evidential databases matches the original Apriori algorithm for binary databases [15]. Since Evidential database represents imprecision and uncertainty, it assimilates fuzzy databases via consonant BBA. The EDMA approach also assimilates other fuzzy Apriori algorithms as [4].

**Algorithm 1** Evidential Data Mining Apriori (EDMA) algorithm

| | |
|---|---|
| **Require:** $\mathcal{EDB}, minsup, PT, Size\_\mathcal{EDB}$ | 14: **function** FREQUENT_ITEMSET($candidate$, |
| **Ensure:** $\mathcal{EIFF}$ | $\quad$ $minsup, PT, Size\_\mathcal{EDB}$) |
| 1: **function** SUPPORT_ESTIMATION($PT$, | 15: $\quad$ $frequent \leftarrow \emptyset$ |
| $\quad$ $I, d$) | 16: $\quad$ **for all** $x$ in $candidate$ **do** |
| 2: $\quad$ $Sup_I \leftarrow 0$ | 17: $\quad\quad$ **if** $Support\_estimation(PT, x, Size\_\mathcal{EDB}) \geq$ |
| 3: $\quad$ **for** j=1 to d **do** | $\quad$ $minsup$ **then** |
| 4: $\quad\quad$ $Sup_{Trans} \leftarrow 1$ | 18: $\quad\quad\quad$ $frequent \leftarrow frequent \cup \{x\}$ |
| 5: $\quad\quad$ **for all** $i \in Pr(j).focal\_element$ | 19: $\quad\quad$ **end if** |
| $\quad$ **do** | 20: $\quad$ **end for** |
| 6: $\quad\quad\quad$ **if** $Pr(j).focal\_element \in I$ | 21: $\quad$ **return** $frequent$ |
| $\quad$ **then** | 22: **end function** |
| 7: $\quad\quad\quad\quad$ $Sup_{Trans} \leftarrow Sup_{Trans} \times$ | 23: $\mathcal{EIFF} \leftarrow \emptyset$ |
| $\quad$ $Pr(j).value$ | 24: $size \leftarrow 1$ |
| 8: $\quad\quad\quad$ **end if** | 25: $candidate \leftarrow candidate\_apriori\_gen(\mathcal{EDB}, size)$ |
| 9: $\quad\quad$ **end for** | 26: **While** ($candidate \neq \emptyset$) |
| 10: $\quad\quad$ $Sup_I \leftarrow Sup_I + Sup_{Trans}$ | 27: $\quad$ $freq \leftarrow Frequent\_itemset (candidate, minsup, PT, Size\_\mathcal{EDB})$ |
| 11: $\quad$ **end for** | 28: $\quad$ $size \leftarrow size + 1$ |
| 12: $\quad$ **return** $\frac{Sup_I}{d}$ | 29: $\quad$ $\mathcal{EIFF} \leftarrow \mathcal{EIFF} \cup freq$ |
| 13: **end function** | 30: $\quad$ $candidate \leftarrow candidate\_apriori\_gen(\mathcal{EDB}, size, freq)$ |
| | 31: **End While** |

**Fig. 1** EDMA algorithm

# 6 Experimentation and Results

In this section, we present how we managed to conduct our experiments and we discuss comparative results.

## 6.1 Evidential Database Construction

No doubt the evidential database is a real life need where opinions are perfectly modeled via BBAs. Despite their real contribution, evidential databases are really hard to find. In [9], tests were conducted on synthetic database. Even in [16], the constructed BBA includes only one evidential attributes. In [10], the authors worked on a simplified naval anti-surface warfare scenario. In the following, we propose a method that allows to construct an evidential database from a numerical dataset. We based our evidential database construction on the ECM [17] clustering approach. It is an FCM-like algorithm based on the concept of credal partition, extending those of hard, fuzzy, and possibilistic ones. To derive such a structure, we minimized the proposed objective function:

$$J_{ECM}(M,V) \triangleq \sum_{i=1}^{d} \sum_{\{j/A_j \neq \emptyset, A_j \subseteq \Omega\}} c_j^{\alpha} m_{ij}^{\beta} dist_{ij}^2 + \sum_{i=1}^{n} \delta^2 m_{i\emptyset}^{\beta} \tag{15}$$

subject to:

$$\sum_{\{j/A_j \neq \emptyset, A_j \subseteq \Omega\}} m_{ij} + m_{i\emptyset} = 1 \quad \forall i = 1, d \tag{16}$$

where $m_{i\emptyset}$ and $m_{ij}$ denote respectively $m_i(\emptyset)$ and $m_i(A_j)$. $M$ is the credal partition $M = (m_1, \ldots, m_d)$ and $V$ is a cluster centers matrix. $c_j^\alpha$ is a weighting coefficient and $dist_{ij}$ is the Euclidean distance. In our case, the parameters $\alpha$, $\beta$ and $\delta$ were fixed to 1, 2 and 10.

In order to obtain evidential databases, this approach was applied on several UCI benchmarks [1]. The studied datasets are summarized on Table 3 in terms of number of instances and attributes. For each dataset, the number of focal elements after ECM application was addressed. The number of focal element is related to the objective function $J_{ECM}$ that was minimized.

**Table 3** Data set characteristics

| Data set | #Instances | #Attributes | #Focal elements |
|---|---|---|---|
| Iris | 150 | 4 | 32 |
| Vertebral Column | 310 | 6 | 64 |
| Diabetes | 767 | 9 | 144 |
| Abalone | 4177 | 9 | 40 |

## 6.2 Comparative Results

We compared the precise support measure integrated into the EDMA algorithm to [9, 10] support metric. As it is shown in Table 4, the $EDMA - Pr$ attribute concerns the introduced precise support and $EDMA - Bel$ refers to the definition given in section 3. For our experimentation, we integrated the ramification support (subsection 3.2) into EDMA algorithm. Since EDMA relies on a table that contains all item's metric values, we created the Belief Table (BT). The Belief Table has the same structure as that of Pr Table (c.f., subsection 4.2) and in which we stored all item's belief. The different approaches were tested on the obtained evidential database as illustrated in subsection 6.1.

**Table 4** Comparative results in terms of the number of frequent pattern number

| Support | Iris | | Diabete | | Vertebral Column | | Abalone | |
|---|---|---|---|---|---|---|---|---|
| | EDMA-Pr | EDMA-Bel | EDMA-Pr | EDMA-Bel | EDMA-Pr | EDMA-Bel | EDMA-Pr | EDMA-Bel |
| 0.9 | 15 | 15 | 319 | 191 | 63 | 63 | 767 | 511 |
| 0.8 | 23 | 15 | 1503 | 319 | 95 | 63 | 767 | 511 |
| 0.7 | 47 | 15 | 5055 | 671 | 415 | 63 | 767 | 511 |
| 0.6 | 95 | 31 | 9074 | 1407 | 799 | 95 | 1919 | 511 |

Table 4 illustrates the number of extracted frequent patterns from the evidential databases. As is it demonstrated, the Precise support extracts more frequent pattern than do the based belief method. This result is expected since the precise metric study all subsets of the superset and considers those having an intersection with the considered itemset. In addition, the number of patterns increases normally as far as the considered *minsup* threshold decreases.

We also conducted performance test on our proposed algorithm which we compared to an exhaustive approach. This approach consists in the Cartesian based algorithm (Cart-Bel in Table 5). The Cartesian algorithm computes all possible BBAs needed for support measure. The complexity of such approach is exponential with respect to the number of focal elements. Indeed, for an evidential database with $k$ attributes each one has $n$ focal elements and $d$ transactions. The arithmetic complexity of a Cartesian product is: $\mathscr{C} = d \times n^k = O(n^k)$.

**Table 5** Comparative results in terms of execution time (seconds)

| Support | Iris | | | Diabete | | | Vertebral Column | | | Abalone | | |
|---|---|---|---|---|---|---|---|---|---|---|---|---|
| | EDMA-Pr | EDMA-Bel | Cart-Bel≈ | EDMA-Pr | EDMA-Bel | Cart-Bel≈ | EDMA-Pr | EDMA-Bel | Cart-Bel≈ | EDMA-Pr | EDMA-Bel | Cart-Bel≈ |
| 0.9 | 0.13 | 0.10 | 96172 | 4.72 | 1.65 | 6.43E+48 | 0.74 | 0.24 | 3.96E+24 | 79.71 | 16.35 | 9.13E+41 |
| 0.8 | 0.13 | 0.11 | 96172 | 175.94 | 3.00 | 6.43E+48 | 1.11 | 0.24 | 3.96E+24 | 75.77 | 16.18 | 9.13E+41 |
| 0.7 | 0.35 | 0.15 | 96172 | 21188 | 12.69 | 6.43E+48 | 15.21 | 0.24 | 3.96E+24 | 77.23 | 16.24 | 9.13E+41 |
| 0.6 | 1.01 | 0.25 | 96172 | 12.21E+4 | 100.56 | 6.43E+48 | 116.32 | 0.33 | 3.96E+24 | 337.87 | 16.21 | 9.13E+41 |

Table 5 illustrates a comparative performance tests between EDMA and Cartesian based algorithm. The proposed algorithm has drastically improved the results. The extraction performance of the EDMA-Bel is better than those of EDMA-Pr. This observation can be explained by the number of extracted patterns. The more frequent candidates are generated, the more time consumed is.

# 7 Conclusion

In this paper, we tackled data mining problem in evidential databases. We focused on evidential itemsets' support estimation. We detailed state of art evidential support metric. To drop the original complexity, we proposed a simplification for their methods by reducing the Cartesian product to a simple belief product. We also introduced a new support measure that brings precision by analyzing deeply the BBA's frame of discernment. The proposed precise measure extracts more hidden frequent patterns than the usual method. The precise measure was applied on an Apriori based algorithm and was tested on evidential databases obtained from transformed datasets. As illustrated in the experimentation section, despite the huge item's number that evidential database contains, EDMA generates all frequent itemsets in a reasonable execution time. This problem can be recovered in future works by tackling compact evidential itemset representation. Indeed, estimating the support exactly from a compact set had never been more challenging. In addition, quality test for the generated frequent patterns is a need. In future work, we plan to study the developpement of a new method to estimate the confidence of evidential associative rules based on our support measure.

# References

1. Frank, A., Asuncion, A.: UCI machine learning repository (2010),
   http://archive.ics.uci.edu/ml
2. Lee, S.: Imprecise and uncertain information in databases: an evidential approach. In: Proceedings of Eighth International Conference on Data Engineering, Tempe, AZ, pp. 614–621 (1992)
3. Zadeh, L.A.: Fuzzy sets. Information and Control 8(3), 338–353 (1965)
4. Hong, T.-P., Kuo, C.-S., Wang, S.-L.: A fuzzy AprioriTid mining algorithm with reduced computational time. Applied Soft Computing 5(1), 1–10 (2004)
5. Chen, Y., Weng, C.: Mining association rules from imprecise ordinal data. Fuzzy Set Syst. 159(4), 460–474 (2008)
6. Lee, Y.-C., Hong, T.-P., Lin, W.-Y.: Mining fuzzy association rules with multiple minimum supports using maximum constraints. In: Negoita, M.G., Howlett, R.J., Jain, L.C. (eds.) KES 2004, Part II. LNCS (LNAI), vol. 3214, pp. 1283–1290. Springer, Heidelberg (2004)
7. Dempster, A.: Upper and lower probabilities induced by multivalued mapping. In: AMS-38 (1967)
8. Shafer, G.: A Mathematical Theory of Evidence. Princeton University Press (1976)
9. Bach Tobji, M.A., Ben Yaghlane, B., Mellouli, K.: Incremental maintenance of frequent itemsets in evidential databases. In: Sossai, C., Chemello, G. (eds.) ECSQARU 2009. LNCS, vol. 5590, pp. 457–468. Springer, Heidelberg (2009)
10. Hewawasam, K.K.R., Premaratne, K., Shyu, M.-L.: Rule mining and classification in a situation assessment application: A belief-theoretic approach for handling data imperfections. Trans. Sys. Man Cyber. Part B 37(6), 1446–1459 (2007)
11. Smets, P.: The Transferable Belief Model and other interpretations of Dempster-Shafer's Model. In: Proceedings of the Sixth Annual Conference on Uncertainty in Artificial Intelligence, UAI 1990, pp. 375–383. MIT, Cambridge (1990)
12. Smets, P.: Belief functions. In: Smets, P., Mamdani, A., Dubois, D., Prade, H. (eds.) Non Standard Logics for Automated Reasoning, pp. 253–286. Academic, London (1988)
13. Smets, P., Kennes, R.: The Transferable Belief Model. Artificial Intelligence 66(2), 191–234 (1994)
14. Hewawasam, K.K.R., Premaratne, K., Shyu, M.-L., Subasingha, S.P.: Rule mining and classification in the presence of feature level and class label ambiguities. In: SPIE 5803, Intelligent Computing: Theory and Applications III, p. 98 (2005)
15. Agrawal, R., Srikant, R.: Fast algorithm for mining association rules. In: Proceedings of International Conference on Very Large Data Bases, VLDB, Santiago de Chile, Chile, pp. 487–499 (1994)
16. Bach Tobji, M.A., Ben Yaghlane, B., Mellouli, K.: Frequent itemset mining from databases including one evidential attribute. In: Greco, S., Lukasiewicz, T. (eds.) SUM 2008. LNCS (LNAI), vol. 5291, pp. 19–32. Springer, Heidelberg (2008)
17. Masson, M.-H., Denœux, T.: ECM: An evidential version of the fuzzy c-means algorithm. Pattern Recognition 41(4), 1384–1397 (2008)

# Automatic Evaluation of the Elastic Modulus of a Capsule Membrane

Thi-Xuan Chu, Anne-Virginie Salsac, Eric Leclerc, and Dominique Barthès-Biesel

**Abstract.** In order to characterize the mechanical properties of bioartificial micro-capsules, we have designed an inverse analysis technique that combines microfluidic experiments with a multi-physical numerical model. The experiments consist of flowing a suspension of capsules in a channel with cross section similar to the particle size. The mechanical properties of the capsule membrane are deduced from the deformed shape by means of a fluid-structure interaction model of the problem. The model couples a boundary integral method to solve for the inner and outer fluid motion with a collocation method to solve for the membrane deformation. An optimization technique is used to find the parameters that provide the same deformation as measured experimentally and deduce an estimation of the membrane properties. The technique is applied on capsules with reticulated protein membranes. We show that it is possible to discriminate the influence of the physico-chemical conditions of fabrication.

## 1 Introduction

Encapsulated liquid droplets enclosed by a thin elastic membrane, are widely found in nature (red blood cells, eggs) and in cosmetic, food or pharmaceutical industry [1]. The deformable membrane that separates the internal and external liquids prevents the diffusion and degradation of the internal substance and controls its release. Capsules are commonly suspended in another liquid. The control of the motion, deformation and potential burst of flowing capsules is a difficult multi-scale and multi-physics problem of biomechanics. Indeed, at the capsule level, it involves a

Thi-Xuan Chu · Anne-Virginie Salsac · Eric Leclerc · Dominique Barthès-Biesel
Biomechanics & Bioengineering Laboratory (BMBI, UMR CNRS 7338),
Université de Technologie de Compiègne, CS60319, 60203 Compiègne, France
e-mail: xuan@itims.edu.vn,
    {anne-virginie.salsac,eric.leclerc,
    dominique.barthes-biesel}@utc.fr
    http://www.utc.fr/bmbi/

V.-N. Huynh et al. (eds.), *Knowledge and Systems Engineering, Volume 2*,
Advances in Intelligent Systems and Computing 245,
DOI: 10.1007/978-3-319-02821-7_34, © Springer International Publishing Switzerland 2014

strong interaction between hydrodynamic forces and elastic forces, which represents a complex fluid-structure coupling problem. At the membrane and macromolecular level, the resistance to external stress depends on the physico-chemical composition of the capsule wall and thus on the chemical reactions that lead to the membrane formation.

The ability to characterize the mechanical properties of the capsule membrane is essential for the design of artificial capsules, but it is a challenging task when the capsules have a small size of order a few tens of micrometers, as is often the case in biomedical applications.

Recently, a new method has been proposed to measure the membrane properties of a capsule population. It consists of flowing a capsule suspension into a cylindrical glass capillary tube with radius comparable to that of the capsules [2, 3]. The experimental set-up allows us to capture the deformed profile and velocity of individual capsules as they flow in the tube. Hydrodynamic forces and boundary confinement lead to a large deformation of the capsules, which can take either a parachute or a slug shape depending on the flow velocity and/or on the relative size between the capsule and tube cross sections. However, this information alone is not sufficient to infer the capsule membrane mechanical properties. First, a sophisticated mechanical model of a flowing capsule is necessary to relate the resulting capsule deformation to the flow conditions. Second, an inverse analysis must be performed in the parameter space of the model to deduce plausible values of the mechanical properties of the capsule membrane.

As an example, we show how the method can be used to determine the mechanical properties of the membrane of capsules and also discriminate between capsules prepared under different physico-chemical conditions.

We first describe the experimental setup, used to generate the confined flow of a capsule suspension in a microchannel, as well as the associated mechanical model. We then detail the procedure through which we identify the capsule mechanical properties. Results are provided for various populations of microcapsules fabricated by creating a two-phase emulsion and reticulating proteins at the interface of the drops.

## 2 Materials and Methods

### 2.1 Experimental Procedure

Microcapsules are prepared using an interfacial cross-linking method [4]. Some of the preparation parameters are varied to obtain membranes with different cross-linking degrees encapsulating liquid droplets with different diameters. Capsules have been produced with a cross-linked membrane made from ovalbumin (Sigma).

The capsules are transferred into glycerin for the measurement. The capsule samples are kept in aqueous suspension after fabrication. For the mechanical evaluation, a volume of 40 $\mu$l of the microcapsule sediment is suspended in 1.8 ml of glycerin, which leads to a 2.2% (w/v) capsule suspension. After mixing by successive pumping in-and-out of a syringe, the suspension is left to rest for 10 min at the room temperature of 23°C to allow the inner water to be replaced by the glycerin solution by osmotic exchange. Rheological measurements of the capsule suspensions at 23°C provided a viscosity value of $\mu = 0.7$ Pa.s.

A syringe pump containing the capsule suspension is connected to the microfluidic system to perfuse a cylindrical glass microchannel with 75 $\mu$m internal diameter of the same order as the capsule diameter.

The capsule motion in the micro-channel is observed with a microscope and recorded with a high-speed camera. The microscope is focused on the axial plane of the channel so that we obtain the cross section of the capsule in its plane of symmetry. The images are recorded in a cross section located at about 200 channel radii from the channel entrance, where the flow is considered to be at steady state [5]. The software ImageJ is used to extract the contour of the capsule manually. We calculate the capsule volume assuming axisymmetry. From the volume, we obtain the initial capsule radius $a$ and size ratio $a/R$, where $R$ is the tube radius. We measure characteristic geometrical quantities of the experimental profile: the total length $L$ and axial length $L_a$ of the capsule along the tube axis, and the area $S$ of the meridional profile (Figure 1$b$). The quantities $L$ and $S$ are obtained averaging the values for the top and bottom half-profiles. Altogether, the contour determination is estimated to lead to a 2% error on the size ratio and on the various geometrical characteristic quantities. From two successive images, we measure the capsule velocity $v$, which varies between 1 and 10 mm/s, depending on the size of the capsule and the flow rate. It is of course impossible to deduce quantitative information about the membrane mechanical properties from the experimental evidence only, unless one has a mechanical model of the process.

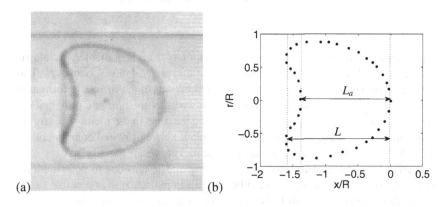

(a)          (b)

**Fig. 1** Initially spherical capsule flowing in a 75 $\mu$m pore: (a) experimental image; (b) capsule contour and definition of geometrical quantities $L$, $L_a$. From [3]

## 2.2 Mechanical Model of the Confined Flow of a Capsule in a Pore

The numerical model initially proposed by [6] and later improved by [5] and [7] is used to determine the flow of a capsule in a cylindrical channel of comparable size. The model assumptions and results are only briefly outlined; more details may be found in [7].

An initially spherical capsule of radius $a$ is filled with an incompressible Newtonian liquid of viscosity $\mu$ and enclosed by an infinitely thin elastic membrane. The capsule is placed in another incompressible Newtonian liquid of viscosity $\mu$, flowing with mean velocity $U$ in a cylindrical channel of radius $R$ (Figure 2). The flow Reynolds number is supposed to be small so that the internal and external fluid motions obey the Stokes equations. Buoyancy effects are negligible. Consequently, a capsule centered on the tube axis has an axisymmetric deformation.

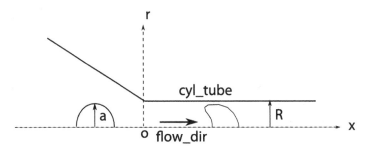

**Fig. 2** Geometry used in the numerical simulation: the undeformed capsule is initially positioned in an hyperbolic entrance, before being flowed into the cylindrical channel. From [3].

The membrane is assumed to consist of an infinitely thin sheet of an hyper-elastic isotropic material with surface shear modulus $G_s$ and area dilation modulus $K_s$. Under the condition of axisymmetry, the principal directions of the surface deformation tensor are along the meridian and parallel curves with corresponding principal extension ratios $\lambda_1$ and $\lambda_2$, respectively. The elastic tensions in the membrane (forces per unit arc length measured in the membrane plane) have principal components $T_1$ and $T_2$ also directed along the meridian and parallel curves, respectively.

Different membrane constitutive laws, relating elastic tensions to deformations, have been proposed to model thin membranes [8]. In particular, it is possible to use laws which are either strain-softening or strain-hardening under large deformation. The law that corresponds to the capsule mechanical behavior must be such that it leads to a constant value of the elastic modulus for large or small deformation. In a previous study of similar ovalbumin capsules, it was shown that a strain-hardening law was not appropriate to describe the capsule membrane, whereas a neo-Hookean (NH) strain-softening law led to constant values of the membrane elastic shear modulus [2]. Consequently, we have used here a neo-Hookean law that assumes that the membrane is an infinitely thin sheet of a three-dimensional isotropic volume-incompressible material. The principal elastic tensions are expressed by [8, 2]:

$$T_1 = \frac{G_s}{\lambda_1 \lambda_2} \left[ \lambda_1^2 - \frac{1}{(\lambda_1 \lambda_2)^2} \right], T_2 = \frac{G_s}{\lambda_1 \lambda_2} \left[ \lambda_2^2 - \frac{1}{(\lambda_1 \lambda_2)^2} \right]. \tag{1}$$

The area dilation modulus is then shown to be $K_s = 3G_s$.

The problem is solved by means of a boundary integral technique, coupled to the Lagrangian tracking of the capsule interface [5, 7].

The model input parameters are the size ratio $a/R$ and the capillary number $Ca = \mu U/G_s$ that measures the ratio between the viscous and elastic forces. The model output data are the capsule deformed profile at steady state and the velocity ratio $v/U$. The model also provides other physical quantities that no current experimental technique can measure, such as the pressure difference $\Delta p$ between the front and back of the capsule, the elastic tension distribution in the capsule membrane and the surface strain energy of the capsule.

The numerical model provides values for the geometrical non-dimensionalized parameters $L/R$, $L_a/R$, $S/R^2$ of the deformed capsule and for the velocity ratio $v/U$ as a function of $a/R$ and $Ca$ (Figure 3). One can define the quantity $L_p = L - L_a$ as a measure of the depth of the parachute. The lengths $L$ and $L_a$ increase with the capillary number (i.e. the deformation) except for small capsules ($a/R < 1$) at low capillary numbers ($Ca < 0.02$) (Figures 3a-b). Figure 3b shows that $L_p$ increases with the capillary number. Figure 3c shows that the meridional area $S/R^2$ does not vary much with the capillary number and depends mainly on the size ratio. The velocity ratio $v/U$, shown in Figure 3d, increases monotonically with the capillary number.

## 2.3 Inverse Analysis Procedure

We have developed an algorithm to automatically perform the inverse analysis of capsule profiles. Tolerances have been defined to account for the uncertainty of the experimental measurements. Depending on the flow conditions, the membrane may appear more or less fuzzy. For ovalbumin capsules this leads to a possible error of 2% on $a/R$, 4% on $L/R$ and $L_a/R$ and 5% on $S/R^2$.

A membrane law is first assumed. For the measured size ratio $a/R$, the algorithm then determines the values of the capillary number $Ca$ for which the experimental and numerical values of $L/R$, $L_a/R$ and $S$ correspond when the tolerances are accounted for. This yields three ensembles of values $\{Ca\}_L, \{Ca\}_{La}$ and $\{Ca\}_S$, the intersection of which is then computed

$$\{Ca\}_{a0} = \{Ca\}_L \bigcap \{Ca\}_{La} \bigcap \{Ca\}_S \tag{2}$$

We then repeat the procedure for $a/R \pm 2\%$, which yields four more intersections of values of $Ca$: $\{Ca\}_{a+1}, \{Ca\}_{a+2}, \{Ca\}_{a-1}, \{Ca\}_{a-2}$. We finally compute the global intersection

$$\{Ca\} = \{Ca\}_{a+1} \bigcap \{Ca\}_{a+2} \bigcap \{Ca\}_{a0} \bigcap \{Ca\}_{a-1} \bigcap \{Ca\}_{a-2} \tag{3}$$

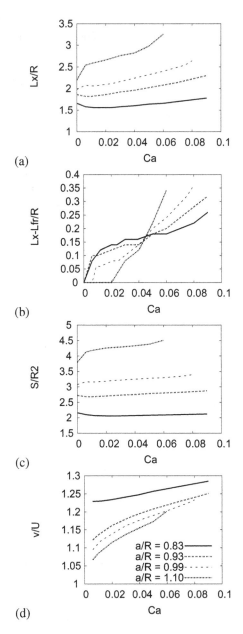

(a)

(b)

(c)

(d)

**Fig. 3** Charts of the results provided by the numerical model: evolution of $L/R$ (a), $L_p/R$ (b), $S/R^2$ (c) and $v/U$ (d) as a function of the capillary number and size ratio. From [3].

If $\{Ca\} = \emptyset$, when we discard the capsule and consider that it cannot be fitted by the model. Otherwise, we compute the mean value of the fitting capillary number as the average of the values $\{Ca\}$. For each value of $Ca \in \{Ca\}$, we calculate the mean fluid velocity $U$ from the capsule velocity $v$ and the velocity ratio $v/U$ of the database. Finally, we calculate the shear modulus that corresponds to each $Ca \in \{Ca\}$ by means of the relation $G_s = \mu U/Ca$.

## 3 Results

As an example, we have characterized ovalbumin capsules with a 60 $\mu$m mean diameter fabricated from a 10% ovalbumin solution at various pH and reticulation times $t_r$ [3]. Typical capsule shapes are illustrated in Figure 4. For a given value of $a/R$, the capsules experience a small deformation with convex front and back at low velocity (or small $Ca$), and larger deformation with convex front and largely concave rear when the velocity (or $Ca$) is increased (Figures 4a-b). Figures 4c-d show the influence of the size ratio $a/R$ for a given capillary number ($Ca = 0.02$).

The automatic inverse analysis procedure has been applied to a capsule population prepared under given pH and reticulation time $t_r$. For instance, the resulting values of $G_s$ are shown for microcapsules fabricated at pH = 6.8 and with two values of $t_r$ in Figure 5. Each cross on the graphs of Figure 5 represents the mean shear modulus obtained for one capsule by means of the inverse analysis. These graphs show that the capsule size ratio does not influence the mechanical properties within the experimental uncertainties. More rigid capsules obtained for larger values of $t_r$ are more difficult to deform: this is why the dispersion in the values of $G_s$ is larger in Figure 5b than in Figure 5a. Indeed, for small deformation, the dependence on $Ca$ of the $L/R$ and $S/R^2$ curves is very small (Figure 3). As a consequence there is a wide interval of values of $Ca$ (and thus of $G_s$) for which the value of, say $L/R$, can be obtained by the inverse analysis, when the capsule does not assume a parachute shape. In conclusion, for a given capsule population, we can define a mean value of $G_s$ which is typically obtained with an error of about 30% or less.

Typical fits between the numerical capsule profiles calculated for the mean $G_s$ value and the experimental profiles are shown for different capsules in Figure 4. The fact that the fit is excellent indicates that we have indeed obtained a plausible value of $G_s$.

The global effect of the physico-chemical parameters used during fabrication, i.e. pH and reticulation time $t_r$, on the capsule membrane is shown in Figure 6. The membrane rigidity (as measured by $G_s$) increases with the reticulation time (Figures 5 and 6). This is due to the fact that the longer the reticulation reaction, the more covalent links are created at the interface. We also find that $G_s$ is essentially insensitive to pH$\leq$ 7, but markedly increases for pH 8. This can be explained by the change of conformation of the interfacial ovalbumin molecules for 7<pH< 7.5 [3]. In conclusion, the measurement of $G_s$ on the capsule scale reflects in part the structure of the membrane on the molecular scale, such as the degree of cross-linking.

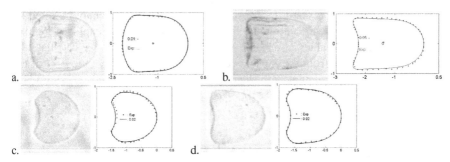

**Fig. 4** Picture and corresponding superposition of the experimental (dotted line) and numerical (continuous line) profiles for ovalbumin capsules fabricated with $t_r = 5$ min flowing down the microchannel: (a) pH 5, $a/R = 1.01$, $Ca = 0.01$ ($v = 1.23$ mm/s); (b) pH 5, $a/R = 1.01$, $Ca = 0.05$ ($v = 3.1$ mm/s); (c) pH 8, $a/R = 0.81$, $Ca = 0.02$ ($v = 1.45$ mm/s); (d) pH 8, $a/R = 0.9$, $Ca = 0.02$ ($v = 2.92$ mm/s)

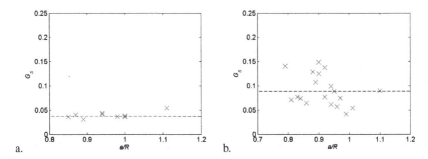

**Fig. 5** Membrane shear modulus $G_s$ of ovalbumin capsules fabricated at pH = 6.8 as a function of the capsule size ratio. The dashed line represents the mean value of $G_s$. (a) $t_r = 5$ min ($G_s = 0.039 \pm 0.005$ N/m), (b) $t_r = 15$ min ($G_s = 0.091 \pm 0.029$ N/m).

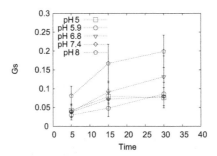

**Fig. 6** Variation of the mean shear modulus of the microcapsule populations as a function of the reaction pH and time of reticulation $t_r$. From [3].

# 4 Conclusion

The present study shows that it is possible to measure the mechanical properties of microcapsules composed of a reticulated protein membrane. A multi-physical model of the motion and deformation of the capsule in a micropore of similar size is necessary to identify the membrane properties from the capsule profiles measured experimentally. A systematic inverse analysis tool has been designed to explore the parameter space of capsule profiles and deduce the most plausible value of the membrane shear elastic modulus. As an illustration, we have applied the technique to capsules with an ovalbumin membrane. We have showed that the shear resistance of an ovalbumin membrane increased with the degree of cross-linking. This indicates that the technique is able to detect changes in the membrane at the microscale. The fact that the characterization technique is non-destructive opens interesting perspectives for online verification processes for industrial capsule fabrication.

**Acknowledgments.** The capsules were fabricated by the team of Dr. Florence Edwards-Lévy, Institut de Chimie Moléculaire de Reims (UMR CNRS 7312), Université de Reims Champagne-Ardenne, Reims, France. This work was supported by the Conseil Régional de Picardie (MODCAP grant).

Dr. Thi-Xuan Chu currently works in the International Training Institute for Materials Science, Hanoi University of Science and Technology (Vietnam).

# References

1. Barthès-Biesel, D.: Modeling the motion of capsules in flow. Current Opinion in Colloid and Interface Science 16, 3–12 (2011)
2. Lefebvre, Y., Leclerc, E., Barthès-Biesel, D., Walter, J., Edwards-Lévy, F.: Flow of artificial microcapsules in microfluidic channels: A method for determining the elastic properties of the membrane. Phys. Fluids 20, 123102 (2008)
3. Chu, T.X., Salsac, A.-V., Leclerc, E., Barthès-Biesel, D., Wurtz, H., Edwards-Lévy, F.: Comparison between measurements of elasticity and free amino group content of ovalbumin microcapsule membranes: discrimination of the cross-linking degree. Int. J. Coll. Interf. Sci. 355, 81–88 (2011)
4. Edwards-Lévy, F., Andry, M.C., Lévy, M.C.: Determination of free amino group content of serum albumin microcapsules using trinitrobenzenesulfonic acid: effect of variations in polycondensation pH. Int. J. Pharm. 96, 85–90 (1993)
5. Diaz, A., Barthès-Biesel, D.: Entrance of a Bioartificial Capsule in a Pore. CMES 3, 321–337 (2002)
6. Quéguiner, C., Barthès-Biesel, D.: Axisymmetric motion of the capsules through cylindrical channels. J. Fluid Mech. 348, 349–376 (1997)
7. Lefebvre, Y., Barthès-Biesel, D.: Motion of a capsule in a cylindrical tube: Effect of membrane pre-stress. J. Fluid Mech. 589, 157–181 (2007)
8. Barthès-Biesel, D., Diaz, A., Dhenin, E.: Effect of constitutive laws for two dimensional membranes on flow-induced capsule deformation. J. Fluid Mech. 460, 211–222 (2002)

9. Lac, E., Barthès-Biesel, D.: Deformation of a capsule in simple shear flow: Effect of membrane prestress. Phys. Fluids 17, 072105 (2005)

10. Sherwood, J.D., Risso, F., Collé-Paillot, F., Edwards-Lévy, F., Lévy, M.-C.: Rates of transport through a capsule membrane to attain Donnan equilibrium. Int. J. Coll. Interf. Sci. 263, 202–212 (2003)

# Recovering the Contralateral Arm Strength Loss Caused by an Induced Jaw Imbalance

Nguyen Van Hoa, Le Minh Hoa, Nguyen Thanh Hai, and Vo Van Toi

**Abstract.** This study brought a more complete picture of the investigated phenomenon regarding how an evoked jaw imbalance caused the loss of arm strength and how to recover this strength. It also suggested that the proposed mechanical model, although simple, was a suitable tool to characterize, optimize and allowed further investigations of this phenomenon, and the emitted hypothesis was plausible. It paves the way to new clinical studies with patients to investigate and manage this kind of temporomandibular joint disorder, and to novel investigations in sport medicine.

## 1 Background

Mouth guards and the likes are effective dental protection devices and also are appropriate investigative tools to study the relationship between temporomandibular joint (TMJ) and other systemic muscles including neck and arm muscles [1, 2, 3, 4, 5, 6, 7, 8]. In a previous study [9] we found that when a subject bit into a spacer using one side of the jaw his/her arm of the contralateral side lost its strength while the strength of the arm of the ipsilateral side remained intact. The statistical study on 34 young and healthy subjects of both genders revealed that. The loss decreased linearly as the thickness of the spacer was increasing. The loss leveled off when the spacer thickness reached about 2mm see Fig.10. In a recent study to determine the coupling between the jaw imbalance and the loss of arm strength we measured the electromyogram (EMG) of muscles of jaw (masseter), neck (trapezoid) and arms (deltoid, brachiodiolis). We found that muscle fatigue and the decrease in muscle contraction level led to the loss of arm strength [10]. Our study is in agreement with reported literature that changes in the dental occlusion height

Nguyen Van Hoa · Le Minh Hoa · Nguyen Thanh Hai · Vo Van Toi
Biomedical Engineering Department,
International University of Vietnam National Universities, Ho Chi Minh City, Vietnam
e-mail: vvtoi@hcmiu.edu.vn

V.-N. Huynh et al. (eds.), *Knowledge and Systems Engineering, Volume 2*,
Advances in Intelligent Systems and Computing 245,
DOI: 10.1007/978-3-319-02821-7_35, © Springer International Publishing Switzerland 2014

affected the strength of different muscles including masseter, temporalis, trapezius, stemocheidomastoid, deltoid, bicep, triceps and latissimus dorsi [11, 12, 13, 14]. Further, other authors reported that a unilateral task performed by one hand induces a contralateral muscle movement on the other hand [15, 16]. These studies suggest that this coupling mechanism has a muscular component i.e., this is a biomechanical issue. An important question now is whether there is a way to bring the arm strength back to its natural value. This question is particular relevant in the investigation of a therapeutic method to manage patients suffered from this kind of TMJ disorder. It is also relevant in sport medicine to help prevent athletes from losing their performance while using the mouth guard.

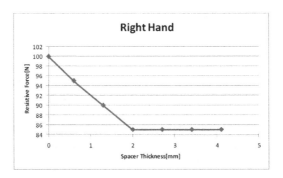

**Fig. 1** (Excerpted from [9] with permission): The variation of the arm strength with respect to the thickness of a bite plate of one subject. The subject resisted against a pull-down force applied on the right arm while firmly biting into a spacer placed on the left side of his jaw. The right arm first gradually lost strength then the loss was leveled when the thickness of the spacer reached about 2mm. The left arm strength remained equal to the initial value when the spacer was not introduced to the left jaw.

In the study presented here we investigated a way to recover the arm strength to its natural value under the jaw imbalance condition.

## 2   Materials and Methods

### 2.1   Establishment of the Hypothesis and the Theoretical Model

Our hypothesis was that there is a coupling system that links the muscles from the jaw to the arms. In the natural situation this system rests on a support that governs the strength of the arm. An action that caused an imbalance in the jaw displaces this support and destabilizes the system. As a consequence, it weakens the strength of the opposite arm. Based on this hypothesis, the mechanism that governs the function of this system can be represented by a mechanical model that consists of a weightless and rigid beam AB which rotates about a fulcrum C located at the center of beam and a constant force spring at each end of the beam. When the system is in action these springs develop the forces $F_1$ and $F_2$, respectively. We assume that physiologically, on healthy people, there is a natural balance between the left and the right parts of the body, therefore:

$$F_1 = F_2 \tag{1}$$

### A. Natural strength of the arm

Fig.1 shows the force diagram that represents the situation when the subject did not bite on any spacer and an external force $F_0$ was applied on one arm until the arm yielded. As a consequence $F_0$ is the natural strength of the arm, $F_1$ and $F_2$ are the forces developed by the aforementioned coupling system against $F_0$. For the sake of consistency let's call A the right side of the subject and B the left side. Hence, in Fig.1 the strength of the subject's right arm was determined.

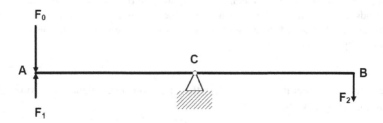

**Fig. 2** The balance of the forces on the model when a maximum external pull-down force $F_0$ applied on one arm (say at A) to make the arm yield. Hence, $F_0$ corresponds to the arm strength when the subject didn't bite into a spacer.

The balance of the torques about C can be described by the following equation:

$$F_0 L = F_1 L + F_2 L \tag{2}$$

Where $L = AC = CB$. Therefore:

$$F_0 = F_1 + F_2 \tag{3}$$

From (1) we deduce that

$$F_0 = 2F_1 = 2F_2 \tag{4}$$

### B. Contralateral arm's strength when the jaw is imbalanced

The fact that the subject bit into a spacer using one side of the jaw (say left) and caused the jaw imbalanced is modeled by the displacement of the fulcrum toward the corresponding side (B) where the spacer was held (Fig.1). The strength of the tested arm shown in Fig.1 therefore changed. For the sake of clarity, we term "ipsilateral" for elements on the side where the spacer is held by the jaw and "contralateral" for elements on the opposite side with respect to the fulcrum.

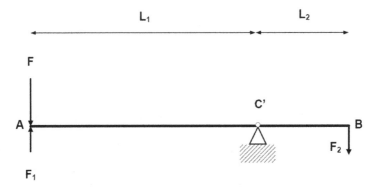

**Fig. 3** The balance of the forces on the model when the subject bites into a spacer using one side of the jaw (say left). In such a case, on the model, the fulcrum was displaced toward the same side (B). Hence, the external force F applied on the arm (say right) until it yields corresponds to the strength of the arm when the subject holds a spacer. It is demonstrated that $F$ is smaller than $F_0$ and is decreasing when the thickness of the spacer is increasing as found experimentally in [9] (see text).

Let F the contralateral arm strength in this situation. The balance of torques about the point C' (Fig.2) yields:

$$F = F_1 + (F_2/n) \tag{5}$$

Where $n = L_1/L_2$.

Because $n > 1$, in comparing (5) with (3), we deduced that $F < F_0$. This finding therefore indicates that when the subject bit into a spacer on one side of the jaw, the strength of his/her contralateral arm was reduced. Importantly, it suggests that the proposed theoretical model well describes the experimental results. Note that $n$ is a function of $d$ and $n > 1$. In a particular case, $n = 1$ when the fulcrum is in the middle of the beam (Fig.1) or $d = 0mm$ then $F = F_0$.

Furthermore, experimental results showed that for $d \leq 2mm$ the contralateral arm strength $F$ decreased linearly [9]. Therefore the behavior of the arm strength with respect to the height d of the spacer can be described as follows:

$$F = \alpha d + \beta \tag{6}$$

Where:
$F$ is the strength of the arm and is a function of $d$
$d$ is the height of the spacer and is a variable
$\alpha$ is the slope of the line, and is constant and negative
$\beta$ is the y-intercept, and is constant and positive.

The force $F_0$ (described in A) is the initial value of $F$. Indeed, from equation (6) when $d = 0$, we find that:

$$F = F_0 = \beta \tag{7}$$

The relationship between $d$ and $n$ can be developed as followed:
From equations (5) and (6) we solve for $n$ in terms of $d$:

$$n = F_2/(\alpha d + \beta - F_1) \tag{8}$$

From (4) and (7) we can deduce:

$$n = \beta/(2\alpha d + \beta) \tag{9}$$

To check the validity of this expression, from (9) we can find that when $d = 0$ then $n = 1$ as it must be.

In summary, based on the concept of the displacement of the fulcrum, the proposed model successfully describes a mechanism that reproduces the fact that an induced imbalance in the jaw caused a loss of the strength of the contralateral arm [9]. The value of spacer thickness d beyond which the loss of arm strength stopped, constitutes a boundary condition of the model. This boundary condition was represented by the horizontal line for $d \geq 2mm$. The model and its limit of validity can be expressed as:

$$\begin{cases} F = \alpha d + \beta & \text{for } 0mm \leq d \leq 2mm \\ F = 2\alpha + \beta & \text{for } d \geq 2mm \end{cases} \tag{I}$$

## C. Deduction from the model

By inspection we found that when an external counter-force called $F_3$ was applied on the ipsilateral side the contralateral force increased. Let's now call $F'$ the strength of the contralateral arm when the imbalance is applied to one side of the jaw and at the same time an external counter-force $F_3$ is applied to the ipsilateral arm (Fig.3) at

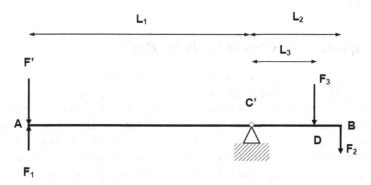

**Fig. 4** When an external counter-force $F_3$ is applied at D and its magnitude is increasing, the force $F'$ will be increasing. Theoretically, $F'$ can reach the value $F_0$ and can even go beyond it. In applying this observation to the individual subject this implies that when the spacer is applied in one side of the jaw, to help the contralateral arm recover its original strength $F_0$ we have to apply an external counter-force $F_3$ on the ipsilateral arm. Theoretically as the magnitude of $F_3$ continues to increase the contralateral arm would become stronger than normal. Note: $F_3$ is a function of the spacer thickness.

D. D is a point between C' and B. The balance of the torques about the fulcrum C' yields:

$$F'L_1 = F_1L_1 + F_2L_2 + F_3L_3 \qquad (10)$$

for $n' = L_2/L_3$, and in taking into account expressions (5), (6) and (9), we obtain:

$$F' = [(2\alpha d + \beta)/(\beta n')]F_3 + (\alpha d + \beta) \qquad (11)$$

Note that:

- $n' \geq 1$, $n'$ depends on the position of $F_3$ and is a function of $d$
- $F' = F$ when $F_3 = 0$; and
- $F' = F_0$ when both $F_3 = 0$ and $d = 0$.

In examining (11) we found that:

- If we replace $F_3 = 0$ (i.e., situation described in B), we find $F' = \alpha d + \beta$ as indicated by (6).
- If we replace $d = 0$ and $F_3 = 0$ (i.e., situation describe in A), we find $F' = F_0 = \beta$ as indicated by (7).

Most importantly, we found that the force F increases linearly with respect to the external counter-force $F_3$. Eq.(11) hence suggested that to make the contralateral arm recover its natural strength, an external counter-force must be applied on the ipsilateral arm. Theoretically, depending on the value of this external force, the contralateral arm would recover its natural strength and become even stronger than normal.

To verify the above theoretical predictions we conducted the experiments as described below.

## 2.2 Experimental Setup to Verify the Predictions

Six subjects, 4 males and 2 females (average age: 22 and standard deviation: 1.26 years old) participated in this investigation. All participants were healthy and showed no musculoskeletal restrictions or diseases. They had a complete dentition, i.e. no premolars or molars were missing in a quadrant. There were no signs of severe malocclusions and no facial malformations. All subjects were free of any medication and didn't report of any type of muscle pain at the time of the experiments. In order to avoid any possible influences on the experimental results neither the experimenters nor the subjects knew the impact of the results on the theoretical model.

Before the experiments each subject filled out a questionnaire, which was kept confidential and included patient's identification, age and gender. His/her height (H) was measured and recorded. The tenets of the Declaration of Helsinki were followed; the local Institutional Review Board approved the study and informed consent was obtained for all subjects.

Two sets of experiments were performed. The first set was to determine the normal arm strength (value $F_0$) of the subject. The experimental protocol was the same as in previous work [9]. The subject was asked to stand with their arms and legs extended in the frontal plan in such a way that the ratio $D/H = 0.25$. The general protocol consisted of asking the subject first to hold the handle of a load which was at rest, i.e., the load did not apply any force on the subject arm. Second, at a warning signal the load was released at once; therefore it generated a pull-down force on the subject's arm. The subject held the load for a maximum of 5 seconds without moving his/her body. If the subject couldn't hold the load for this period, the load was removed and lightened incrementally. If the subject could hold the load for this period, the load was removed and added incrementally. Third, the subject relaxed for five minutes and the measurements restarted. In the end, the maximum load the subject could hold was recorded as the strength of his/her arm.

The second set of experiments was to determine the effect of the force $F_3$, applied on the ipsilateral arm, on the recovery force F' of the contralateral arm. To this end, the subject bit into a 2mm thick spacer, made out of a firm material, using the teeth of one side of the jaw. A weight of a determined value was hung on the wrist of the ipsilateral arm. The strength of the contralateral arm was then measured as mentioned above.

## 3   Results

Fig.4 shows the variation of the strength of the contralateral arm ($F'$) versus the external force applied on the ipsilateral arm ($F_3$) when the subjects bit onto a spacer of 2mm thick using one side of the jaw. The average values obtained on 6 subjects were plotted with the corresponding standard deviation values. We found that $F'$ increased linearly with respect to $F_3$ then leveled off when $F_3$ was about a half of $F'$. The maximum value that $F'$ could reach corresponded to the natural strength ($F_0$) of the subject. We also found that between female and male subjects except the strength of female was lower than male there was no difference between the recovery behavior and rate among them. Among the male subjects we found an exception (Fig.5 - right).

Fig.5 (left) shows the results of a typical subject (NVH). Fig.5 (right) shows an exception: the recovery of this subject (DNH) was immediate. The result clearly shows that the natural strength of this subject could not be exceeded as $F_3$ continued to increase. To determine whether we can induce an "extra" strength to the subject otherwise, we performed a pilot experiment. We found that even the subject did not bite onto a spacer, a force $F_3$ applied on one arm did not induce any extra strength to the other arm.

To compare quantitatively between the experimental and theoretical results we took the experimental data of subject NVH shown in Fig.5 (left) to determine the parameters in equation (11). Thus, for this specific subject, the theoretical recovery

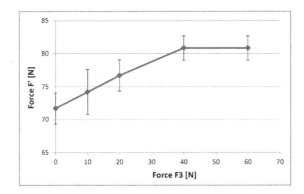

**Fig. 5** Average and standard deviation (vertical bars) values of the variation of the strength $F'$ (in Newtons) with respect to the counter force $F_3$ (in Newtons) across all six subjects. Each subject held a spacer of 2mm thickness in his/her left jaw, $F_3$ was applied on the left arm and $F'$ was measured on the right arm. We observed that when $F_3$ increased the arm strength gradually reached the original value $F_0$ then leveled off

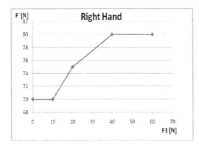

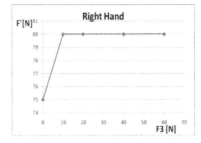

**Fig. 6** Variation of the strength $F'$ (in Newtons) with respect to the counter-force $F_3$ (in Newtons) of two subjects NVH (left) and DNH (right). Both subjects held a spacer of 2mm thickness in their left jaws, $F_3$ was applied on the left arms and $F'$ was measured on the right arms. We observed that when $F_3$ increased the arm strength increased to reach the original value then leveled off. Although subject b reached the natural strength faster this value couldn't be exceeded

strength of the contralateral arm $F'$ with respect to the external force $F_3$ applied on the ipsilateral arm is:

$$F' = 0.25F_3 + 70 \tag{12}$$

The equation of the rising portion of the behavior of this subject shown on Fig.5 (left), determined by the linear fitting, is:

$$F' = 0.27F_3 + 69 \tag{13}$$

Fig.6 shows the experimental and theoretical behaviors of $F'$ with respect to $F_3$ as shown in Fig.5 (left) and represented by equation (12), respectively. We found that with the exception at one point that corresponds to $F_3 = 10N$, they matched quite well.

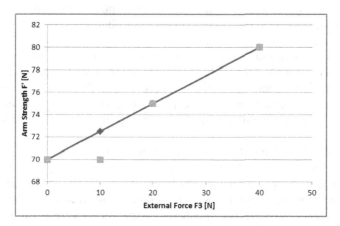

**Fig. 7** Variations of the arm strength $F'$ of the contralateral arm with respect to the counter force $F_3$ applied on the ipsilateral arm when a spacer of 2mm thick applied to one side of the jaw. The solid line represents the behavior predicted from the theoretical model, the squares were the values obtained from the experiments. We found that except one point (at $F_3 = 10N$), the experimental and theoretical results match well

In summary the obtained experimental results mainly showed that:

1. For a specific spacer thickness the recovery strength of the contralateral arm increases linearly with respect to the external force $F_3$ applied on the ipsilateral arm as predicted by the theoretical model. The theoretical results obtained from Eq.11 match well with the experimental results (Fig.6).
2. The natural strength of the contralateral arm $F_0$ cannot be exceeded (the highest plateaus in Fig.4 and 5). This value, represented by a horizontal line $F' = \beta$, constitutes a limit of the validity of (11). In other words, the recovery strength $F'$ of the contralateral arm with respect to the external force $F_3$ applied on the ipsilateral arm and its validity can be expressed as:

$$\begin{cases} F' = [(2\alpha d + \beta)/(\beta n')]F_3 + (\alpha d + \beta) \text{ for } F_3 < (-\alpha d\beta n')/(2\alpha d + \beta) \text{ and } 2 > d > 0 \\ F' = \beta \text{ for other values of } F_3 \text{ and } d \end{cases} \tag{II}$$

## 4 Discussion

In this study, we emitted a hypothesis that there is a coupling system that links the muscles of the jaw to those of the arms. In the natural situation this system rests on a support that governs the strength of the arm. The forces at both ends of the system are equal. An action that caused an imbalance in the jaw displaces this support and destabilizes the system. As a consequence, it weakens the arm strength on one side.

Based on this hypothesis, we developed a mechanical model similar to a seesaw that consists of a weightless and rigid beam which rotates about a fulcrum C located at the center of beam and two constant force springs placed at the ends of the beam. Displacing the fulcrum toward one end of the beam diminishes the force of the other end.

A mathematical model derived from this lumped model and expressed by the set of Eq.(I) successfully described the linear loss of the contralateral arm strength as found in the experiments [9] and took into account the limit of the dependency of this loss to the spacer thickness. This expression therefore governs the behavior of the loss of the contralateral arm strength with respect to the degree of imbalance of the jaw.

Based on the mechanical model we predicted that it is possible to recover the strength of the contralateral arm to its natural strength by applying an external counter-force on the ipsilateral arm. The theoretical model also suggested that there is a linear relationship between this counter-force and the recovery of the contralateral arm strength. To verify this prediction, experiments were performed with 6 healthy subjects. The experimental results clearly confirmed this prediction. In addition, they showed that when the counter-force reached a value about a half of the natural strength the contralateral arm fully recovered its strength. Further increase of the counter-force did not improve the strength of the contralateral arm. This finding allowed us to set the limit to the recovery mathematical model as expressed by the set of Eq.(II).

Although Eq.(I) governs the loss of the contralateral arm strength and Eq.(II) governs the recovery of its strength, there is a certain relationship between them. Indeed in the Eq.(II) if the counter force is zero Eq.(II) is reduce to Eq.(I). This fact shows that the mechanical model although simple is suitable for describing the phenomenon and is a valuable tool for further investigations.

The heart of the model is based on the concept of the moving fulcrum. This investigation showed that this concept is tangible. It is interesting to correlate this concept to the anatomical or physiological facts to understand its nature and mechanism.

In a pilot study, we applied the counter-force at the ipsilateral shoulder instead at the wrist of the subject and found similar results. This finding is particular important because it would pave the way to develop a practical therapeutic device to help true sufferers from this temporomandibular disorder to recover their natural strength. The device could be designed in the form of a jacket which has an appropriate weight that continuously exerts a force on the ipsilateral shoulder of the patient. It is interesting to further explore this idea.

The present investigation helped describe more comprehensively the reported phenomenon as follows: "An imbalance of the jaw evoked by biting unilaterally a firm spacer in one side of the jaw caused the contralateral arm of the subject to lose its strength. The loss decreased linearly as the thickness of the spacer was increasing. The loss leveled off when the spacer thickness reached about 2mm. By applying a counter-force which was about one half of the natural strength on the ipsilateral arm, the contralateral arm fully recovered its strength. This strength could not be exceeded even with a higher counter-force." It is interesting to extend this study in clinics with patients suffered from similar temporomandibular disorders or loss of teeth on one side of the jaw.

## 5  Conclusions

The present investigation revealed that:

1. The subjects can recover the strength of their contralateral arms weakened by an evoked imbalance of the jaw induced unilaterally. The recovery occurred when a counter-force was applied on the ipsilateral arm. The recovered strength increased linearly as the counter-force was increasing and reached its natural value when the counter-force was about a half of this value. We also found that further increase of the counter-force did not improve the contralateral arm strength. These findings allowed a better understanding of this phenomenon and suggested the design of a practical device which may help the recovery. It is interesting to extend this investigation to sufferers from this kind of temporo-mandibular disorder.
2. The heart of the model is based on the concept of the moving fulcrum. It is interesting to correlate this concept to the anatomical or physiological facts to understand its nature and mechanism.
3. The good fit between the theoretical and experimental data proves that although the model is simple, it constitutes an appropriate tool to investigate this phenomenon.

Further investigations of the effects of other parameters in our model would help to even better understand this phenomenon.

**Acknowledgments.** This work is supported by grant No. 106.99-2010.11 from the Vietnam National Foundation for Science and Technology Development (NAFOSTED). We would like to thank Benjamin Kelley, Dao Tien Tuan, and Ngo Thanh Hoan for their comments and assistance.

## References

1. Linderholm, H., Lindqvist, B., Ringqvist, M., Wennström, A.: Isometric bite force and its relation to body build and general muscle force. Acta Odont Scand 29, 563–568 (1971)
2. Bakke, M.: Mandibular elevator muscles: physiology, action, and effect of dental occlusion. Scand. J. Dent. Res. 101, 314–331 (1993)
3. Gelb, H., Mehta, N.R., Forgione, A.G.: The relationship between jaw posture and muscular strength in sports dentistry: a reappraisal. Cranio. 14(4), 320–325 (1996)
4. Abdallah, E.F., Mehta, N.R., Forgione, A.G., Clark, R.E.: Affecting upper extremity strength by changing maxillo-mandibular vertical dimension in deep bite subjects. Cranio 22(4), 268–275 (2004)

5. Raadsheer, M.C., van Eijden, T.M.G.J., van Ginkel, F.C., Prahl-Andersen, B.: Human jaw muscle strength and size in relation to limb muscle strength and size. Eur. J. Oral Sci. 112, 398–405 (2004)

6. Matsuo, K., Mudd, J., Jerome, V., Kopelman, N., Atlas, R.: Duration of the second stage of labor while wearing a dental support device: A pilot study. Obstet Gynaecol. Res. 35(4), 672–678 (2009)

7. Garner, D.P., McDivitt, E.: Effects of Mouthpiece Use on Airway Openings and Lactate Leve in Healthy College Males. Compedium 30(2), 9–13 (2009)

8. Cuccia, A., Caradonna, C.: The relationship between the stomatognathic system and body posture. Clinics 64(1), 61–66 (2009)

9. Hòa, L.M., Huân D.N., Thao, N.H., Hoàn, N.T., Khoa, T.Q.D., Tâm, N.H.M., Van Toi, V.: Relationship between Dental Occlusion and Arm Strength. In: Van Toi, V., Khoa, T.Q.D. (eds.) BME 2010. IFMBE Proceedings, vol. 27, pp. 266–269. Springer, Heidelberg (2010)

10. Dang, K., Le, H., Nguyen, H., Toi, V.V.: Analyzing surface EMG signals to determine relationship between jaw imbalance and arm strength loss. BioMedical Engineering On-Line 11, 55 (2012), doi:10.1186/1475-925X-11-55

11. Tsukimura, N.: Study on the relation between the stomatognathic system condition influences of postural changes in vertical maxillomandibular relation on the back strength. J. Japan Prostodont Soc. 36, 705–719 (1992)

12. Forgione, A.G., Mehta, N.R., McQuade, C.F., Westcott, W.L.: Strength and bite, Part 2: Testing isometric strength using a MORA set to a functional criterion. Cranio. 10(1), 13–20 (1992)

13. Van Spronsen, P.H., Weijs, W.A., Van Ginkel, F.C., Prahl-Andersen, B.: Jaw muscle orientation and moment arms of long-face and normal adults. J. Dent. Res. 75, 1372–1380 (1996)

14. Chakfa, A.M., Mehta, N.R., Forgione, A.G., Al-Badawi, E.A., Lobo, S.L., Zawawi, K.H.: The effect of step-wise increases in vertical dimension of occlusion on isometric strength of cervical flexors and deltoid muscles in nonsymptomatic females. Cranio. 20(4), 264–273 (2002)

15. Post, M., Bayrak, S., Kernell, D., Zijdewind, I.: Contralateral muscle activity and fatigue in the human first dorsal interosseous muscle. J. Appl. Physio. 105(1), 70–82 (2008)

16. Shinohara, M., Keenan, K.G., Roger, M., Enoka, R.M.: Contralateral activity in a homologous hand muscle during voluntary contractions is greater in old adults. J. Appl. Physiol. 94, 966–974 (2003)

# Estimation of Patient Specific Lumbar Spine Muscle Forces Using Multi-physical Musculoskeletal Model and Dynamic MRI

Tien Tuan Dao, Philippe Pouletaut, Fabrice Charleux, Áron Lazáry, Peter Eltes, Peter Pal Varga, and Marie Christine Ho Ba Tho

**Abstract.** Trunk muscle forces are of great interest in the diagnosis and treatment of low back pain diseases. Musculoskeletal modeling is often used to estimate muscle forces using optimization principle. Available parameterized multibody lumbar spine models used generic geometries and literature-based values leading to inaccurate muscle architecture and muscle forces not reliable for a specific case. In this present study, a multi-physical musculoskeletal model of the lumbar spine was developed from medical imaging to estimate patient specific trunk muscle forces with lumbar spine range of motions derived from dynamic MRI data in supine position. As results, a 3D patient specific musculoskeletal model was developed with 126 muscle fascicles. Maximal estimated forces of all muscle groups range from 3 to 40 N for hyperlordosis motion. The higher muscle forces were estimated in iliocostalis lumborum pars lumborum. This study has demonstrated that patient specific modeling is essential for clinical analysis of lumbar spine.

## 1 Introduction

Lumbar spine is one of the most important load-sharing structures of the human body. In addition to its protective function for vital internal organs, this structure allows internal and external loads to be transmitted through back muscles, intervertebral joints, ligaments and discs in a coordinative manner [1]. Lumbar spine muscles play an essential role in the generation of spinal motions. The coordination of these

Tien Tuan Dao · Philippe Pouletaut · Marie Christine Ho Ba Tho
UTC CNRS UMR 7338, Biomechanics and Bioengineering (BMBI),
University of Technology of Compiègne, BP 20529, 60205 Compiègne cedex, France

Fabrice Charleux
ACRIM-Polyclinique St Côme, BP 70409 - 60204 Compiègne Cedex, France

Áron Lazáry · Peter Eltes · Peter Pal Varga
National Center for Spinal Disorders - Buda Health Center - Nagy Jeno u. 8,
1126 Budapest - Hungary

V.-N. Huynh et al. (eds.), *Knowledge and Systems Engineering, Volume 2,*   411
Advances in Intelligent Systems and Computing 245,
DOI: 10.1007/978-3-319-02821-7_36, © Springer International Publishing Switzerland 2014

muscles contributes to the stability of the trunk and the whole body under internal and external loadings. In fact, the understanding of this complex mechanical behavior of the lumbar spine structures plays an important role in the diagnosis of low back pain as well as in the prescription of appropriate and optimal functional rehabilitation treatment planning [2], [3]. For these purposes, musculoskeletal modeling of the lumbar spine is commonly used to provide information (e.g. tissue stress, muscle force or joint loading) inside the lumbar spine structures and to determine how the mechanical behavior of lumbar spine works in normal and abnormal cases.

A number of in silico deformable lumbar spine models have been developed to study the in vivo and in vitro tissue stress under internal and external loading conditions [4], [5], [6], [7]. Medical imaging techniques such as Magnetic Resonance Imaging (MRI) or Computed Tomography (CT) and finite element method allows lumbar spine model to be developed and analyzed in a subject or patient specific manner. Individualized geometrical and mechanical properties of the lumbar spine structures are commonly used in these models [4]. Moreover, at the cell-to-tissue level, micromechanical models with osmo-poro-visco-hyper-elastic discs (OVED) based on constituent content and their material properties have been developed recently to get a better understanding of the normal and abnormal mechanical behaviour of the intervertebral disc [8],[9]. However, despite their bio-fidelity character, this modeling approach is very time-consuming and requires advanced modeling knowledge (e.g. formulation of constitutive laws of the biological tissues). Besides, rigid multi-bodies modeling is an alternative solution to study the complex mechanical behavior of the lumbar spine [10], [11]. In particular, this approach allows trunk muscle forces to be estimated leading to provide objective criteria for the evaluation of functional rehabilitation.

In the literature, in silico rigid multi-bodies models of the lumbar spine ranging from basic free body diagram to 3D musculoskeletal model [11], [12], [13] have been developed. These 3D lumbar spine multi-body models showed a variation of the number of muscles under consideration. For example, LifeMOD/ADAMS model includes 6 erector spinae muscle fascicles [14], AnyBody model includes 154 muscle fascicles [11]. Enhanced AnyBody model has 214 muscle fascicles [13]. The most physiologically detailed musculoskeletal model is the current OpenSIM lumbar spine model including 238 muscle fascicles [12]. Moreover, passive stiffness structures (Intervertebral Disc (IVD)) and intra-abdominal pressure (IAP) activation were also taken into consideration [13]. Some models were used to analyze the postural effect and stabilities or to optimize the correction of spinal deformities [15]. Other models were developed to be shared and reused to investigate a range of research questions [11],[12]. These models used optimization to solve the redundancy problem (i.e. number of muscles contributed into a specific motion is greater than the number of physical law equations describing this motion) for the muscle force estimation [10],[16]. Hill-based models ranging from simplest to full versions were used to describe the trunk muscle behavior. The simplest version deals with the consideration of only maximal muscle force [11], [13], [15] in the optimization algorithm to estimate muscle forces. There is no Force-Length-Velocity relationship due to the lack of intrinsic back muscle-tendon properties such as muscle moment arm or tendon lack length. Recently, full Hill-based muscle model was integrated into 3D musculoskeletal model for the better description of the simulation muscle behavior [12]. Moreover, these models were simulated under static and dynamic loading conditions. For dynamic activities, 3D motion capture technique was used to simulate large displacement such as flexion/extension. Electromyography (EMG) signals were also used to describe the trunk muscle recruitment patterns [17]. According to our knowledge, most of these models used generic or parameterized geometries and literature-based properties (e.g. muscle and joint properties, body segment inertial parameters). Moreover, one could note that these models have not been used for clinical applications. In the framework of the application of in silico models for personalized medicine purpose, patient specific data should be derived from medical imaging to provide accurate and reliable data for clinical decision-making purpose [16].

The objective of this paper was to develop a patient specific 3D musculoskeletal model of the lumbar spine using medical imaging to estimate the patient specific trunk muscle forces during hyperlordosis motion derived from dynamic MRI data in supine position.

## 2   Materials and Methods

### 2.1   Medical Imaging Protocols

A Computed Tomography (CT) protocol was developed to acquire the anatomical lumbar spine images of one patient (female: 60 year old, 65kg body mass, 160cm height, 25.39 kg/m Body Mass Index (BMI)) at the National Center for Spinal Disorders (Budapest, Hungary). The patient signed an informed consent agreement. The acquisition was performed using a Hitachi CT machine and consisted in 344 slices of thickness 1.25 mm. The pixel resolution is 0.6 mm x 0.6 mm and the field of view 31 cm x 31 cm (512 x 512 pixels).

A dynamic MRI protocol was developed at the Polyclinique Saint Cme (Compigne, France) to acquire the range of hyperlordosis motion of the lumbar spine in supine posture. This protocol was developed for a 1.5T MRI GE machine with a 20-mm slice thickness. A sagittal dynamic FIESTA (fast imaging employing steady state acquisition) sequence was used. The repetition time and echo time are set up as 3.689ms and 1.152ms respectively. The acquisition matrix is 512x512 pixels. The FOV is set up as 360x360mm. Seven normal subjects (males: 38.3  10.5 year old, 78.8  11.2 kg body mass, 1.75  0.1 m height, 25.8  4.7 kg/m Body Mass Index (BMI)) participated into the data acquisition. They signed informed consent agreements. A position change strategy was set up as follows: each position change corresponds to each change of MRI machine sound and the subject keeps this position during next MRI machine sound time to have an image with the best quality. This procedure is repeated until the subject returned to initial position from his (her) maximal hyperlordosis position (Fig. 1). The total acquisition time of one measure is around 50 second.

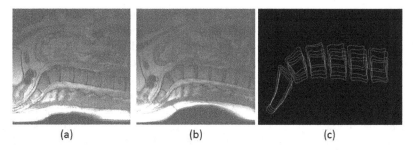

**Fig. 1** Dynamic MRI data: initial (a) and maximal (b) hyperlordosis positions and contour-based reconstruction of hyperlordosis motion (c) of one subject

## 2.2 Image Processing: Musculoskeletal Geometries and Inertial Properties

Our patient specific lumbar spine model has 7 rigid segments: T12, L1 to L5, and pelvis and sacrum. Based on the 2D CT-based slice-by-slice anatomical images, bony segments (all vertebrae and pelvis) were segmented from surrounding tissues using ScanIP module (Simpleware, UK). A threshold-based segmentation method was applied with a value of 231 Hounsfield units (Fig. 2a and Fig. 2b). Then, each image was verified and cleaned to ensure that the segmented pixels belong to the tissue of interest as well as to avoid unnecessary segmented pixels respectively.

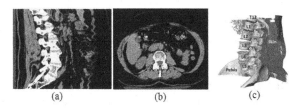

**Fig. 2** Segmented skin (in red) and bony vertebral segments (in light yellow) of the lumbar spine in sagittal (a) and axial (b) planes and 3D reconstructed model (c) of the lumbar spine structures: T12, L1 to L5, pelvis and sacrum, and back skin structures

Based on the 2D segmented slices, marching cube algorithm was used to reconstruct the 3D Stereolithography (STL)-based geometries of each vertebra and back skin (Fig. 2c). This process was performed using ScanIP module (Simpleware, UK).

In addition to the patient specific 3D geometries of all bony segments, their body segment inertial parameters were also computed. Anatomical axes using the parallel axis theorem were used to determine individualized segmental mass, positions of mass center and moments of inertia properties of all body segments [16].

## 2.3  Musculoskeletal Model: Joint and Muscle Modeling

Each intervertebral joint is modeled as a 3DOF (Degrees Of Freedom) OpenSIM custom joint [18] with a fixed location of instantaneous axes of rotation (i.e. center of rotation) [12]. The location of instantaneous axes of rotation (IAR) of each intervertebral joint is determined using the radiography-based estimation provided by Pearcy and Bogduk [12], [19]. International Society of Biomechanics (ISB) recommendation on the definition of spine joint coordinate system was used [20]. First, the joint between L5 and Sacrum & Pelvis segments was modeled with 3 coordinates: flexion-extension, lateral bending and axial rotation. Its respective range of motion was derived from dynamic MRI data for a hyperlordosis motion in supine position. Then, kinematic constrains were applied using available OpenSIM spatial transformations and functions (e.g. linear function and piecewise linear function) to create and constrain the motion of all other lumbar spine joints rang from L5/L4 to T12/L1 [12].

Based on our reconstructed bony geometries, our patient specific model includes 126 muscle fascicles consisting of 4 muscle groups: erector spinae (iliocostalis lumborum pars lumborum (8 fascicles) and iliocostalis thoracis pars lumborum (10 fascicles)), quadratus lumborum (36 fascicles), psoas major (22 fascicles), multifidus (spinous process (40 fascicles), entry level (10 fascicles)). The origin-insertion points of each muscle are defined using anatomical landmarks [12]. These anatomical landmarks are added using a virtual palpation procedure. Normal behavior of lumbar spine muscles is assumed. Thus, all modeling properties (e.g. physiological cross-sectional area, maximum isometric force, muscle and muscle-tendon lengths, sarcomere length, optimal fiber length, pennation angle and tendon slack length) of each muscle are based on the OpenSIM lumbar spine default values [12].

A Hill-based rheological muscle model [21] was used to describe the muscle behavior (Fig. 3). This model was adjusted using age-related muscle mechanics [12]. Hill-based model has 3 behavior curves such as the tendon Force-Length curve, muscle Force-Length curve, and muscle Force-Velocity curve.

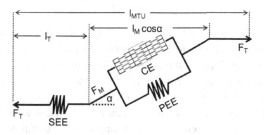

**Fig. 3** Graphical representation of a Hill-based model

## 2.4 Dynamic MRI: Maximal Hyperlordosis Range of Motion in Supine Position

As described above, the maximal range of L5/S1 joint is needed to simulate the hyperlodosis motion. This range of values was quantified at the L5/S1 level for all acquired healthy subjects at the latest MRI image when each subject reaches his or her maximal hyperlordosis position. Then, maximal hyperlordosis L5/S1 angle connected by the middle lines of L5 and S1 were measured (Fig. 4).

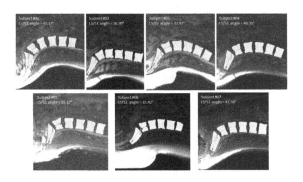

**Fig. 4** Maximal hyperlordosis L5/S1 angles of all subjects

## 2.5 Multi-body Dynamics Using Inverse Method

Based on the motion data, inverse dynamics principle aims to estimate the joint torque of different intervertebral joint levels. The equations of motion [10] describing the dynamics of a musculoskeletal system are expressed as follows:

$$(M(q)\ddot{q} + C(q,\dot{q}) + G(q) + T_{MT}) = 0 \tag{1}$$

$$(T_{MT}) = R(q)F^{MT} \tag{2}$$

Where q is the joint angles set for n joints; $M(q)$ is the system mass (n x n) matrix; $C(q,\dot{q})$ is the centrifugal and coriolis loading (n x 1) matrix; $G(q)$ is the gravitational loading (n x 1) matrix; $T_{MT}$ is the muscular joint torques (n x 1) matrix; $R(q)$ is the muscle moment arms (n x m) matrix; $F^{MT}$ is the muscle-tendon force (m x 1) matrix; and E is external forces (e.g. ground reaction forces). The muscle moment arms matrix could be computed using the principle of virtual work [10], [12]. Thus, the moment arm of muscle j with respect to joint axis i is computed as follows:

$$(R_{ij}(q)) = -\frac{\partial L_j(q)}{\partial q_i} \tag{3}$$

where $L_j(q)$ is the length of muscle j. Thus, it is necessary to use accurate definition of origin and insertion points of each muscle.

When the joint kinematics data (q) and external forces (E) are available, one could compute the muscle joint torques during motion by using the following inverse dynamics equation:

$$(T_{MT}) = M(q)\ddot{q} + C(q,\dot{q}) + G(q) + E \tag{4}$$

### 2.6  Muscle Force Estimation Using Static Optimization

By using the computed net joint torque, muscle forces could be estimated using static optimization for each lumbar spine muscle fascicle [16]. From mathematical point of view, the muscle force estimation problem is undetermined because the number of unknown muscle forces is greater than the number of equations for all joints $(m \succ n)$. Thus, to solve this problem, one of the most used approaches is the inverse dynamics-based static optimization. The constitutive equations to minimize the objective function (i.e. $F^{ojb}$ could be a mono-objective or multi-objectives function) are expressed as follows:

$$\left(F^{ojb}\right) = \left(\sum_{j=1}^{126} a_j t_i\right)^2 \tag{5}$$

subject to

$$\left(R(q)F^{MT}\right) = T_{MT} \quad \text{with } 0 \le F^{MT} \le F_M^0 . \tag{6}$$

where $F_M^0$ is the peak isometric muscle force; $a_j$ is the muscle activation level and $t_i$ is the simulation time.

### 2.7  Model Analysis

The development of our patient specific model and all simulations were performed on a personal computer (Dell Precision T3500, 2.8 Hz, 3GB Random-Access Memory (RAM)) using Visual C++ 2010 (Microsoft, USA) and OpenSIM 2.4 engine [18] respectively. Patient specific muscle fascicle forces are computed and reported. These forces were also compared to those extracted from literature.

Data post-processing was performed using Matlab R2012a (The Matworks, Inc., Natick, MA).

## 3  Computational Results

### 3.1  Patient Specific 3D Musculoskeletal Lumbar Spine Model

Our patient specific lumbar spine model including all vertebral segments, pelvis and sacrum, intervertebral joint and muscle fascicles is illustrated in Fig. 3. This model consists of 4 modeled muscle groups including 126 muscle fascicles.

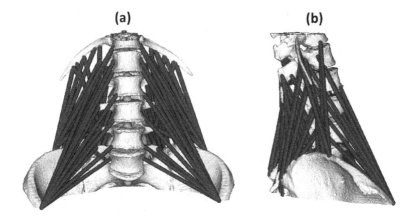

**Fig. 5** Our patient specific lumbar spine model: frontal view (a) and lateral view (b)

## 3.2 Maximal Range of Hyperlordosis Motion in Supine Position

Maximal range of hyperlordosis motion in supine posture of all seven subjects is depicted in Table 1. The mean L5/S1 angle for all subjects is around $41.8° \pm 8.1°$.

**Table 1** Maximal range of hyperlordosis motion of all subjects at the L5/S1 joint

| Subject | L5/S1 angle (°) |
|---------|-----------------|
| 01      | 41.37           |
| 02      | 30.70           |
| 03      | 37.92           |
| 04      | 44.35           |
| 05      | 55.17           |
| 06      | 35.42           |
| 07      | 47.58           |

## 3.3 Patient Specific Trunk Muscle Forces

All 126 muscle fascicle forces were estimated using inverse dynamics and static optimization algorithms. The computing time of hyperlordosis (extension-like motion) motion is around $15 \pm 1$ min. The total forces all muscle groups for hyperlordosis motion are illustrated in Fig. 6. The comparisons between our muscle force range of values and those extracted from literature are depicted in Table 1. Maximal estimated forces of all muscle groups range from 3 to 40N. The higher muscle forces were estimated in iliocostalis lumborum pars lumborum. Moreover, all muscle group forces are decreasing from supine initial position to around 40-degrees hyperlordosis extended position, except for psoas major muscles, which are increasing slightly.

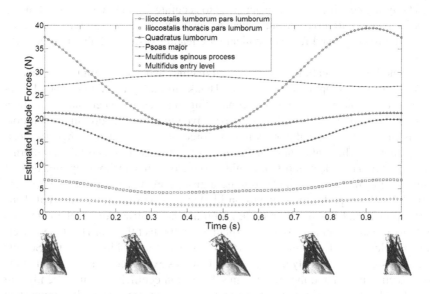

**Fig. 6** Estimated muscle forces for supine hyperlordosis (extension-like) motion

**Table 2** Maximal estimated trunk muscle forces (N): ILpL = iliocostalis lumborum pars lumborum, ITpL = iliocostalis thoracis pars lumborum, QL = quadratus lumborum, PS = psoas major, MSP = multifidus spinous process, MEL = multifidus entry level

| Studies | ILpL | ITpL | QL | PS | MSP | MEL |
|---|---|---|---|---|---|---|
| Han et al. 2012 [13] | 15 | 15 | 13 | 70 | 11 | 11 |
| Schultz et al. 1982[22] | 0 | | | | | |
| This study | 40 | 7 | 21 | 29 | 19 | 3 |

# 4 Discussion

Musculoskeletal disorder patients have specific musculoskeletal geometries and behaviors ranging out of normality [16]. To provide accurate intrinsic information of how the biomechanical loading and pathophysiological processes impact on the muscle behavior of each patient, personalized musculoskeletal model needs to be developed [23]. In this study, a patient specific model derived from medical imaging was developed to estimate muscle forces using inverse dynamics and static optimization algorithms. In fact, individualized geometries and their body segment inertial parameters were created and computed. It is well known that these properties relates strongly to the kinetic results (e.g. joint moment) [16]. Consequently, estimated muscle forces are influenced indirectly by this effect. Thus, individualized body segment inertial parameters lead to accurate joint moments and estimated muscle forces. Furthermore, individualized geometries allow accurate muscle architecture to be defined for each patient case leading also to reliable muscle force estimation.

Muscle force estimation using inverse dynamics and static optimization has become a customized approach [10], [16]. Thanks to the optimization principle, undetermined problem (i.e. number of contracting muscles is greater than the number of motion equations at all joints) could be solved to provide approximate solution. In this study, patient specific trunk muscle forces were estimated for hyperlordosis flexed motion. The higher muscle forces were estimated in iliocostalis lumborum pars lumborum muscle group. Our results are concordant with those reported by Han et al. 2012 [13]. Thus, all modeled muscle groups are active and they contributed actively on the hyperlordosis (extension-like) motion. However, Schultz et al. 1982 [22] reported that iliocostalis muscle group has no contribution on the extension motion. The disagreement could be explained by the fact that our Hill-type muscle model includes the force-length and force-velocity relationships while the Schultz et al.'s model [22] uses only maximal muscle forces expressed by the product of the cross-sectional area and the specific muscle tension coefficient. Consequently, our model could provide more accurate estimated forces. In fact, our results showed that muscle rheological model needs include complex behaviors such as force-length and force-velocity relationship to estimate accurately muscle forces. Such muscle force values are of great interest for performing objective diagnosis and appropriate treatment prescription of lumbar spine disorders. Moreover, these muscle forces could be used to define or calibrate boundary and loading conditions for finite element model of the lumbar spine [5].

One of possible improvements of the current lumbar spine model is the use of individualized mechanical muscle properties (e.g. muscle stiffness) to improve the accuracy of estimated forces. For this topic, in our previous study, muscle shear modulus property derived from MR elastography was used to adjust force-length curve aiming to provide more intrinsic information for the accurate estimation of healthy and pathological muscle forces [25]. For the joint modeling, there is a lack of consensus about the use of right joint type (e.g. ball-and-socket or spherical or custom joints). Consequently, new efforts need to be investigated to develop new individualized joint modeling approach. For the muscle force estimation, effort will be needed to develop an individualized objective function for each patient case. Furthermore, neural control needs to be included into patient specific musculoskeletal model to develop patient specific neuro-musculoskeletal model describing better the relationship between neural control commands and muscle intrinsic behavior [25].

# 5 Conclusions

A patient specific musculoskeletal model of the lumbar spine was developed. Patient specific trunk muscle forces during hyperlordosis (extension-like) motion in supine posture were estimated using dynamic MRI data. This study has demonstrated that patient specific modeling is essential for clinical analysis of lumbar spine.

**Acknowledgments.** The research leading to these results has been funded from the European Union Seventh Framework Programme (FP7/2007-2013) under grant agreement n°269909.

# References

1. Adams, M.A.: Biomechanics of back pain. Acupunct. Med. 22(4), 178–188 (2004)
2. Norris, C.M.: Spinal Stabilisation: Limiting Factors to End-range Motion in the Lumbar Spine. Physiotherapy 81(2), 64–72 (1995)
3. Haynes, W.: New strategies in the treatment and rehabilitation of the lumbar spine. Journal of Bodywork and Movement Therapies 7(2), 117–130 (2003)
4. Périé, D., Sales De Gauzy, J., Ho Ba Tho, M.C.: Biomechanical evaluation of Cheneau-Toulouse-Munster brace in the treatment of scoliosis using optimisation approach and finite element method. Med. Biol. Eng. Comput. 40(3), 296–301 (2002)
5. Noailly, J., Wilke, H.J., Planell, J.A., Lacroix, D.: How does the geometry affect the internal biomechanics of a lumbar spine bi-segment finite element model? Consequences on the validation process. Journal of Biomechanics 40(11), 2414–2425 (2007)
6. Schmidt, H., Shirazi-Adl, A., Galbusera, F., Wilke, H.J.: Response analysis of the lumbar spine during regular daily activities-A finite element analysis. Journal of Biomechanics 43(10), 1849–1856 (2010)
7. Schmidt, H., Reitmaier, S.: Is the ovine intervertebral disc a small human one?: A finite element model study. Journal of the Mechanical Behavior of Biomedical Materials 17, 229–241 (2013)
8. Schroeder, Y., Sivan, S., Wilson, W., Merkher, Y., Huyghe, J., Maroudas, A., Baaijens, F.P.T.: Are disc pressure, stress, and osmolarity affected by intra and extra fibrillar fluid exchange? Journal of Orthopaedic Research 25, 1317–1324 (2007)
9. Alicia, R.J., Chun-Yuh, H., Wei, Y.G.: Effect of endplate calcification and mechanical deformation on the distribution of glucose in intervertebral disc: a 3D finite element study. Computer Methods in Biomechanics and Biomedical Engineering 14(2), 195–204 (2011)
10. Erdemir, A., McLean, S., Herzog, W., van den Bogert, A.J.: Model-based estimation of muscle forces exerted during movements. Clin. Biomech. 22(2), 131–154 (2007)
11. de Zee, M., Hansen, L., Wong, C., Rasmussen, J., Simonsen, E.B.: A generic detailed rigid-body lumbar spine model. Journal of Biomechanics 40(6), 1219–1227 (2007)
12. Christophy, M., Faruk Senan, N.A., Lotz, J.C., O'Reilly, O.M.: A musculoskeletal model for the lumbar spine. Biomech. Model Mechanobiol. 11(1-2), 19–34 (2012)
13. Han, K.S., Zander, T., Taylor, W.R., Rohlmann, A.: An enhanced and validated generic thoraco-lumbar spine model for prediction of muscle forces. Medical Engineering and Physics 34(6), 709–716 (2012)
14. Huynh, K.T., Gibson, I., Lu, W.F., Jagdish, B.N.: Simulating dynamics of thoracolumbar spine derived from LifeMOD under haptic forces. World Academy of Science, Engineering and Technology 64, 278–285 (2010)

15. Stokes, I.A.F., Gardner-Morse, M.: Lumbar spine maximum efforts and muscle recruitment patterns predicted by a model with multijoint muscles and flexible joints. J. Biomech. 28(2), 173–186 (1995)

16. Dao, T.T., Marin, F., Pouletaut, P., Aufaure, P., Charleux, F., Ho Ba Tho, M.C.: Estimation of Accuracy of Patient Specific Musculoskeletal Modeling: Case Study on a Post-Polio Residual Paralysis Subject. Computer Method in Biomechanics and Biomedical Engineering 15 (7), 745–751 (2012)

17. Gagnon, D., Arjmand, N., Plamondon, A., Shirazi-Adl, A., Lariviére, C.: An improved multi-joint EMG-assisted optimization approach to estimate joint and muscle forces in a musculoskeletal model of the lumbar spine. Journal of Biomechanics 44(8), 1521–1529 (2011)

18. Delp, S.L., Anderson, F.C., Arnold, A.S., Loan, P., Habib, A., John, C.T., Guendelman, E., Thelen, D.G.: OpenSim: Open-source Software to Create and Analyze Dynamic Simulations of Movement. IEEE Transactions on Biomedical Engineering 54(11), 1940–1950 (2007)

19. Pearcy, M.J., Bogduk, N.: Instantaneous axes of rotation of the lumbar intervertebral joints. Spine 13, 1033–1041 (1998)

20. Wu, G., Siegler, S., Allard, P., Kirtley, C., Leardini, A., Rosenbaum, D., Whittle, M., D'Lima, D.D., Cristofolini, L., Witte, H., Schmid, O., Stokes, I.: ISB recommendation on definitions of joint coordinate system of various joints for the reporting of human joint motion–part I: ankle, hip, and spine. J. Biomech. 35(4), 543–548 (2002)

21. Hill, A.V.: The heat of shortening and dynamics constants of muscles. Proc. R. Soc. Lond. B 126(843), 136–195 (1938)

22. Schultz, A., Andersson, G., Ortengren, R., Haderspeck, K., Nachemson, A.: Loads on the lumbar spine. Validation of a biomechanical analysis by measurements of intradiscal pressures and myoelectric signals. J. Bone Joint Surg. Am. 64, 713–720 (1982)

23. Blemker, S.S., Asakawa, D.S., Gold, G.E., Delp, S.L.: Image-based musculskeletal modeling: Applications, advances, and future opportunities. Journal of Magnetic Resonance Imaging 25, 441–451 (2007)

24. Bensamoun, S.F., Dao, T.T., Charleux, F., Ho Ba Tho, M.C.: Calculation of in vivo muscle forces derived from MR elastography. Journal of Biomechanics 45(1), S489 (2012)

25. Buchanan, T.S., Lloyd, D.G., Manal, K., Besier, T.F.: Neuromusculoskeletal modeling: estimation of muscle forces and joint moments and movements from measurements of neural command. Journal of Applied Biomechanics 20(4), 367–395 (2004)

# Subject Specific Modeling of the Muscle Activation: Application to the Facial Mimics

Marie Christine Ho Ba Tho, Tien Tuan Dao, Sabine Bensamoun, Stéphanie Dakpe, Bernard Devauchelle, and Mohamed Rachik

**Abstract.** Facial muscle activation information is of interest for the simulation of the facial mimics. In the present study, three positions (smile, pronunciation of sound 'Pou' and 'O') describing specific motions of the facial mimics were acquired using MRI. Finite element (FE) simulations of these three facial muscle activation behaviors were performed on a specific muscle zygomaticus major (ZM) one of the most relevant in facial mimics. Numerical results were compared qualitatively and quantitatively with those derived from MRI images. The MRI-based average displacements of the ZM muscle are $4 \pm 2$mm, $4.5 \pm 1.4$ mm and $6 \pm 3$mm for 'Smile', 'Pou' and 'O' positions respectively. The FE-based average displacements of the ZM muscle are $1.9 \pm 0.8$ mm $2 \pm 1$ mm and $2.8 \pm 1.1$ mm for 'Smile', 'Pou' and 'O' positions respectively. This present study shows the development of a methodology for subject specific modeling with confrontation between experimental and numerical results for simulation of facial mimics.

## 1 Introduction

Facial expressions or motions are basic behaviors of the human face. Due to the accident or musculoskeletal face disorders, these behaviors could be altered. The understanding of facial mimics is needed to recover these behaviors. To achieve this

Marie Christine Ho Ba Tho · Tien Tuan Dao · Sabine Bensamoun
UTC CNRS UMR 7338, Biomechanics and Bioengineering (BMBI),
University of Technology of Compiègne, BP 20529, 60205 Compiègne cedex, France

Stéphanie Dakpe · Bernard Devauchelle
CHU Amiens- Service Chirurgie Maxillo-Faciale, Stomatologie, Amiens, France

Mohamed Rachik
UTC UMR 7337 Roberval, University of Technology of Compiègne, BP 20529,
60205 Compigne cedex, France

V.-N. Huynh et al. (eds.), *Knowledge and Systems Engineering, Volume 2,*                     423
Advances in Intelligent Systems and Computing 245,
DOI: 10.1007/978-3-319-02821-7_37, © Springer International Publishing Switzerland 2014

goal, the modeling of facial muscles and their impact on the motion of the facial skin needs to be investigated. This provides useful quantitative and objective criterias for the 3D evaluation of facial expressions or motions.

Facial muscle activation information is of great interest for the accurate simulation of facial mimics leading to provide objective criteria in clinical rehabilitation treatment of facial disfigurement. As current clinical routine practices, the evaluation of facial mimics has been based only on the subjective palpation and observations of the clinicians leading to inappropriate treatment prescription. Thus, numerical simulation could be used to provide more intrinsic internal information of the facial muscle contraction and coordination mechanism for an objective evaluation [1]. Rigid-body biomechanical models have been used to study the mechanical behavior of facial motions such as mastication with temporomandibular joint [2], [3], [4]. However, this modeling approach cannot simulate accurately the muscle behavior due to modeling limitations and assumptions [5]. Biomechanics deformable model is an alternative solution to improve the accuracy of the numerical simulation. Some studies aimed to simulate facial expressions or motions due to the muscle contraction using biomechanical advanced muscle constitutive models [6], [7], [8], [9]. However, these models are based on generic geometries and there is a lack of objective data for the simulation set up as well as for the model evaluation in a clinical context.

The objective of this present study was to develop a subject specific methodology with confrontation between experimental and simulation results of the facial muscle activation behaviors. Magnetic Resonance Imaging (MRI) technique has been used for the acquisition of specific facial positions reflecting specific motions. Then, the information of facial muscle activation behavior was used as input data for the finite element model (geometry, elongation and shortening of the muscle for each position). Indeed, the anatomical variability and mechanical characteristics of the facial muscles needed to be considered for direct clinical application [1].

## 2  Materials and Methods

### 2.1  MRI Protocol

A muscle-specific 3Tesla MRI protocol (3DFSPGR Sagittal T1 sequence, FOV = 24x24 $mm^2$, slice thickness = 1.6mm, acquisition time = 7 seconds) was developed at the CHU Amiens to acquire anatomical images of the facial soft tissues (fat and skin) and the zygomaticus major (ZM) muscle at 4 different positions (neutral, smile, pronunciation of sound 'O', pronunciation of sound 'Pou') of a healthy subject (female, 24 yo, 1.5 m, 57 kg) as illustrated in Fig. 1.

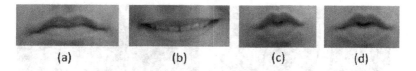

**Fig. 1** Different positions of the face: neutral (a), smile (b), pronunciation of sound 'O' (c) and pronunciation of sound 'Pou' (d)

## 2.2   Image Processing and Reconstruction

The facial muscle tissues (levator labii superioris, orbicularis oris, levator anguli oris (caninus), zygomaticus major, depressor anguli oris (triangularis), musculus mentalis, depressor labii inferioris (or quadratus labii inferioris) were segmented using manual segmentation with the ScanIP module (Simpleware, UK) (Fig. 2). Other facial soft tissues (fat and skin) were segmented using semi-automatic segmentation. Due to the complexity of the facial soft tissue morphologies, an experienced clinician performed these segmentations to ensure each segmented pixel belongs to the related facial soft tissues (Fig. 2). Then, marching cube algorithm was used to reconstruct the 3D geometries of the face and muscle models. The result was saved in STL format for further processing.

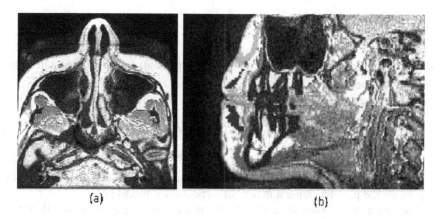

**Fig. 2** Segmented MRI images of the facial muscles: axial (A) and Sagittal (B) planes

## 2.3   Meshing and Muscle Activation Behavior Information

The 3D geometries of the facial muscle tissues were generated and smoothed using a 'home-made' process [10], [11]. Then the muscle mean-line length was computed to quantify the elongation or the contraction level of the facial muscles (Fig. 3).

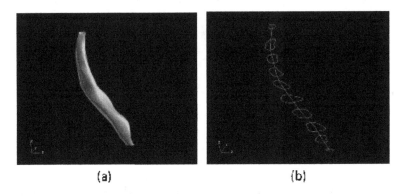

**Fig. 3** Reconstructed model (a) and muscle mean-line length (b) of the ZM

## 2.4 Muscle Constitutive Modeling

In this present study, facial muscle is modelled as a transversely isotropic hypere-lastic material [12].

The strain energy function is expressed as follows:

$$(U) = U_1(\overline{I}_1^B) + U_f(\overline{\lambda}_f, \xi^{CE}) + U_J(J) \tag{1}$$

$$\left(U_1(\overline{I}_1^B)\right) = a\left\{expb\left[\overline{I}_1^B - 3\right] - 1\right\} \tag{2}$$

where $U_1(\overline{I}_1^B)$ is the isotropic strain part; a, b are constants and $\overline{I}_1^B$ is the first invariant of the left Cauchy-Green deformation tensor $B = J^{-\frac{2}{3}}FF^T$.

The strain energy associated with muscle fiber behaviors is expressed as follows:

$$\left(U_f(\overline{\lambda}_f, \xi^{CE})\right) = U_{PE}(\overline{\lambda}_f) + U_{SE}(\overline{\lambda}_f, \xi^{CE}) \tag{3}$$

where $\overline{\lambda}_f$ is elongation of the muscle fiber, $\xi^{CE}$ is the contraction amplitude re-flecting the muscle activation level, CE represents the contractile element and PE represents the parallel element in the Hill-type muscle model [12].

The strain energy associated with the volume change is expressed as follows:

$$(U_J(J)) = \frac{1}{D}(J - 1)^2 \tag{4}$$

where D is a constant, J = det(F) is the gradient deformation tensor.

From the strain energy density function (01), the Cauchy stress function is de-duced as follows:

$$(\sigma) = \frac{2}{J}dev\left[\overline{B}\frac{\partial(U_1 + U_f)}{\partial\overline{B}}\right] + \frac{\partial U_J}{\partial J}I \tag{5}$$

where dev is the deviator and I is the order-2 tensor.

Then after the derivation, the stress tensor could be expressed as follows:

$$(\sigma) = \frac{1}{J}\left[U_1'\left(2\overline{B} - \frac{2}{3}\overline{I}_1^B I\right) + U_f'\left(\overline{\lambda}_f n \otimes n - \frac{1}{3}\overline{\lambda}_f I\right)\right] + U_J'I \tag{6}$$

where $n = \frac{J^{-\frac{2}{3}}FN}{\overline{\lambda}_f}$ is the vector related to the fiber direction defined in a non-deformed configuration.

The derivations of $U_1', U_f', U_J'$ are expressed as follows:

$$\left(U_1'\right) = ba\left\{exp\left[\overline{I}_1^B - 3\right]\right\} \tag{7}$$

$$\left(U_{PE}'\right) = \left\{4\sigma_0(\overline{\lambda}_f - 1)^2\right\} if \overline{\lambda}_f \succ 1 \tag{8}$$

$$\left(U_{SE}'\right) = 0.1\sigma_0\left\{exp\left[100(\overline{\lambda}_f - 1 - \xi^{CE})\right] - 1\right\} \tag{9}$$

$$\left(U_J'\right) = \frac{2}{D}(J-1) \tag{10}$$

It is important to note that $U_{PE}' = 0$ when $\overline{\lambda}_f \leq 1$. All simulation coefficients were set up as literature values $a = 8.2110^{-4}$ MPa, b=1.79, $\sigma_0 = 0.6688$ MPa and $D = 10^{-3}(MPa)^{-1}$. It is important to note that the values of $\xi^{CE}$ were set up using the information derived from MRI data of the facial muscle in different positions.

## 2.5 Finite Element Simulation and Analysis

In the present study, only the behavior of the ZM muscle was simulated and analyzed. The meshed model of the ZM muscle includes 9228 nodes and 5557 tetrahedral elements. The proposed muscle behavior was implemented using a subroutine VUMAT of Abaqus 6.11.

The simulation of ZM muscles activity reflecting the facial mimics expressions were performed. The displacement data will be used for the confrontation between MRI-based data and FE-based results.

# 3 Computational Results

## 3.1 Geometrical Models

3D geometrical models of the face and muscles are illustrated in Fig. 4.

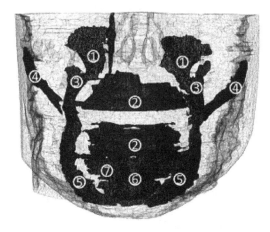

**Fig. 4** Reconstructed face and muscles: levator labii superioris (1), orbicularis oris (2), levator anguli oris (3), zygomaticus major (4), depressor anguli oris (5), musculus mentalis (6), depressor labii inferioris (7)

## 3.2  MRI-Based Muscle Activation Information

The muscle length and strain of the zygomaticus major are depicted in Table 1. The activation patterns of zygomaticus major in neutral position versus smile, pronunciations of sound 'Pou' and 'O' are illustrated in Fig. 5, Fig. 6 and Fig. 7 respectively.

**Table 1** Muscle length and strain of the zygomaticus major for all analyzed positions

| Position | Neutral | Smile | Pou | O |
|---|---|---|---|---|
| L (mm) | 52 | 48 | 57 | 64 |
| $\frac{\Delta L}{L}$ (%) | 0 | -6.82 | 10.4 | 24 |

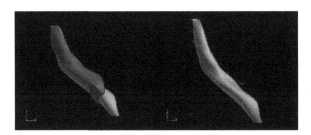

**Fig. 5** MRI-based muscle activation: neutral (blue color) vs. smile (purple color) positions

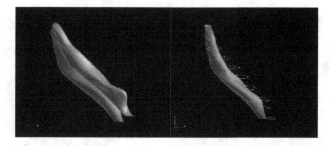

**Fig. 6** MRI-based muscle activation: neutral (blue color) vs. 'Pou' (yellow color) positions

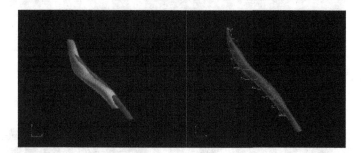

**Fig. 7** MRI-based muscle activation: neutral (blue color) vs. 'O' (red color) positions

## 3.3 Finite Element Simulations

The finite element simulation of the ZM in three expressions (smile, pronunciation of sound 'Pou' and 'O') are illustrated in Fig. 8, Fig. 9 and Fig. 10 respectively.

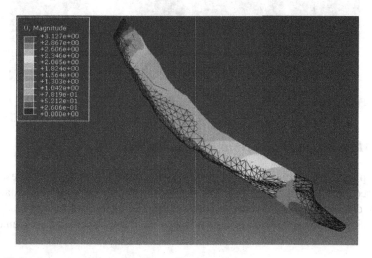

**Fig. 8** Finite element muscle activation: smile expression

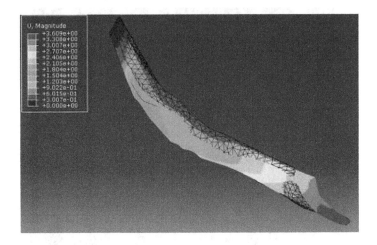

**Fig. 9** Finite element muscle activation: 'Pou' expression

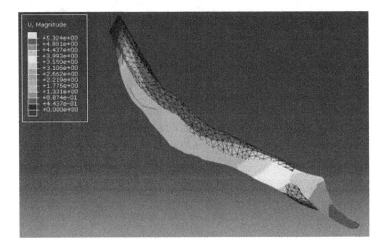

**Fig. 10** Finite element muscle activation: 'O' expression

## 3.4  MRI-Based Data versus FE-Based Results

The displacements between the neutral position and each position of interest were compared and reported in Table 2. The displacement derived from MRI data is the average displacement of all geometrical points on the surfaces of neutral model and each position of interest. The FE-based displacement is the average displacement of nodes.

**Table 2** MRI-based data vs. FE-based simulation results

| Parameters | Smile $\xi^{CE} = -6.8$ | Pou $\xi^{CE} = 10.4$ | O $\xi^{CE} = 24$ |
|---|---|---|---|
| MRI-based displacement (mm) | $4 \pm 2$ | $4.5 \pm 1.4$ | $6 \pm 3$ |
| FE-based displacement (mm) | $1.9 \pm 0.8$ | $2 \pm 1$ | $2.8 \pm 1.1$ |

# 4 Discussion

Medical imaging allows subject specific or patient specific geometries to be acquired. However, these image-based geometries showed irregular defects for complex and small tissues such as facial muscle or cartilage models [14]. Thanks to our specific procedure, image-based surface model is smoothed and then, related mesh can be generated in a flexible manner. In fact, the number of elements can be reduced leading to reduce computing time. In addition, modifications can be done directly during mesh generation process to adapt to a specific simulation purpose such as FE analysis of facial mimics.

In our present study, the confrontation between MRI-based data and finite element simulation result revealed that the numerical results are underestimated by the factor 2. Moreover, the best simulation was obtained for the smile expression.

Advanced muscle constitutive models have been used for biomechanical analysis [6], [7], [8], [9], [15], [16], [17], [18], [19], [20] as well as for computer simulation and animation [21], [22]. Muscle was commonly modeled as a transversely isotropic hyperelastic material. These studies showed that this material is appropriate for describing the skeletal muscle behavior. However, experimental data need to be acquired in a subject specific manner to improve the bio-fidelity of numerical model. For the geometrical properties, medical imaging becomes customized technique to acquire specific geometries of the subject under investigation. However, other intrinsic properties such as muscle mechanical and activation properties are commonly set up using experience-based hypotheses or under numerical convergence requirement. Consequently, new effort needs to be investigated to introduce these properties derived from advanced experimental techniques such as Diffusion Tensor MRI [23], [24] or Magnetic Resonance Elastography (MRE) [25], [26] into numerical model leading to accurate simulation results.

In this present study, only one activated muscle (zygomaticus major) was simulated and analyzed. It is well-known that more muscles contribute to the facial motions. Other facial muscles will be investigated to have a more complete model leading to the simulation of muscle coordination mechanism of facial mimics. Finally, the identification of in vivo mechanical properties of facial muscles via the MRE will be performed to create a model with individualized geometrical and mechanical properties via medical imaging techniques [1].

# 5 Conclusions

This present study introduces a methodology for the confrontation between experimental and numerical results of the facial muscle activation behavior leading to the accurate simulation of the facial mimics.

**Acknowledgments.** This work was carried out in the framework of the Equipex FIGURES supported by the French Government, through the program 'Investments for the future' managed by the National Agency for Research (ANR-10-EQPX-01-01). This present study has been funded by the Picardie Region (SIMOVI project).

# References

1. Ho Ba Tho, M.C.: Bone and joints modelling with individualised geometric and mechanical properties derived from medical images. Computer Mechanics and Engineering Sciences 4(3&4), 489–496 (2003)
2. Langenbach, G.E.J., Hannam, A.G.: The role of passive muscle tensions in a three-dimensional dynamic model of the human jaw. Archives of Oral Biology 44(7), 557–573 (1999)
3. May, B., Saha, S., Saltzman, M.: A three-dimensional mathematical model of temporomandibular joint loading. Clinical Biomechanics 16, 489–495 (2001)
4. Shi, J., Curtis, N., Fitton, L.C., O'Higgins, P., Fagan, M.J.: Developing a musculoskeletal model of the primate skull: Predicting muscle activations, bite force, and joint reaction forces using multibody dynamics analysis and advanced optimisation methods. Journal of Theoretical Biology 310, 21–30 (2012)
5. Dao, T.T.: Modeling of Musculoskeletal System of the Lower Limbs: Biomechanical Model vs. Meta Model (Knowledge-based Model). PhD Thesis, University of Technology of Compigne, 1–194 (2009)
6. Chabanas, M., Luboz, V., Payan, Y.: Patient specific Finite Element model of the face soft tissue for computer-assisted maxillofacial surgery. Medical Image Analysis 7(2), 131–151 (2003)
7. Röhrle, O., Pullan, A.J.: Three-dimensional finite element modelling of muscle forces during mastication. Journal of Biomechanics 40(15), 3363–3372 (2007)
8. Hung, A., Mithraratne, K., Sagar, M., Hunter, P.: Multilayer Soft Tissue Continuum Model: Towards Realistic Simulation of Facial Expressions. World Academy of Science, Engineering and Technology 54, 134–138 (2009)
9. Nazari, M., Perrier, P., Chabanas, M., Payan, Y.: Simulation of dynamic orofacial movements using a constitutive law varying with muscle activation. Computer Methods in Biomechanics & Biomedical Engineering 13(4), 469–548 (2010)
10. Dao, T.T., Pouletaut, P., Goebel, J.C., Pinzano, A., Gillet, P., Ho Ba Tho, M.C.: Differential Geometrical Transformation Applied to Point Clouds Generation from Image-based Surface Model. Proceedings of International Society of Biomechanics Congress, 531 (2011)
11. Bideau, N., Dao, T.T., Charleux, F., Aufaure, P., Ho Ba Tho, M.C., Rassineux, A.: Dveloppement d'une chaine de calcul par ments finis partir de donnes issues de l'imagerie mdicale. In: Proceedings of 10me Colloque National en Calcul des Structures, CSMA, pp. 1–6 (2011)

12. Martins, J.A.C., Pires, E.B., Salvado, R., Dinis, P.B.: A numerical model of passive and active behavior of skeletal muscles. Comput. Methods Appl. Mech. Engineering 151, 419–433 (1998)

13. Dao, T.T., Marin, F., Pouletaut, P., Aufaure, P., Charleux, F., Ho Ba Tho, M.C.: Estimation of Accuracy of Patient Specific Musculoskeletal Modeling: Case Study on a Post-Polio Residual Paralysis Subject. Computer Method in Biomechanics and Biomedical Engineering 15(7), 745–751 (2012)

14. Dao, T.T., Pouletaut, P., Goebel, J.C., Pinzano, A., Gillet, P., Ho Ba Tho, M.C.: In vivo characterization of morphological properties and contact areas of the rat cartilage derived from high-resolution MRI. Biomedical Engineering and Research 32(3), 204–213 (2011)

15. Blemker, S.S., Asakawa, D.S., Gold, G.E., Delp, S.L.: Image-based musculskeletal modeling: Applications, advances, and future opportunities. Journal of Magnetic Resonance Imaging 25, 441–451 (2007)

16. Yucesoy, C.A., Koopman, B.H.F.J.M., Huijing, P.A., Grootenboer, H.J.: Three-dimensional finite element modeling of skeletal muscle using a two-domain approach: linked fiber-matrix mesh model. Journal of Biomechanics 35, 1253–1262 (2002)

17. Fernandez, J.W., Buist, M.L., Nickerson, D.P., Hunter, P.J.: Modelling the passive and nerve activated response of the rectus femoris muscle to a flexion loading: A finite element framework. Medical Engineering & Physics 27, 862–870 (2005)

18. Blemker, S.S., Pinsky, P.M., Delp, S.L.: A 3D model of muscle reveals the causes of nonuniform strains in the biceps brachii. Journal of Biomechanics 38, 657–665 (2005)

19. Tang, C.Y., Zhang, G., Tsui, C.P.: A 3D skeletal muscle model coupled with active contraction of muscle fibres and hyperelastic behavior. Journal of Biomechanics 42, 865–872 (2009)

20. Lu, Y.T., Zhu, H.X., Richmond, S., Middleton, J.: A visco-hyperelastic model for skeletal muscle tissue under high strain rates. Journal of Biomechanics 43, 2629–2632 (2010)

21. Sifakis, E., Selle, A., Robinson-Mosher, A., Fedkiw, R.: Simulating Speech with a Physics-Based Facial Muscle Model. In: ACM SIGGRAPH/Eurographics Symposium on Computer Animation (SCA), pp. 1–10 (2006)

22. Sifakis, E., Neverov, I., Fedkiw, R.: Automatic Determination of Facial Muscle Activations from Sparse Motion Capture Marker Data. ACM Transactions on Graphics 24, 417–425 (2005)

23. Shinagawa, H., Murano, E.Z., Zhuo, J., Landman, B., Gullapalli, R.P., Prince, J.L., Stone, M.: Effect of oral appliances on genioglossus muscle tonicity seen with diffusion tensor imaging: A pilot study. Oral Surgery, Oral Medicine, Oral Pathology, Oral Radiology, and Endodontology 107(3), e57–e63 (2009)

24. McMillan, A.B., Shi, D., Pratt, S.J.P., Lovering, R.M.: Diffusion Tensor MRI to Assess Damage in Healthy and Dystrophic Skeletal Muscle after Lengthening Contractions. Journal of Biomedicine and Biotechnology, 1–10 (2011)

25. Bensamoun, S., Ringleb, S.I., Littrell, L., Chen, Q., Brennan, M., Ehman, R.L., An, K.N.: Determination of thigh muscle stiffness using magnetic resonance elastography. J. Magn. Reson. Imaging 22, 242–247 (2006)

26. Ringleb, S.I., Bensamoun, S., Chen, Q., Manduca, A., Ehman, R.L., An, K.N.: Applications of Magnetic Resonance Elastography to Healthy and Pathologic Skeletal Muscle. J. Magn. Reson. Imaging 25(2), 301–309 (2007)

# Ultrasound Wave Propagation in a Stochastic Cortical Bone Plate

Salah Naili, Vu-Hieu Nguyen, Mai-Ba Vu, Christophe Desceliers,
and Christian Soize

**Abstract.** Ultrasonic guided-wave technologies are powerful nondestructive testing techniques to characterize bone material. This work aims to evaluate the effect due to spatial heterogeneity of bone material properties on its ultrasound response using axial transmission technique. A probabilistic model is introduced to describe the mechanical behavior of bone material. The numerical results focused on studying of FAS (First Arriving Velocity) showing that this quantity strongly depends on the dispersion induced by statistical fluctuations of stochastic elasticity field.

## 1 Introduction

It is well-known that cortical bone is a highly complex composite material formed by a hierarchical and multiscale constituents. Due to the fluctuation of pore distribution and physical properties of mineralization of bone tissues, cortical bone at the vascular scale is a heterogeneous and random medium. These factors would not be neglected when performing diagnostics of cortical bone. One of the most usual techniques used for diagnostics of long bone is known as ultrasonic pulsed through-transmission or axial transmission technique (ATT). This technique, which measures the wave velocity in the bone longitudinal direction, has been shown particularly suitable to predict mechanical as well as geometrical characteristics of the bone [10]. In the past, most models of ATT considered cortical long bone as a medium with homogeneous properties along its longitudinal direction. Hence, homogeneous or functionally graded material properties have been used to model the cortical bone plate [2, 7, 12, 13]. In practice, the exploitation of measured signal

Salah Naili · Vu-Hieu Nguyen · Mai-Ba Vu
Université Paris Est, Laboratoire Modélisation et Simulation Multi-Echelle
(MSME UMR 8208 CNRS), 61 avenue du Général de Gaulle 94010 Créil, France

Christophe Desceliers · Christian Soize
Université Paris Est, Laboratoire Modélisation et Simulation Multi-Echelle
(MSME UMR 8208 CNRS), 5, Boulevard Descartes, 77454 Marne-la-Vallée, France

V.-N. Huynh et al. (eds.), *Knowledge and Systems Engineering, Volume 2,*   435
Advances in Intelligent Systems and Computing 245,
DOI: 10.1007/978-3-319-02821-7_38, © Springer International Publishing Switzerland 2014

data naturally needs to also take into account the uncertainty of material character-istics. However, most of parametric studies of wave propagation in bone are mainly limited to deterministic media.

Some studies have recently been carried out to investigate the influence of random properties of cortical bone in the context of ultrasound characterization. A probabilistic model based on the maximum entropy principle has been con-structed for considering cortical bone plate as homogeneous or multi-layered media [11, 3, 5]. It has been shown that a simplified mechanical model with an additional stochastic modeling of bone elasticity properties are able to represent the *in vivo* measurements in the statistical sense. However, in these works, material inhomo-geneities along longitudinal direction of bone have always been neglected.

This paper presents a stochastic model to consider the random fluctuation of ma-terial properties in both radial and longitudinal directions in a cortical bone plate. Section 2 presents the description of the problem and the numerical method to com-pute the time domain solution of the acoustic response of a heterogeneous bone plate coupled with fluid. Next, we provides a procedure to construct a parametric model for the elasticity properties of bone plate. Then, Section 3 shows a numerical test to study the sensitivity of $V_{FAS}$ due to dispersion of the stochastic elasticity field in the bone. Last, Section 4 gives some conclusions of this work.

## 2 Modeling of Transient Wave Propagation in a Heterogeneous Bone Immersed in Fluid

### 2.1 Geometrical Configuration

Let $\mathbf{R}(O; \mathbf{e}_1, \mathbf{e}_2, \mathbf{e}_3)$ be a reference Cartesian frame where $(\mathbf{e}_1, \mathbf{e}_2, \mathbf{e}_3)$ is an orthonor-mal basis. Figure 1 shows a common configuration for modeling an ultrasound axial transmission test. It consists of an anisotropic elastic layer sandwiched between two idealized acoustic fluids. The bone material is assumed to be heterogeneous in the plane $(\mathbf{e}_1, \mathbf{e}_2)$ but homogeneous along $\mathbf{e}_3$ direction. An acoustic line source, which is parallel to $\mathbf{e}_3$, produces an excitation at $\mathbf{x}^s = (x_1^s, x_2^s)$ inside the upper fluid layer. Hence, the problem may be reduced into a two-dimensional plane strain problem in the $(O; \mathbf{e}_1, \mathbf{e}_2)$ plane. In Figure 1, the infinite bone layer occupying the domain $\Omega^b$ with a constant thickness $h$ ($\Omega^b = \{x_1 \in \mathbb{R}, 0 \geq x_2 \geq -h\}$). This bone plate is loaded on its upper and lower surfaces by two fluid halfspaces which represents the soft tissues. The upper fluid domain is denoted by $\Omega_1^f$ ($\Omega_1^f = \{x_1 \in \mathbb{R}, x_2 \geq 0\}$) and the lower one is denoted by $\Omega_2^f$ ($\Omega_2^f = \{x_1 \in \mathbb{R}, x_2 \leq -h\}$). The interfaces between the bone ($\Omega^b$) and the fluids ($\Omega_1^f$ and $\Omega_2^f$) are denoted by $\Gamma_1^{bf}$ and $\Gamma_2^{bf}$, respectively (see Fig. 1). In that follows, the whole of domain is denoted by $\Omega = \Omega_1^f \cup \Omega^b \cup \Omega_2^f$.

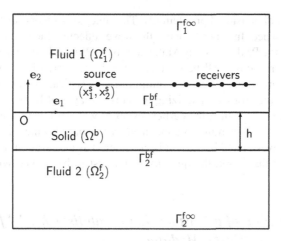

**Fig. 1** Geometrical configuration of the trilayer model for ultrasound axial transmission test

## 2.2 Governing Equations and Numerical Methods

For the sake of simplifying, the described trilayer system, which consists of a heterogeneous anisotropic elastic layer $\Omega^b$ and two homogeneous inviscid acoustic fluid layers, will be all modeled as an elastic medium. At each point $\mathbf{x}$ in $\Omega$, the constitutive law equation is expressed by

$$\boldsymbol{\sigma} = \bar{\mathbb{c}}\,\boldsymbol{\varepsilon}, \quad \text{with } \boldsymbol{\varepsilon} = \frac{1}{2}(\operatorname{grad}\mathbf{u} + (\operatorname{grad}\mathbf{u})^T), \tag{1}$$

where $\boldsymbol{\sigma}(\mathbf{x},t)$ and $\boldsymbol{\varepsilon}(\mathbf{x},t)$ denote the stress and strain tensors, respectively; $\mathbf{u}(\mathbf{x},t)$ denotes the displacement vector; $\mathbb{c}(\mathbf{x})$ is the fourth-order elasticity tensor of the fluid or solid depending on $\mathbf{x}$: $\bar{\mathbb{c}}(\mathbf{x}) = \mathbb{c}^f$ for $\mathbf{x} \in \Omega_1^f \cup \Omega_2^f$ and $\bar{\mathbb{c}}(\mathbf{x}) = \mathbb{c}(\mathbf{x})$ for $\mathbf{x} \in \Omega^b$. The elasticity tensor of the fluid $\mathbb{c}^f$, which is homogeneous in $\Omega_1^f \cup \Omega_2^f$, is as an isotropic elastic tensor without shear modulus. The elasticity tensor of the solid $\mathbb{c}$ depends on $\mathbf{x}$ and is determined by 6 elastic constants. Note that in the next section, the probabilistic model only concerns the tensor of elasticity of the domain $\Omega^b$.

By neglecting the body force field, the equation of motion in domain $\Omega$ reads

$$\operatorname{div}\boldsymbol{\sigma} = \rho\,\ddot{\mathbf{u}}, \quad \forall \mathbf{x} \in \Omega, \tag{2}$$

where $\rho(\mathbf{x})$ is the mass density and div designates the divergence operator. The domain $\Omega$ is at rest for $t < 0$. Note that no interface condition is required for this elastoacoustic problem because both of fluid and bone domains are modeled as elastic media.

An explicit time domain finite difference scheme, which is based on a staggered grid formulation for the velocity and stress components, has been implemented [9] to solve the two-dimensional problem (2). The scheme is second-order in time and fourth-order in space. In order to avoid the wave reflected due to the boundaries at finite distances, the PML Perfectly Matched Layer (PM) technique has been is used.

The acoustic response will be captured at a linear array of receivers as shown in Fig. 1. This emitter and receiver configuration is typical one in ultrasonic axial transmission devices for characterizing the cortical layer of bone. When using the axial transmission technique, the earliest event or wavelet (also called by First Arriving Signal, FAS) of the multicomponent signal recorded at the receivers has been the most often investigated because it is considered as a relevant index of bone status [18, 1, 8]. In this work, the quantity of interest to be studied will be the FAS velocity.

### 2.3 Construction of a Parametric Probabilistic Model for an Anisotropic Elastic Medium

Parametric probabilistic models are built by modeling the local physical properties of the medium. In the present study, only the uncertainty of elasticity tensor $\mathbb{C}(\mathbf{x})$, of which the mean value is defined by $\mathbb{c}(\mathbf{x})$, will be considered. At each point $\mathbf{x} \in \Omega^b$, tensor $\mathbb{C}$ depends on 21 independent random variables. In practice, it is impossible to have a sufficient large set of experimental data, especially for bone material, to estimate the probability distribution of the random elasticity tensor. Moreover, the random spatial variation of the elastic property should be considered. To overcome this difficulty, we use Soize's model whereby the probability distribution is built by full-filling the maximum entropy principle. Using this principle, the stochastic elasticity tensor $\mathbb{C}(\mathbf{x})$ may be parameterized by *its mean values* $\mathbb{c}(\mathbf{x})$ *via* its matrix representation, and by a *minimal set of essential parameters* which consists of only 4 parameters: one scalar dispersion level $\delta$ and one vector $\boldsymbol{\lambda}$ that contains three spatial correlation lengths. The detail of essential steps to estimate the random field $\mathbb{C}(\mathbf{x}; \mathbb{c}, \delta, \boldsymbol{\lambda})$ at every point $\mathbf{x} \in \Omega^b$ may be found in [16, 17] .

## 3  Numerical Results

This section will present some illustrative results on studying the ultrasound wave propagation through a random 4mm-thickness cortical bone layer. For this simulation, the pressure source acts on a spatial length of 0.75 mm which is horizontally placed in the fluid at 2 mm from the upper interface of the solid layer. The excitation signal is a Gaussian with a center frequency $f_c = 1$ MHz [4]. The responses will be calculated at 14 receivers are regularly spaced with a pitch of 0.8 mm and a distance of emitter to closest receiver equal to 11 mm.

Fluid and Solid Material Properties

Both fluid layers, which represent the soft tissues and bone marrow, are considered to be an inviscid water with a mass density $\rho_f = 1\,000$ kg.m$^{-3}$ and a bulk modulus $K_f = 2.25$ GPa.

The mean model of the bone material is assumed to be a transversely isotropic medium. This behavior has been experimentally shown by different authors [6, 14, 15] to be a realistic approximation of cortical bone properties. This example uses data given by [6] who measured the homogenized bone properties by performing tensile and torsional tests with a mechanical testing system on 18 different human femoral bone specimens. The components of the mean elasticity tensor are given by: $a_{11} = 23.05$ GPa, $a_{22} = a_{33} = 15.08$ GPa, $a_{12} = a_{13} = 8.72$ GPa, $a_{44} = 3.3$ GPa, $a_{55} = a_{66} = 4.7$ GPa.

Parameters for the Uncertain Elasticity Model

As discussed before, this study is restricted to only consider the uncertain heterogeneous in the plane $(\mathbf{x}_1, \mathbf{x}_2)$. Thus only the dispersion $\delta$ and two correlation lengths $(\lambda_1, \lambda_2)$ need to be introduced to control the random variation of elasticity field. For this example, a correlation length $\lambda_1 = \lambda_2 = 2 \times 10^{-4}$ $m$, which may be seen as a typical center-to-center distance between osteons in cortical bone, is used. Two different values of the dispersion $\delta = 0.1$ and $\delta = 0.3$ will be investigated. Figure 2 shows a map illustrating the spatial variation of $A_{ij}$.

Numerical Parameters for the Finite-Difference Solver

The finite difference simulation is performed for each of 800 Monte-Carlo realizations. A rectangular domain with size $0.03 \times 0.01$ m is used to be able to contain all details of the model, $i.e$ three layers with the source and receivers. The grid steps in both $x_1$- and $x_2$-direction are chosen to be identical: $\Delta x_1 = \Delta x_2 = 2 \times 10^{-5}$ m to ensure that grid intervals are smaller than about 1/8 of the smallest wavelength in whole domains. As a result, the grid has $1501 \times 501$ points (about $1.4 \times 10^6$ degrees of freedom). The time step is chosen by using CFL stability condition $\Delta t < \alpha \Delta / c$, where $\alpha$ is a constant, $\Delta$ is the smallest space interval and $c$ is the maximum wave velocity in the domain. For this study, the time step size is fixed by $\Delta t = 10^{-9}$ s.

Results and Discussions

Fig. 3 (left) presents the $V_{FAS}$ obtained for all of 800 Monte-Carlo realizations. Two levels of dispersion $\delta = 1$ and $\delta = 0.3$ have been considered. It can be seen that when $\delta = 0.1$, the values of $V_{FAS}$ oscillate around the one corresponding to the mean model which is shown as the red continuous line ($V_{FAS}^{mean} = 3606$ m.s$^{-1}$). When the dispersion is higher ($\delta = 0.3$), the measured values of $V_{FAS}$ are globally decrease. It may be explained by the fact that when the heterogeneity of the medium is greater, the scattering phenomenon becomes more significant and may modify the

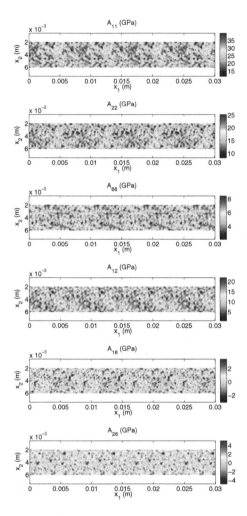

**Fig. 2** Spatial variation of the elasticity tensor components $A_{ij}$

global wave velocity in the domain. In addition, we obtained a greater dispersion of $V_{FAS}$ is more important in the case $\delta = 0.3$ than the one in the case $\delta = 0.1$.

In Fig. 3 (right), the probability density functions of $V_{FAS}$ are computed for both cases. One may state that the probability density function of $V_{FAS}$ strongly depends on the dispersion $\delta$. It means that in the practice, neglecting the uncertain heterogeneity in bone may lead to a poor prediction of the mean value of mechanical properties by using $V_{FAS}$ as a index.

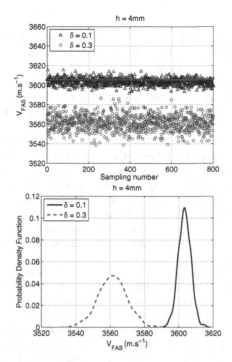

**Fig. 3** (left) Values of $V_{FAS}$ evaluated at each realization; the solid line is obtained for the homogeneous material properties; (right) probability density functions of $V_{FAS}$

## 4  Conclusion

The heterogeneity of bone's mechanical properties is significant and need to be considered when studying the sensitivity of ultrasound response in cortical long bone. As the statistical data on real bone material is hardly found, a parametric probabilistic method, which is based on the maximum entropy principle, has been used to generate an optimal probabilistic model for taking into account the uncertainties of bone elasticity. A explicit FDTD solver has been developed for simulating the wave propagation in a transversely isotropic heterogeneous medium in the time domain. It has been shown that the FAS velocity is very sensitive to dispersion of the bone's elasticity tensor in statistic sense. Detailed study need to be carried out and will be presented in a forthcoming paper.

Although the uncertain elastic model of the plate presented in this paper was developed to study behavior of cortical long bone, this procedure is of wider interest and may be applied to the characterization of other materials.

# References

1. Barkmann, R., Kantorovich, E., Singal, C., Hans, D., Genant, H.K., Heller, M., Gluer, C.C.: A new method for quantitative ultrasound measurements at multiple skeletal sites. J. Clin. Densitometry 3, 1–7 (2000)
2. Bossy, E., Talmant, M., Defontaine, M., Patat, F., Laugier, P.: Bidirectional axial transmission can improve accuracy and precision of ultrasonic velocity measurement in cortical bone: a validation on test materials. IEEE Transactions on Ultrasonics, Ferroelectrics and Frequency Control 51(1), 71–79 (2004)
3. Desceliers, C., Soize, C., Grimal, M., Talmant, Q., Naili, S.: Determination of the random anisotropic elasticity layer using transient wave propagation in a fluidsolid multilayer: model and experiments. J. Acoust. Soc. Am. 125(4), 2027–2034 (2008a)
4. Desceliers, C., Soize, C., Grimal, Q., Haat, G., Naili, S.: Three dimensional transient elastic waves in multilayer semi-infinite media solved by a time-space-spectral numerical method. Wave Motion 45(4), 383–399 (2008b)
5. Desceliers, C., Soize, C., Naili, S., Haiat, G.: Probabilistic model of the human cortical bone with mechanical alterations in ultrasonic range. Mechanical Systems and Signal Processing 32, 170–177 (2012)
6. Dong, X.N., Guo, X.E.: The dependence of transversely isotropic elasticity of human femoral cortical bone on porosity. J. Biomech. 37(8), 1281–1287 (2004)
7. Haiat, G., Naili, S., Grimal, Q., Talmant, M., Desceliers, C., Soize, C.: Influence of a gradient of material properties on ultrasonic wave propagation in cortical bone: Application to axial transmission. J. Acoust. Soc. Am. 125(6), 4043–4052 (2009)
8. Hans, D., Srivastav, S.K., Singal, C., Barkmann, R., Njeh, C.F., Kantorovich, E., Gluer, C.C., Genant, H.K.: Does combining the results from multiple bone sites measured by a new quantitative ultrasound device improve discrimination of hip fracture? J. Bone Miner. Res. 14(4), 644–651 (1999)
9. Levander, A.R.: Fourth-order finite-difference P-SV seismograms. Geophysics 53(11), 1425–1436 (1988)
10. Lowet, G., Van der Perre, G.: Ultrasound velocity measurements in long bones: measurement method and simulation of ultrasound wave propagation. Journal of Biomechanics 29, 1255–1262 (1996)
11. Macocco, K., Grimal, Q., Naili, S., Soize, C.: Elastoacoustic model with uncertain mechanical properties for ultrasonic wave velocity prediction; application to cortical bone evaluation. Journal of the Acoustical Society of America 119(2), 729–740 (2006)
12. Naili, S., Vu, M.-B., Grimal, Q., Talmant, M., Desceliers, C., Soize, C., Haiat, G.: Influence of viscoelastic and viscous absorption on ultrasonic wave propagation in cortical bone: Application to axial transmission. J. AcousT. Soc. Am. 127(4), 2622–2634 (2010)
13. Nguyen, V.-H., Naili, S.: Ultrasonic wave propagation in viscoelastic cortical bone plate coupled with fluids: a spectral finite element study. Computer Methods in Biomechanics and Biomedical Engineering (2012)
14. Reilly, D.T., Burnstein, A.H.: The mechanical properties of cortical bone. J. Bone Joint Surg. Am. 56, 1001–1022 (1974)
15. Rho, J.Y.: An ultrasonic method for measuring the elastic properties of human tibial cortical and cancellous bone. Ultrasonics 34(8), 777–783 (1996)
16. Soize, C.: Random matrix theory for modeling uncertainties in computational mechanics. Computer Methods in Applied Mechanics and Engineering 194, 1333–1366 (2005)

17. Soize, C.: Non-Gaussian positive-definite matrix-valued random fields for elliptic stochastic partial differential operators. Computer Methods in Applied Mechanics and Engineering 195, 26–64 (2006)
18. Stegman, M.R., Heaney, R.P., Travers-Gustafson, D., Leist, J.: Cortical ultrasound velocity as an indicator of bone status. Osteoporos. Int. 5(5), 349–533 (1995)

# Author Index